Mathematics: Theory & Applications

Editors

Richard V. Kadison
Isidore M. Singer

Bernhelm Booß-Bavnbek
Krzysztof P. Wojciechowski

Elliptic Boundary Problems for Dirac Operators

Birkhäuser
Boston • Basel • Berlin

1993

Bernhelm Booß-Bavnbek
IMFUFA
Roskilde University
4000 Roskilde
Denmark

Krzysztof P. Wojciechowski
Department of Mathematics
IUPUI
Indianapolis, IN 46202
USA

Library of Congress Cataloging In-Publication Data
Booss-Bavnbek, Bernhelm, 1941-
Elliptic boundary problems for Dirac operators / Bernhelm Booss
-Bavnbek, Krzysztof P. Wojciechowski
p. cm. -- (Mathematics)
Includes bibliographical references and index.
ISBN 0-8176-3681-1 (H : acid-free). -- ISBN 3-7643-3681-1 (H : acid-free)
1. Differential equations, Elliptic. 2. Boundary value problems.
3. Dirac equation. I. Wojciechowski, Krzysztof P., 1953- .
II. Title. III. Series: Mathematics (Boston, Mass.)
QA377.B686 1993 93-22006
515'.353--dc20 CIP

Birkhäuser

ISBN 0-8176-3681-1
ISBN 3-7643-3681-1

Typeset by authors in AMS-TEX
Printed and bound by Quinn-Woodbine, Woodbine, NJ
Printed in the U.S.A.

9 8 7 6 5 4 3 2 1

Contents

Preface

Elliptic boundary problems have enjoyed interest recently, especially among C^*-algebraists and mathematical physicists who want to understand single aspects of the theory, such as the behaviour of Dirac operators and their solution spaces in the case of a non-trivial boundary.

However, the theory of elliptic boundary problems by far has not achieved the same status as the theory of elliptic operators on closed (compact, without boundary) manifolds. The latter is nowadays recognized by many as a mathematical work of art and a very useful technical tool with applications to a multitude of mathematical contexts. Therefore, the theory of elliptic operators on closed manifolds is well-known not only to a small group of specialists in partial differential equations, but also to a broad range of researchers who have specialized in other mathematical topics.

Why is the theory of elliptic boundary problems, compared to that on closed manifolds, still lagging behind in popularity? Admittedly, from an analytical point of view, it is a jigsaw puzzle which has more pieces than does the elliptic theory on closed manifolds. But that is not the only reason. Having worked with the topology, geometry, and analysis of elliptic boundary problems for many years, the authors feel that the existing books on elliptic boundary problems, excellent as they are, present the theory in too great a generality and by means of too intricate a machinery for our intended readership: graduate students in, for example, geometry, topology or mathematical physics, and researchers from fields other than partial differential equations. They will need insight into specific situations, which involve manifolds with boundary and their Dirac operators. Our aim is to give these readers a book that presents the absolute minimum of the necessary machinery and illustrates how the methods work with interesting examples.

To follow our presentation, the reader should be familiar with the basic calculus of pseudo-differential operators on closed manifolds. Knowledge of the first pages of P. Gilkey's book [1984; pp. 1–36] (new edition in preparation) is quite satisfactory for our purpose.

Our Chapters 17 and 22–26 furthermore presuppose knowledge of the index theory of elliptic operators on closed manifolds. Once again Gilkey's book provides the reader with all needed material. As an auxiliary reference, we also refer to Booß & Bleecker [1985].

Analytically, the heart of our book consists of deep results proved long ago by A. Calderón and R. Seeley: in our book Theorem 12.4, Theorem 13.1, Corollary 13.7, and Theorem 13.8. Our presentation closely follows, with some simplifications from elementary constructions developed in our previous work, the original work of Seeley (see Seeley [1966], [1969] and Palais [1965a; Seeley's contribution to Chapter 17]). Inevitably, some proofs are relatively long and complicated, but still we hope that the aforementioned simplifications make the theory of elliptic boundary problems for Dirac operators accessible to a larger audience. In fact, even the theory seems to benefit from those simplifications. They lead, for example, to recognizing that the Cauchy data spaces of a Dirac operator and its adjoint are orthogonal after Clifford multiplication. This is crucial to almost all index calculations presented in the text.

The text itself consists of 26 chapters, divided into 3 parts. The first part — Chapters 1-10 — gives a detailed exposition of Dirac operators acting on sections of bundles of Clifford modules. In the first 7 chapters, we closely follow standard references like Atiyah, Bott & Shapiro [1964], Lawson & Michelsohn [1989], and especially Gilkey [1989] and Branson & Gilkey [1992a]. It is worth mentioning that Chapter 3 distinguishes three different classes of operators. We say that A is an *operator of Dirac type*, if the principal symbol of A^2 is defined by the metric tensor; A is a *generalized Dirac operator*, if A is the composition of a connection operator and Clifford multiplication; and we call A a *compatible Dirac operator*, or in short, a *Dirac operator*, if the connection used to define A is compatible with Clifford multiplication. For simplicity, all results in the book are proved only for Dirac operators, but they also hold for operators of Dirac type. In fact, except for certain index formulas, they hold even for arbitrary linear elliptic operators of first order. The reason why some of our index formulas do not hold for all first-order elliptic operators is that they are computed by reducing the index problem to a problem on the boundary. That works when the *unique continuation property* (or *uniqueness of the Cauchy problem*—UCP) for the solutions holds. For general elliptic operators, the lack of UCP implies that not all solutions have non-trivial traces on the boundary. That would lead to an additional term in our index formulas.

Chapter 8 proves the simplest among different variants of the unique continuation principle for Dirac operators. Let v be a solution of the equation $Av = 0$, where A is a Dirac operator on a connected manifold. Assume that $v = 0$ on some open set; then $v = 0$ everywhere. The proof closely follows the beautiful exposition in F. Treves [1980], which he based on ideas due to Calderón. Typically, we apply the UCP to a solution v of $Av = 0$ with $v = 0$ on a closed submanifold of codimension 1, so that v vanishes everywhere. The result allows the reducing of index computations for elliptic boundary problems or elliptic transmission problems (*general linear conjugation problems*) to calculations on the boundary.

Chapter 9 shows that any Dirac operator on a compact manifold with boundary extends to a Dirac operator on a closed manifold. Moreover, the constructed extension is an invertible operator. That greatly simplifies the proof of the Calderón-Seeley theorem, the crucial analytical result of our book (Theorem 12.4).

Chapter 10 applies the glueing construction of Chapter 9 to the relative index theorem and to the excision property for Dirac operators.

The second part of the book — Chapters 11-17 — presents some basic analytical and topological tools that enable us to analyse the nature of the index of elliptic boundary problems. In Chapter 11, we investigate Sobolev spaces on manifolds with boundary. Let v be a section on a compact manifold X with boundary Y and let v belong to the Sobolev space H^s. Then the main result is that v, provided $s > \frac{1}{2}$, has a well-defined *trace* $v|_Y$, which is an element of the Sobolev space $H^{s-\frac{1}{2}}$.

The crucial analytical result is contained in Chapter 12 : Let A be a Dirac operator on a compact manifold X with boundary Y. First we construct a *Poisson type operator* K_+, mapping sections on Y to (smooth) solutions of A in $X \setminus Y$. Then we obtain the *Calderón operator* $\mathcal{P}_+$ by restricting $K_+(h)$ to the boundary Y, where h is a section on Y. We show that $\mathcal{P}_+$ is a projection onto the space $H(A) := \{v|_Y \mid Av = 0 \text{ in } X \setminus Y\}$ of *Cauchy data* of the operator A. Moreover, $\mathcal{P}_+$ is a pseudo-differential operator and K_+ provides an isomorphism of $H(A)$ with the space of solutions. In the modern literature, the operators $\mathcal{P}_+$ and K_+ are often constructed modulo smoothing operators. This leads only to approximate projections and approximate isomorphisms, but it simplifies the construction and is still sufficient to prove the regularity theorems and develop the index theory. However, this is the point where usually extra machinery appears, and generality is achieved at the expense of precision and detail. Therefore we chose a different path. The most important properties of K_+ and $\mathcal{P}_+$ are collected in Theorem 12.4.

The proof of Theorem 12.4 is rounded off in Chapter 13, which, among other results, proves the existence of a well-defined trace on the boundary for any solution of $Av = 0$. Chapter 11 showed that v has a well-defined trace, if v belongs to the Sobolev space H^s with $s > \frac{1}{2}$. Here we show that for any s, an element v of H^s with $Av = 0$ in $X \setminus Y$ has a well-defined trace in $H^{s-\frac{1}{2}}$ on Y.

Chapter 14 investigates another operator, which lives on the boundary and is induced by the *tangential part* B of A. B is a total Dirac operator, and in particular it is self-adjoint. Then, for any real number a, the *spectral projection* $\mathrm{P}_{\geq a}$ of B onto the interval $[a, \infty)$ is a pseudo-differential operator which differs from $\mathcal{P}_+$ by a compact operator. For the operator A, these spectral projections define specific boundary conditions first considered by Atiyah, Patodi & Singer in [1975]. We discuss generalized Atiyah-Patodi-Singer problems in the third part of the book.

In Chapter 15, we switch to topological tools and introduce the *Grassmannian* Gr, which is the natural object for studying the index theory of elliptic boundary problems. The Grassmannian is the space of all pseudo-differential projections, which differ from $\mathcal{P}_+$ by a compact operator. The idea of studying the infinite-dimensional Grassmannian is due to B. Bojarski [1979]. We investigated the topological structure of Gr and found applications to the cutting and pasting of the index of elliptic operators in a series of papers (see Wojciechowski [1981], [1985a], [1986], and Booß & Wojciechowski [1985b], [1986]; more recently, in the context of spectral estimates, Booß & Wojciechowski [1989] and Douglas & Wojciechowski [1991; Appendix B]). The Grassmannian appeared in other contexts in the works of other authors (see, for example, Sato, Miwa & Jimbo [1979], Sato [1981], Segal & Wilson [1985], and Pressley & Segal [1986]). Chapter 15 studies the topological structure of Gr and shows that, for $P \in \mathrm{Gr}$, the operator $P\mathcal{P}_+$ (acting from the range of $\mathcal{P}_+$ to the range of P) is a Fredholm operator and that its index $\mathbf{i}(P, \mathcal{P}_+)$ distinguishes the connected components of Gr. For P, $\mathcal{P}_+$ orthogonal, the integer $\mathbf{i}(P, \mathcal{P}_+)$ can also be interpreted as the spectral flow of a corresponding family $\{(2P - \mathrm{Id}) + 2t(\mathcal{P}_+ - P)\}$ of self-adjoint operators.

That is why Chapter 16 studies the topology of the non-trivial component $\widehat{\mathcal{F}}_*$ of the space of *self-adjoint Fredholm operators*. It is well known that $\widehat{\mathcal{F}}_*$ is a classifying space for the functor K^{-1} (see Atiyah & Singer [1969]). To avoid the original and quite complicated proof, we present instead a computation of the homotopy groups of that space. In particular, the first homotopy group is equal to the integers. This is all we need to know in this book.

Chapter 17 is devoted to the study of the *spectral flow* of a family $\{B_t\}_{t \in S^1}$ which is just the class of $\{B_t\}$ in $\widehat{\mathcal{F}}_*$. It is a nice feature in that it has a straightforward analytical interpretation: the number of the eigenvalues of B_t, which change their sign from $-$ to $+$ as t goes along S^1, minus the number of eigenvalues which change their sign from $+$ to $-$. We discuss various topological formulas, giving the spectral flow in case $\{B_t\}$ is a family of elliptic operators of non-negative order. Such families arise naturally, when elliptic boundary problems or elliptic transmission problems induce specific parametrizations of the tangential parts of Dirac operators in a collar of the boundary, resp. the partitioning submanifold.

That is explained in the third part of the book — Chapters 18–26 — which presents the basic theory of elliptic boundary problems and other applications of the machinery that we have developed. Chapter 18, following the beautiful lectures of Seeley [1969], introduces the concept of an *elliptic boundary condition* for a Dirac operator A. It is a pseudo-differential operator R (of order 0), which satisfies a specific compatibility condition with the operator A. Following Seeley, we furthermore assume a technical condition, namely that the range of the operator R is closed in suitable Sobolev spaces. This last condition implies, by an argument from M. Birman's and A. Solomyak's work [1982] on *elliptic fans*, that the orthogonal projections onto the range and the kernel of the operator R are pseudo-differential operators. One consequence is that the *boundary integral* $R\mathcal{P}_+ : H(A) \to \text{range}(R)$ is a Fredholm operator (Corollary 18.15). In that way it brings the index theory of elliptic boundary problems back to the Grassmannian Gr.

Chapter 19 investigates the unbounded operator $A_R : \operatorname{dom} A_R \to L^2$ — the *L^2-realization* of A under the boundary condition $R(v|_Y) = 0$ — and shows that the space of solutions of A_R is finite-dimensional (Corollary 19.2). Moreover, Chapter 19 proves that any solution of A_R is smooth up to the boundary and shows the range of the operator A_R to be closed.

The structure of the index of the operator A_R is discussed in Chapter 20. First we show that A_R is a closed operator in L^2 and then we construct the adjoint operator. A nice consequence of Seeley's aforementioned technical condition is the explicit form obtained for the adjoint elliptic boundary condition Q which gives $(A_R)^* = (A^*)_Q$. That way around we show that $A_R : \operatorname{dom} A_R \to L^2$ is a Fredholm operator with kernel and cokernel consisting of smooth sections. We reduce the index computation of a global elliptic boundary problem to computing the index of the composition of two projections over the boundary which differ by a compact operator:

$$
\begin{aligned}
(0.1) \quad \operatorname{index} A_R &= \dim\ker A_R - \dim\ker(A^*)_Q \\
&= \operatorname{index}\{R\mathcal{P}_+ : \operatorname{range}\mathcal{P}_+ \longrightarrow \operatorname{range} R\} = \mathbf{i}(R, \mathcal{P}_+)\,.
\end{aligned}
$$

Then we explain the relations between the indices of the realization A_R and the full mapping pair (A, R) in the case of *local elliptic boundary conditions* i.e. when the principal symbol of the boundary condition can be written as the lifting of an isomorphism of the original vector bundles to the cotangent sphere bundle.

It is worth mentioning that the boundary conditions we call *local elliptic* are called *elliptic* or *Šapiro-Lopatinskiĭ* conditions in the classical literature on partial differential equations. Those boundary problems which we, in accordance with the terminology of global analysis, call *global elliptic* or simply *elliptic* were either not treated or are a subclass of what is called *injectively elliptic* in certain parts of the literature (see e.g. Grubb [1990]). Notice also that *local* elliptic boundary problems, for which the designation elliptic used to be reserved, play an important but secondary role for Dirac operators where they only appear in odd dimensions or for systems and transmission problems.

Chapter 21 uses formula (0.1) to prove the *Agranovič-Dynin formula* $\operatorname{index} A_{R_1} - \operatorname{index} A_{R_2} = \operatorname{index} R_1\mathcal{P}_+(R_2)^*$. Next, the Agranovič-Dynin formula is used to prove the *cobordism theorem* for Dirac operators, which says that the index of a Dirac operator, which bounds, is equal to 0. Historically, the theorem was the basis of the first proof of the Atiyah-Singer index theorem (see Atiyah & Singer [1963] and also Palais [1965a]). That supports the perspective, elaborated in the rest of the book, of looking at the seemingly more elementary index theory of elliptic operators on closed manifolds through the glasses of the seemingly more intricate, but in reality more fundamental, index theory of elliptic boundary problems.

The *Atiyah-Patodi-Singer index theorem*, the index formula for the operator $A_{\mathrm{P}_{\geq}}$, is investigated in Chapters 22 and 23. Recall that $\mathrm{P}_{\geq} := \mathrm{P}_{\geq 0}$ denotes the spectral projection of the tangential part B of A onto the interval $[0, +\infty)$. The operator $A_{\mathrm{P}_{\geq}}$ finds so many applications in geometry, topology, and number theory that it deserves a separate treatment. (For one approach, aimed specifically at manifolds with corners, see Melrose [1992]). We only prove the index formula for $A_{\mathrm{P}_{\geq}}$. In doing so, we follow the original paper by Atiyah, Patodi & Singer [1975], beginning with the Atiyah-Bott formula

$$\operatorname{index} A_{\mathrm{P}_{\geq}} = \operatorname{tr}\exp(-t\,(A_{\mathrm{P}_{\geq}})^*\,A_{\mathrm{P}_{\geq}}) - \operatorname{tr}\exp(-t\,A_{\mathrm{P}_{\geq}}\,(A_{\mathrm{P}_{\geq}})^*).$$

We obtain the corresponding heat kernels by glueing the heat kernels on the closed double $\widetilde{X}$ of X with the heat kernels of the corresponding operators on the cylinder $Y \times [0, +\infty)$. Therefore, in the formula we have a contribution from the interior, a contribution from the cylinder, plus the *error* term. We take the interior contribution from the local index theorem for Dirac operators, but compute the cylinder contribution. The only problem left is to show that the error disappears, as t approaches 0. Here we use *Duhamel's principle*, which is explained in detail in Chapter 22C. We could have used the finite propagation speed property, but we chose the more classical approach still used in recent literature. Chapter 22 is rounded off with an L^2-reformulation of the results and a discussion of a *three-dimensional example* treated before by Taubes [1990] and Yoshida [1991].

Having proved the Atiyah-Patodi-Singer formula, we use it together with the Agranovič-Dynin formula in Chapter 23 to obtain an index formula for A_P, where P is an arbitrary element of the Grassmannian, and to discuss the *additivity* of the index of Dirac operators with spectral boundary conditions. This allows the writing of the signature of a $4k$-dimensional compact oriented manifold with boundary as the true index of an elliptic boundary problem - without the error term coming from the kernel of the tangential part of the signature operator.

In Chapter 24, we continue discussing additivity problems and the index of Dirac operators on *partitioned* manifolds. We prove a theorem conjectured by Bojarski [1979]: Let M be a closed manifold and Y a submanifold of codimension 1, which divides M into two parts $M = X_+ \cup X_-$, where $X_+ \cap X_- = Y$. Let A^+ denote the restriction of A to X_+. Then we have the formula

$$\begin{aligned}\operatorname{index} A &= \operatorname{index}\{(\mathrm{Id} - \mathcal{P}_-)\mathcal{P}_+ : H_+(A) \to H_-(A)^\perp\}\\ &= \operatorname{index} G\mathcal{P}_+G^*\mathcal{P}_+\,,\end{aligned}$$

where $\mathcal{P}_+$ denotes the Calderón projector of A^+ and G denotes Clifford multiplication by the unit normal vector. We also discuss a formulation of this theorem in the language of Fredholm pairs and then proceed to *the general linear conjugation problem.* We are looking for couples (s_+, s_-), which are solutions of the following problem on M

$$
\begin{aligned}
A^+ s_+ &= 0 && \text{in } X_+ \\
A^- s_- &= 0 && \text{in } X_- \qquad (0.2)\\
\text{and} \quad s_- &= T s_+ && \text{on } Y\ ,
\end{aligned}
$$

where T is a suitable bundle isomorphism. Bojarski's theorem constitutes the special case of $T = G$. We prove the formula

$$\operatorname{index}_{\mathrm{LCP}}(A, T) = \operatorname{index}_{\mathrm{LCP}}(A, G) + \operatorname{index} \mathcal{P}_+ T \mathcal{P}_+,$$

under a certain consistency assumption on the isomorphism T, where $\operatorname{index}_{\mathrm{LCP}}(A, T)$ denotes the index of the general linear conjugation problem (0.2).

Chapter 25 investigates the underlying geometrical problem of surgery — *cutting and pasting* — of Dirac operators, that is, we cut M along Y and glue it differently by using a diffeomorphism and a corresponding automorphism of the bundle of Clifford modules. Assuming consistency of the data on Y, we obtain a new Dirac operator and come up with a formula for the difference of the indices.

The final Chapter 26 illustrates those results, using a simple, but multifaceted example. We investigate the case of $M = S^2$, which is a perfect illustration of all the phenomena discussed in this book. It might be a good idea to start reading this book with a quick look at this chapter.

Many colleagues and students have helped to improve this book, and we heartily thank them all. Our emphasis on Dirac operators and the related geometrical concepts owes much to remarks by Peter Gilkey. Special thanks also to Ron Douglas, who suggested writing this book, to Ulrich Bunke, Jerry Kaminker, Slawomir Klimek, and Anders Madsen for continued and inspiring discussions of many details, to Ingrid Jensen for drawing the figures, to Birthe Wolter for typing, and to Krystyna and Władek Uscinowicz for providing us with housing in Szczecin for our monthly meetings some years ago when we began our joint work.

List of Notations

erfc, complementary error function, 227
E^{Φ}, bundle after cutting and pasting 277
$\widehat{\mathrm{Ell}}$, $\widehat{\mathrm{Ell}}_{\sigma}$, elliptic self-adjoint operators of positive order, 142, 153
$\mathcal{E}\ell\ell^{\times}$, special group of invertible elliptic operators, 112
$\mathcal{E}\ell\ell^{\times}{}_{P_+}$, subgroup of P_+-invariant invertible elliptic operators, 116
$\mathcal{E}$, continuous extension operator of codimension one, 72
$\mathcal{E}(t,x,y)$, heat kernel, 59
$\mathcal{E}$, $\mathcal{E}_*$, heat kernels on manifolds with boundary, 214, 237
$\mathcal{E}_{\mathrm{c}}$, $\mathcal{E}_{\mathrm{c}\,*}$, heat kernels on the semi-infinite cylinder, 223
$\mathcal{E}_{\mathrm{d}}$, $\mathcal{E}_{\mathrm{d}\,*}$, heat kernels on the closed double, 231
η_B, eta-function, 229

$\widehat{F}^{\infty}$, deformation retract of $\widehat{\mathcal{F}}_*$, 140
$\mathrm{Fred}^2(H)$, $\mathrm{Fred}^2{}_{H_+}(H)$, spaces of Fredholm pairs of subspaces, 265
$\mathcal{F}$, space of Fredholm operators, 127
$\widehat{\mathcal{F}}$, $\widehat{\mathcal{F}}_{\pm}$, $\widehat{\mathcal{F}}_*$, spaces of self-adjoint Fredholm operators, 128 ff
φ, Thom isomorphism, 150

g, genus of Riemann surface, 249
G, Green's form, 24
$\mathcal{G}$, group of units of Calkin algebra, 129
$\widehat{\mathcal{G}}$, self-adjoint units of Calkin algebra, 129
$\widehat{G}$, $\widehat{G}_{\pm}$, $\widehat{G}_*$, unitary retracts, 132
GL_c, GL^{∞}, special groups of invertible operators, 113
$GL_{\mathcal{P}_+}$, $\mathcal{P}_+$-invariant bounded invertible operators, 266
Gr_{p_+}, Grassmannian of pseudo-differential projections, 111
Gr_{P_+}, connected component of the Grassmannian which contains P_+, 116
$\mathrm{Gr}(H)$, $\overline{\mathrm{Gr}(\mathcal{P}_+)}$, full Grassmannian, 133
γ, γ_{ε}, γ_0, standard trace operators, 68, 95, 180
Γ, Gamma-function, 229
Γ_5, Gamma-five matrix, 8
$\Gamma^{\mu}_{\nu\sigma}$, Christoffel symbols, 15

h, $h_{\pm}$, kernel dimensions, 242, 245
H^k, higher Hopf bundle, 283
H^*, $H^{\mathrm{even}}_{\mathrm{comp}}$, cohomology functors, 272
$H^s(\mathbf{R}^n)$, $H^s(X)$, $H^s(Y)$, H^s_{loc}, Sobolev spaces, 67, 216
$H_{\pm}(A)$, $H_{\pm}(A,s)$, spaces of Cauchy data, 76
$\mathcal{H}_{\pm}$, spaces of limiting values, 244

index$(\cdot,\cdot)$, index of Fredholm pair, 263
$\mathbf{i}(P_2,P_1)$, virtual codimension, 119
$\mathcal{I}_R$, orthogonal projection onto the range of R, 166

J, reflection operator, 72

str, symmetric trace, 59
$S^{\pm}$, spinors of even/odd chirality, 40
$\widetilde{S^{\pm}}$, bundles of extended spinors, 53
$\not{S}_{\mathbf{C}}$, associated bundle over spin manifold, 37
σ_1, σ_2, σ_3, Pauli matrices, 33
σ_B, principal symbol of operator B, 19
$[\sigma_B]$, symbol class, 147
$[\sigma'_B]$, stable symbol class, 148

t-index, topological index, 150
tr, trace of operator, 60
T^{Φ}, correcting operator after cutting and pasting, 279
τ, Todd class, 150

$\mathcal{U}$, $\mathcal{U}_{\mathcal{P}_+}$, spaces of unitary operators, 133

$\widetilde{X}$, closed double, 51
X_∞, non-compact Riemannian elongation, 243
$\chi(M)$, Euler characteristic, 280

PART I

Clifford Algebras and Dirac Operators

1. Clifford Algebras and Clifford Modules

We study bundles over a point, recalling the definition of the Clifford algebra $C\ell(V, q)$ of a real vector space V of dimension m equipped with a positive definite inner product q; the $\mathbb{Z}_2$-grading of Clifford algebras is shown, followed by an introduction of complex representations of Clifford algebras and the concept of complex $C\ell(V, q)$-modules and of Clifford multiplication; the isomorphism classes of irreducible $C\ell(V, q)$-modules are studied.

We begin by fixing the notation: Let V be a real vector space of finite dimension m equipped with a symmetric positive definite bilinear form q. Think of V as the tangent space of a Riemannian manifold in a given point endowed with the corresponding Euclidean metric q. For most of the following results the positivity is not needed and we may take $V = \mathbf{R}^4$, e.g. the Minkowski space quadratic form instead of the Euclidean. Denote the *Clifford algebra* of (V, q) by $\mathrm{C}\ell(V, q)$ or in short, $\mathrm{C}\ell(V)$. It is the associative algebra with unit generated by the elements of V (and the identity 1) under the relation

$$v \cdot w + w \cdot v = -2q(v, w) \cdot 1, \quad v,\ w \in V. \tag{1.1}$$

Assume that $\mathrm{C}\ell(V, q)$ is not generated by any proper subspace of V.

Remarks 1.1. (a) We have the following reformulation of our definition. Let

$$\mathcal{T}(V) := \mathbf{R} \oplus V \oplus (V \otimes V) \oplus \ldots \oplus \otimes^k V \oplus \ldots \tag{1.2}$$

denote the complete tensor algebra of V, and define $\mathcal{I}_q(V)$ to be the two-sided ideal in $\mathcal{T}(V)$ generated by all elements of the form $v \otimes v + q(v, v)1$, where v runs through V (or — equivalently — generated by all elements of the form $v \otimes w + w \otimes v + 2q(v, w)$, where v and w run through V). Then we have

$$\mathrm{C}\ell(V, q) = \mathcal{T}(V)/\mathcal{I}_q(V). \tag{1.3}$$

(b) The canonical map $V \to \mathrm{C}\ell(V, q)$, given by the composition

$$V \hookrightarrow \mathcal{T}(V) \to \mathrm{C}\ell(V, q),$$

is an injection. Moreover, if $\{v_1, \ldots, v_m\}$ is a basis for V, then 1 together with the products $\{v_1^{i_1} \cdot \ldots \cdot v_m^{i_m}\}$ (with $i_k = 0$ or $= 1$) form a basis for $C\ell(V,q)$; hence in particular $\dim C\ell(V,q) = 2^m$. For further relations between the Clifford algebra $C\ell(V,q)$ and the exterior algebra ΛV of V, see Example 1.4 below.

(c) The Clifford algebra $C\ell(V,q)$ is up to isomorphisms characterized by the following properties:

(1) $C\ell(V,q)$ is a real algebra with unit;
(2) V is contained in $C\ell(V,q)$ and generates $C\ell(V,q)$;
(3) $v_i \cdot v_j + v_j \cdot v_i = -2\delta_{ij}$ for a chosen q-orthonormal basis for V;
(4) $C\ell(V,q)$ is not generated by any proper subspace of V.

(d) An equivalent characterization of $C\ell(V,q)$ is obtained by the following *universal property*: $C\ell(V,q)$ is the unique associative (real) algebra with the property that any linear map f of V into an associative (real) algebra $\mathcal{A}$ with the unit extends uniquely to an algebra homomorphism $\tilde{f} : C\ell(V,q) \to \mathcal{A}$, if f satisfies

(1.4) $$f(v) \cdot f(v) = -q(v,v)1 \qquad \text{for all } v \in V.$$

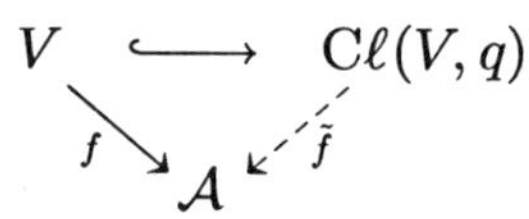

Fig. 1.1 The universal property of $C\ell(V, q)$

(e) Following (b) we see that $C\ell(V,q)$ decomposes into the direct sum

(1.5) $$C\ell(V,q) = C\ell^+(V,q) \oplus C\ell^-(V,q),$$

where $C\ell^+(V,q)$ and $C\ell^-(V,q)$ are the subspaces spanned by those elements of the basis which have an even, resp. odd number of exponents i_k equal to 1. The $\mathbf{Z}_2$-grading can also be obtained from (d), namely, by extending the linear map $\alpha(v) := -v$ from V into $C\ell(V,q)$ to an automorphism

$$\tilde{\alpha} : C\ell(V,q) \to C\ell(V,q)$$

which in fact is an involution ($\tilde{\alpha}^2 = Id$). Then $C\ell^{\pm}(V,q)$ are the ± 1 eigenspaces of $\tilde{\alpha}$. Clearly $C\ell^+(V,q)$ is a subalgebra, and we have that

$$C\ell^- \cdot C\ell^- \subset C\ell^+ \quad \text{and} \quad C\ell^+ \cdot C\ell^- = C\ell^- \cdot C\ell^+ \subset C\ell^-.$$

(f) Let V' be a subspace of V of codimension 1, let $q' := q|_{V' \times V'}$ denote its inner product, and choose $v_n \perp V'$. Then the correspondence

$$v^+ + v^- \mapsto v^+ + v^- \cdot v_n, \qquad v^{\pm} \in C\ell^{\pm}(V',q')$$

is a Clifford algebra isomorphism of $C\ell(V',q')$ with $C\ell^+(V,q)$.

Recall the concept of Clifford modules and Clifford multiplication.

Definition 1.2. (a) A $\mathbf{C}$-representation of the Clifford algebra $C\ell(V,q)$ is an $\mathbf{R}$-algebra homomorphism

$$\mathbf{c} : C\ell(V,q) \to \mathrm{Hom}_{\mathbf{C}}(W,W) \tag{1.6}$$

into the algebra of $\mathbf{C}$-linear transformations of a finite-dimensional complex vector space W. Any vector space W together with a $\mathbf{C}$-representation of $C\ell(V,q)$ in $\mathrm{Hom}_{\mathbf{C}}(W,W)$ is called a (complex) $C\ell(V,q)$-*module* and we shall use the terms $\mathbf{C}$-representation of $C\ell(V,q)$ and $C\ell(V,q)$-*module* synonymously. We shall refer to the corresponding action

$$\varphi \cdot w := \mathbf{c}(\varphi)(w) \quad \text{for } \varphi \in C\ell(V,q) \text{ and } w \in W \tag{1.7}$$

as *Clifford multiplication* of w by φ.
(b) A $\mathbf{C}$-representation $\mathbf{c} : C\ell(V,q) \to \mathrm{Hom}_{\mathbf{C}}(W,W)$ is called *reducible*, if the vector space W can be written as a non-trivial direct sum

$$W = W_1 \oplus W_2$$

such that $\mathbf{c}(\varphi)(W_j) \subset W_j$ for $j = 1, 2$ and for all $\varphi \in C\ell(V,q)$. In this case one can write $\mathbf{c} = \mathbf{c}_1 \oplus \mathbf{c}_2$, where $\mathbf{c}_j(\varphi) := \mathbf{c}(\varphi)|_{W_j}$ for $j = 1, 2$. A representation will be called *irreducible*, if there are no proper invariant subspaces.
(c) Two representations:

$$\mathbf{c}_j : C\ell(V,q) \to \mathrm{Hom}_{\mathbf{C}}(W_j,W_j), \qquad j = 1, 2\,,$$

are said to be *equivalent*, if there exists a $\mathbf{C}$-linear isomorphism $F : W_1 \to W_2$ such that $F\,\mathbf{c}_1(\varphi)F^{-1} = \mathbf{c}_2(\varphi)$ for all $\varphi \in C\ell(V,q)$.

The following result about the isomorphism classes of irreducible $C\ell(V,q)$-modules is usually derived from general properties of simple (matrix) algebras. Here is a direct proof.

Lemma 1.3. *Let V be a real vector space of dimension m equipped with a positive definite inner product. Then every (complex) $C\ell(V)$-module decomposes uniquely as a sum of irreducibles. More precisely:*
(a) If $m = 2n$, then there is (up to isomorphisms) exactly one irreducible $C\ell(V)$-module Δ and we have $\dim \Delta = 2^n$.

(b) If $m = 2n+1$, then there are (up to isomorphisms) exactly two irreducible $C\ell(V)$-modules $\Delta_\pm$ and we have $\dim \Delta_\pm = 2^n$. Moreover, $\mathbf{c}(\tau) = \pm 1$ on $\Delta_\pm$ for any normalized orientation τ of V.

Note. Any orthonormal basis $\{v_\nu\}$ of V defines a *normalized orientation*

$$\tau = i^{[(m+1)/2]}\, v_1 \cdot \ldots \cdot v_m \tag{1.8}$$

of V. Here [-] denotes the greatest integer function; the normalization is chosen so that $\tau \cdot \tau = 1$; moreover, τ commutes with V in $C\ell(V)$ if m is odd, and anti-commutes if m is even.

Proof of Lemma 1.3. Begin with the case $m = 2n$ and fix an orthonormal basis $\{v_1, \ldots, v_{2n}\}$ for V. Consider a (complex) $C\ell(V)$-module W and the related representation $\mathbf{c} : C\ell(V) \to \mathrm{Hom}_{\mathbf{C}}(W, W)$. Let

$$\mathbf{c}_\mu := \mathbf{c}(v_\mu) \quad \text{for } 1 \le \mu \le m$$

and

$$\alpha_\nu := i\, \mathbf{c}_{2\nu-1}\, \mathbf{c}_{2\nu} \quad \text{for } 1 \le \nu \le n\,. \tag{1.9}$$

One sees that the $\{\alpha_\nu\}$ are a commuting family of involutions in $\mathrm{Hom}(W, W)$ inducing a decomposition of W into complementary ± 1 eigenspaces of α_ν

$$W = W_\nu^+ \oplus W_\nu^-\,. \tag{1.10}$$

If $\varepsilon = (\varepsilon_1, \ldots, \varepsilon_n)$ is a choice of signs, we have

$$W = \oplus_\varepsilon W(\varepsilon)\,, \tag{1.11}$$

where

$$\begin{aligned} W(\varepsilon) &:= \{w \in W \mid \alpha_\nu w = \varepsilon_\nu w \text{ for } 1 \le \nu \le n\} \\ &= W_1^{\varepsilon_1} \cap \cdots \cap W_n^{\varepsilon_n} \end{aligned} \tag{1.12}$$

are the simultaneous eigenspaces and the sum in (1.11) is taken over all 2^n sign combinations.

Let $\varepsilon_0 := (+, \ldots, +)$. Then multiplication by $\mathbf{c}(v(\varepsilon))$ defines isomorphisms between $W(\varepsilon_0)$ and $W(\varepsilon)$, where

$$v(\varepsilon) := \prod_{\{\kappa \mid \varepsilon_\kappa = -\}} v_{2\kappa}\,.$$

The reason is that $\mathbf{c}(v(\varepsilon))$ commutes with α_ν if $\varepsilon_\nu = +$, and anti-commutes with α_ν if $\varepsilon_\nu = -$. Consequently we get from (1.11)

$$\dim W = 2^n \cdot \dim W(\varepsilon_0)\,. \tag{1.13}$$

Let $\{\varphi_1, \dots, \varphi_a\}$ be a basis for $W(\varepsilon_0)$. For any $\sigma \in \{1, \dots, a\}$ we shall define an irreducible $C\ell(V)$-module W_σ such that $W = \oplus_\sigma W_\sigma$ and all W_σ are mutually isomorphic. Set

$$W_\sigma := \mathcal{L}_{\mathbf{C}}\{\mathbf{c}(v(\varepsilon))\varphi_\sigma\}_\varepsilon\,,$$

where $\mathcal{L}_{\mathbf{C}}$ denotes the linear span. This space is invariant under $\{\mathbf{c}(v(\varepsilon)), \alpha_\nu\}_{\varepsilon,\nu}$. Since these elements generate $C\ell(V)$ as an algebra, the complex vector space W_σ is a $C\ell(V)$-module and is actually irreducible by construction. This yields the decomposition $W = \oplus W_\sigma$.

Furthermore, the map

$$\mathbf{c}(v(\varepsilon))\varphi_1 \mapsto \mathbf{c}(v(\varepsilon))\varphi_\sigma$$

extends to a $C\ell(V)$-module isomorphism between W_1 and W_σ for any σ; so $\Delta := W_1$ and $\dim \Delta = 2^n$ because of (1.13). This ends the proof of (a).

Now let $m = 2n+1$. Recall that the normalized orientation τ is central in $C\ell(V)$ and that $\tau^2 = 1$; hence $W = W^+ \oplus W^-$ decomposes into the ± 1 eigenspaces of $\mathbf{c}(\tau)$. Then $W^\pm$ are $C\ell(V')$-modules, where $V' := \mathcal{L}_{\mathbf{C}}\{v_1, \dots, v_{m-1}\}$. Now use (a) to decompose $W^\pm$ into the direct sum of a^+, resp. a^- irreducible $C\ell(V')$-modules Δ'. Since $\{V', \tau\}$ generates $C\ell(V)$ as an algebra, we obtain

$$W = (\Delta' \oplus \cdots \oplus \Delta') \oplus (\Delta' \oplus \cdots \oplus \Delta')$$

as a $C\ell(V)$-module, where the first sum consists of a^+ copies of Δ' (where the representation is provided by $\mathbf{c}$) and the second of a^- copies (where the representation is provided by $-\mathbf{c}$). Then (b) follows.

Notice that, for m even, there is only one irreducible $C\ell(V)$-module and that $(\Delta, \mathbf{c})$ and $(\Delta, -\mathbf{c})$ are isomorphic. Since $\mathbf{c}(\tau)$ anti-commutes with $\mathbf{c}(V)$, $\mathbf{c}(\tau)$ provides the isomorphism. If m is odd, then the two irreducible $C\ell(V)$-modules $\Delta_+ := (\Delta', \mathbf{c})$ and $\Delta_- := (\Delta', -\mathbf{c})$ are not isomorphic. □

Examples 1.4. (a) Our first example of a Clifford module is the (complexified) Clifford algebra $C\ell(V)$ itself.
(b) As a second example, recognize also that the exterior algebra ΛV (omit the complexification) is a $C\ell(V)$-module. Begin by specifying how V acts on ΛV. Let $\{v_1, \ldots, v_m\}$ be an orthonormal basis for V and let $\{v_1^*, \ldots, v_m^*\}$ denote its dual basis, i.e. $v_\mu^*(v_\nu) = \delta_{\mu\nu}$. Consider the corresponding elements of $\mathrm{Hom}(\Lambda V, \Lambda V)$, namely,

$$\begin{aligned} \mathrm{ext}_\mu &= \mathrm{ext}(v_\mu) := && \text{exterior multiplication by the vector } v_\mu \\ \mathrm{int}_\mu &= \mathrm{int}(v_\mu^*) := && \text{interior multiplication (contraction)} \\ & && \text{by the covector } v_\mu^*. \end{aligned}$$

Then $\mathrm{ext}_\mu \mathrm{int}_\nu + \mathrm{int}_\nu \mathrm{ext}_\mu = \delta_{\mu\nu}$, which implies that $\mathbf{c}^\mu := \mathrm{ext}_\mu - \mathrm{int}_\mu$ satisfies the defining relation (1.1) for the generators of $C\ell(V)$. Hence the action $\alpha \mapsto \mathbf{c}^\mu(\alpha)$, for given v_μ, defines a linear map $\mathbf{c}$ of V into $\mathrm{Hom}(\Lambda V, \Lambda V)$ which satisfies the Clifford condition (1.4) and uniquely extends across the whole of $C\ell(V)$ by the universal property (see Remark 1.1.d above).
(c) The two representations

$$\mathrm{id} : C\ell(V) \to \mathrm{Hom}(C\ell(V), C\ell(V)) \quad \text{and} \quad \mathbf{c} : C\ell(V) \to \mathrm{Hom}(\Lambda V, \Lambda V)$$

are equivalent under the isomorphism

$$\begin{aligned} F : C\ell(V) &\longrightarrow \Lambda V \\ \mathbf{c}^I \cdot 1 &\mapsto \mathrm{ext}_I(1), \end{aligned}$$

where I denotes a multiindex. Notice that F maps the unit $1 \in C\ell(V)$ onto the identity $1 \in \Lambda^0 V$ and that

$$\Gamma_5^2 = 1$$

when $\Gamma_5 := i^{[(m+1)/2]}\, \mathbf{c}^1 \cdots \mathbf{c}^m$ denotes the *Gamma-five matrix*, as in physics terminology. Considered as an element of (the complexified) Clifford algebra $C\ell(V)$, it is equal to the normalized orientation introduced in (1.8).
(d) If $m = 2n$ is even, the representation of $C\ell(V)$ in $\mathrm{Hom}(\Lambda V, \Lambda V)$ provides an alternative construction of the irreducible $C\ell(V)$-module Δ introduced in Lemma 1.3. Let

$$\wp^\mu := \mathbf{c}^\mu + i\, \mathbf{c}^{n+\mu} \qquad \text{and} \qquad \overline{\wp}^\mu := \mathbf{c}^\mu - i\, \mathbf{c}^{n+\mu}$$

for $\mu = 1, \ldots, n$ and define $\Delta := \mathcal{L}_{\mathbf{C}}\{\wp^I \wp_5\}_I$ with

$$\wp_5 := \overline{\wp}^1 \overline{\wp}^2 \cdots \overline{\wp}^n \,.$$

Observe that

$$\wp^\mu \wp^\nu + \wp^\nu \wp^\mu = 0, \;\; \overline{\wp}^\mu \, \overline{\wp}^\nu + \overline{\wp}^\nu \overline{\wp}^\mu = 0, \;\text{ and }\; \wp^\mu \, \overline{\wp}^\nu + \overline{\wp}^\nu \wp^\mu = -4\delta_{\mu\nu} \,.$$

This shows that $\{\wp^I \overline{\wp}^J\}_{I,J}$ form a basis of the (complexified) $C\ell(V)$ and that $\Delta = \mathcal{L}_{\mathbf{C}}\{\wp^I \wp_5\}$. Hence $\Delta \cong \Lambda_{\mathbf{C}}(\wp^1, \ldots, \wp^n)$ as vector space, so $\dim_{\mathbf{C}} \Delta = 2^n$, and $C\ell(V) \xrightarrow{\cong} \mathrm{Hom}(\Delta, \Delta)$, i.e. $C\ell(V)$ is simple and Δ irreducible.

2. Clifford Bundles and Compatible Connections

We show how the natural operations for vector spaces with quadratic forms carry over to vector bundles with metrics. For a Riemannian manifold X (with or without boundary), we obtain the Clifford bundle $C\ell(X) := C\ell(TX,g)$. We show that there exists a connection D for any bundle S of complex left modules over $C\ell(X)$ which is compatible with Clifford multiplication and extends the Riemannian connection on X to S.

There is a well-known principle by which the natural operations on vector spaces carry over canonically to vector bundles. In the same way the natural operations for vector spaces with quadratic forms carry over to vector bundles with metrics.

Definition 2.1. Let X be a compact Riemannian manifold (with or without a smooth boundary).
(a) The *Clifford bundle* $\mathrm{C}\ell(X)$ of X is a bundle over X, where the fibre over a point $x \in X$ consists of the Clifford algebra $\mathrm{C}\ell(TX_x, g_x)$; here g_x denotes the Riemannian metric for the real tangent vector space TX_x. The fibrewise multiplication in $\mathrm{C}\ell(X)$ gives an algebra structure to the space of sections of $\mathrm{C}\ell(X)$; the decomposition

$$\mathrm{C}\ell(X) = \mathrm{C}\ell^+(X) \oplus \mathrm{C}\ell^-(X) \tag{2.1}$$

is inherited from the plus-minus decomposition of the algebras.
(b) A *representation* of $\mathrm{C}\ell(X)$ (or synonymously a $\mathrm{C}\ell(X)$-*module* or a *left Clifford multiplication* of module sections by sections in the Clifford bundle) is a fibre preserving algebra morphism

$$\mathbf{c} : \mathrm{C}\ell(X) \to \mathrm{Hom}(S, S)\,,$$

where S is a smooth complex vector bundle over X.

Without loss of generality one can always assume that S is endowed with a Hermitian structure $\langle \ldots, \ldots \rangle$ which makes Clifford multiplication by tangent vectors skew-adjoint. This follows from the next lemma.

Lemma 2.2. *Let X be a compact Riemannian manifold (with or without a smooth boundary) of dimension m, S a complex vector bundle over X, and $\mathbf{c} : TX \to \operatorname{Hom}(S, S)$ a bundle map with*

$$\mathbf{c}(v)^2 = -|v|^2 \operatorname{Id} : S_x \to S_x$$

for $v \in TX_x$. Then there exists a Hermitian Clifford-skew-adjoint metric on S so that at each $x \in X$ we have

$$\langle \mathbf{c}(v)s, s' \rangle = - \langle s, \mathbf{c}(v)s' \rangle$$

for all s, $s' \in S_x$ and all vectors $v \in TX_x$, i.e. $\mathbf{c}(v)^* = -\mathbf{c}(v)$.

Note. Lemma 2.2 provides the nice property that Clifford multiplication in a $C\ell(X)$-module by unit vectors of TX is unitary, i.e. at each $x \in X$

$$\langle v \cdot s, v \cdot s' \rangle = \left\langle s, -v^2 s' \right\rangle = \langle s, s' \rangle \tag{2.2}$$

for all s, $s' \in S_x$ and all *unit* vectors $v \in (TX)_x$.

Proof. One proves the lemma in two steps: first locally, then globally using a partition of unity. Since step 2 is canonical, we shall concentrate on step 1. Assume that the dimension of X is even, $m = 2n$. (Then the case $m = 2n+1$ is also solved by the structural splitting, as in the second part of the proof of Lemma 1.3 above.) For fixed $x \in X$ choose an orthonormal basis $v_1, \ldots, v_{2n}$ of TX_x. With the arguments and notations of the proof of Lemma 1.3 we obtain a decomposition $S_x = \oplus_\varepsilon S_x(\varepsilon)$ of the fibre S_x into 2^n 'eigenspaces' $S_x(\varepsilon)$ and canonical isomorphisms

$$\mathbf{c}(v(\varepsilon)) : S_x(\varepsilon_0) \xrightarrow{\cong} S_x(\varepsilon),$$

where $\varepsilon = (\varepsilon_1, \ldots, \varepsilon_n)$ is a choice of signs, $\varepsilon_0 = (+, \ldots, +)$, and $v(\varepsilon) \in C\ell(TX_x)$ the product of those basis vectors $v_{2\kappa}$ for which the sign $\varepsilon_\kappa = -$.

Choose any Hermitian metric on $S_x(\varepsilon_0)$ and expand the metric on all $S_x(\varepsilon)$ by the isomorphism $\mathbf{c}(v(\varepsilon))$. This gives a Hermitian metric on S_x satisfying the following two conditions:

(1) $S_x(\varepsilon) \perp S_x(\varepsilon')$ for all sign choices $\varepsilon \neq \varepsilon'$;
(2) $S_x(\varepsilon) \cong S_x(\varepsilon_0)$ is a unitary isomorphism.

We write $\mathbf{c}_j := \mathbf{c}(v_j)$ and check that $\mathbf{c}_j^* = -\mathbf{c}_j$. Indeed, let $s, s' \in S_x$,

$$s = \sum_{\varepsilon} s_\varepsilon \quad , \qquad\qquad s' = \sum_{\varepsilon} s'_\varepsilon \,.$$

Then one has

$$\begin{aligned}
\langle \mathbf{c}_j\, s, s' \rangle_{S_x} &= \langle \sum_{\varepsilon} \mathbf{c}_j\, s_\varepsilon, \sum_{\varepsilon} s'_\varepsilon \rangle_{S_x} = \sum_{\varepsilon} \langle \mathbf{c}_j\, s_\varepsilon, s'_{\varepsilon^j} \rangle_{S_x(\varepsilon^j)} \\
&= \sum_{\varepsilon} \langle \mathbf{c}(v(\varepsilon^j))^{-1}\, \mathbf{c}_j\, s_\varepsilon, \mathbf{c}(v(\varepsilon^j))^{-1} s'_{\varepsilon^j} \rangle_{S_x(\varepsilon_0)} \\
&= \sum_{\varepsilon} \langle s_\varepsilon, -\mathbf{c}_j\, s'_{\varepsilon^j} \rangle_{S_x(\varepsilon)} \\
&= -\langle s, \mathbf{c}_j\, s' \rangle_{S_x} \,.
\end{aligned}$$

Here

$$\mathbf{c}_j : S_x(\varepsilon) \xrightarrow{\cong} S_x(\varepsilon^j)$$

for all $j = 1, \ldots, 2n$ and sign choices ε, where

$$\varepsilon^j := (\varepsilon_1, \ldots, \hat{\varepsilon}_{[(j+1)/2]}, \ldots, \varepsilon_n)$$

with

$$\hat{\varepsilon}_{[(j+1)/2]} := \begin{cases} + & \text{if } \varepsilon_{[(j+1)/2]} = - \\ - & \text{else}\,. \end{cases}$$

It is easiest to check the transformations for the component s_{ε_0} of s in $S_x(\varepsilon_0)$ and the corresponding component $s'_{\varepsilon_0^j}$ of s' in $S_x(\varepsilon_0^j)$, since here by definition

$$-\mathbf{c}_j : S_x(\varepsilon_0^j) \to S_x(\varepsilon_0)$$

is the inverse of $\mathbf{c}_j$ - both for j even and odd. From $\mathbf{c}_j^* = -\mathbf{c}_j$ follows $\mathbf{c}^* = -\mathbf{c}$. □

The non-trivial aspects of Clifford modules are related to the choice of a suitable connection which provides the geometry of the Dirac operator.

Definition 2.3. Let X be a compact Riemannian manifold (with or without a smooth boundary).
(a) Let k denote the field of real or complex numbers and let E be a smooth k-vector bundle of fibre dimension N over X. A *connection*

(*'covariant derivative'*) on E is a k-linear first order partial differential operator

$$D : C^\infty(X;E) \longrightarrow C^\infty(X;T^*X \otimes E)$$
$$s \mapsto Ds$$

such that

$$(2.3)\quad D(fs) = df \otimes s + f Ds \qquad \text{for } f \in C^\infty(X) \text{ and } s \in C^\infty(X;E)\,.$$

Given a smooth vector field $v : X \to TX$ on X, we thus obtain a map

$$D_v : C^\infty(X;E) \to C^\infty(X;E)$$

called the covariant derivative with respect to v. At a given point $x \in X$, $(D_v s)_x := \langle v, Ds\rangle_x$ only depends on v_x and on the values of s in a neighbourhood of x.
(b) A connection D on a Riemannian (or Hermitian) bundle E will be called *Leibnizian* (sometimes also *Riemannian*), if it satisfies the inner product rule

$$(2.4)\qquad v \cdot \langle s, s'\rangle \;=\; \langle D_v s, s'\rangle \;+\; \langle s, D_v s'\rangle$$

for all $v \in TX$ and s, $s' \in C^\infty(X;E)$, where $\langle \cdot, \cdot \rangle$ denotes the inner product in the fibres of E and '$v\cdot$' the derivation of functions in v direction.
(c) A connection D on a $\mathrm{C}\ell(X)$-module S is defined to be *compatible* with the $\mathrm{C}\ell(X)$-module structure, if it is Leibnizian with respect to a suitable Hermitian structure on S, and $D\,\mathbf{c} = 0$. Here $\mathbf{c}$ is considered to be the composition

$$C^\infty(X;TX) \hookrightarrow C^\infty(X;\mathrm{C}\ell(X)) \to \mathrm{Hom}(S,S) = C^\infty(X;S \otimes S^*)$$

and D is lifted to C^∞-sections in the bundle $TX^* \otimes (S \otimes S^*)$ by

$$(D\,\mathbf{c})(v)(s) := D(\mathbf{c}(v)s) - \mathbf{c}(D^g v)s - \mathbf{c}(v)(Ds)\,,$$

where $v \in C^\infty(X;TX)$, $s \in C^\infty(X;S)$, and D^g denotes the Levi-Civita connection on TX for the Riemannian metric g. (When we extend the Levi-Civita connection on the whole Clifford bundle $\mathrm{C}\ell(X)$ in the standard way, $D\,\mathbf{c}$ is defined in the same way for $v \in C^\infty(X;\mathrm{C}\ell(X))$.) Hence our requirement $D\,\mathbf{c} = 0$ means that the connection D on S is a *module derivation* which extends the Levi-Civita connection D^g on X to S:

$$D(\mathbf{c}(v)s) = \mathbf{c}(D^g v)s + \mathbf{c}(v)(Ds)\,.$$

It is quite natural to demand a connection to be compatible, since $D\,\mathbf{c} = 0$ means that the Clifford multiplication acts *parallel* in all fibres S_x. The most elementary example of a compatible connection is the natural covariant differentiation in the exterior algebra

$$\nabla : C^\infty(X; \Lambda(T^*X)) \longrightarrow C^\infty(X; T^*X \otimes \Lambda(T^*X))$$

of a Riemannian manifold X which by definition is compatible with the Clifford multiplication introduced in Example 1.4 and which extends the Levi-Civita connection of X. The next lemma is the first ingredient in establishing the existence of a compatible connection D for any $\mathrm{C}\ell(X)$-module S.

Lemma 2.4. *Let X be a Riemannian manifold of dimension m; let $\{v_\nu\}$ be a local orthonormal frame for TX over a contractible open set $U \subset X$, and let S be a (complex) $\mathrm{C}\ell(X)$-module. Then we can choose a local frame for S over U so that the matrices $\mathbf{c}(v_\nu)$ are constant.*

Proof. The orthonormal framing of $TX|_U$ provides the trivializations

$$TX|_U = \mathbf{R}^m \times U \quad \text{and} \quad \mathrm{C}\ell(X)|_U = \mathrm{C}\ell(\mathbf{R}^m) \times U\,.$$

Let the eigenspaces $S(\varepsilon)$ be defined pointwise as in (1.12) above in the proof of Lemma 1.3. Because of (1.13), these spaces have constant fibre dimension $a := 2^{-[m/2]} \cdot \dim S$ and define smooth vector subbundles of S. Since U is contractible, the bundle $S(\varepsilon_0)$ (with the choices of signs named as in the proof of Lemma 1.3) is trivial over U; thus we can find a smooth frame $\{\varphi_1, \ldots, \varphi_a\}$ for $S(\varepsilon_0)$ over U. This permits us to repeat the construction of Lemma 1.3 to show:

$$(2.5)\quad D|_U = \begin{cases} (\overbrace{\Delta \oplus \cdots \oplus \Delta}^{a \text{ times}}) \times U & \text{if } m \text{ is even} \\ (\overbrace{\Delta_+ \oplus \cdots \oplus \Delta_+}^{a_+ \text{ times}} \oplus \overbrace{\Delta_- \oplus \cdots \oplus \Delta_-}^{a_- \text{ times}}) \times U & \text{if } m \text{ is odd.} \end{cases}$$

This shows that the Clifford module structure locally is a product. □

Now we shall prove the existence of compatible connections. Let S be a $\mathrm{C}\ell(X)$-module with Clifford multiplication $\mathbf{c} : \mathrm{C}\ell(X) \to$

for all $\varphi \in C^\infty(X; C\ell(X))$ and all $s \in C^\infty(X; S)$, where $\cdot$ denotes Clifford multiplication and D^g the canonical extension of the Levi-Civita connection on $C\ell(X)$. Moreover, to any compatible connection on S, which is built on local spin connections, there belongs a Hermitian metric on S for which Clifford multiplication becomes skew-adjoint and hence satisfies condition (2.2) above.

Remark 2.6. The delicacy of the preceding proposition is due to three circumstances:

(1) The decomposition of a given $C\ell(X)$-module S in a direct sum of irreducible $C\ell(X)$-modules as in Lemma 1.3 is only locally possible and in general not globally possible. Actually, these irreducible bundles exist globally, if and only if X admits a spin structure. But there are famous topological obstructions (the non-vanishing of the first two Stiefel-Whitney classes) which prevent some X from admitting a spin structure. In general the spin connections D^S, defined above in (2.8) and (2.9), are not globally defined either.

(2) Nevertheless Proposition 2.5.b ascertains the global existence of a compatible connection for any $C\ell(X)$-module S, whether X is spin or not.

(3) Locally this compatible connection has a standard form as the following proposition shows. The global uniqueness is lost by patching the compatible local connections together.

Proposition 2.7. *Let D be a compatible connection on S and let $U \subset X$ be a contractible open set. Choose a local orthonormal frame for $TX|_U$ and use Lemma 2.4 to decompose*

$$S|_U = \begin{cases} \Delta(U) \otimes V_0 & \textit{if } m \textit{ is even} \\ \Delta_+(U) \otimes V_+ \oplus \Delta_-(U) \otimes V_- & \textit{if } m \textit{ is odd} \end{cases}$$

with suitable trivial bundles V_0, V_+, V_-. Then there exist unique connections D^0, D^+, D^- on these bundles so that

$$D_U = \begin{cases} D^S \otimes 1 \oplus 1 \otimes D^0 & \textit{if } m \textit{ is even} \\ \{D^{+,S} \otimes 1 \oplus 1 \otimes D^+\} \oplus \{D^{-,S} \otimes 1 \oplus 1 \otimes D^-\} & \textit{if } m \textit{ is odd.} \end{cases}$$

Proof. We suppose m even; the case m odd is analogous. Let ω be the connection 1-form of the given connection D and let $\omega^S \otimes 1$ be the connection 1-form of the spin connection D^S defined in (2.8), (2.9). Since both D and D^S are compatible, $\omega - (w^S \otimes 1)$ commutes with the Clifford module structure, hence

$$\omega - (\omega^S \otimes 1) = 1 \otimes \theta \quad \text{for some } \theta \in \operatorname{End} V_0 \,.$$

Let θ define a connection D^0, then $D = D^S \otimes 1 \oplus 1 \otimes D^0$. □

3. Dirac Operators

We define a canonical first order differential operator A : $C^\infty(X;S) \to C^\infty(X;S)$, called the Dirac operator of S. Next we find the principal symbols of A and A^2 and show that A is formally self-adjoint with an explicit Green's formula.

There are many different concepts of a Dirac operator: Atiyah-Singer (or classical Dirac) operators and twisted Atiyah-Singer (or twisted Dirac) operators on spin manifolds; operators of Dirac type, generalized Dirac operators, and compatible Dirac operators on arbitrary Riemannian manifolds; full and split (odd-parity) Dirac operators, boundary Dirac operators, and so on. As we shall see, each definition has its own merits and range of application.

Let X be a smooth manifold (with or without boundary) of dimension m and let S be a complex vector bundle over X. Recall the definition of the *principal symbol* of a differential operator $P : C^\infty(X;S) \to C^\infty(X;S)$ of order $k \geq 1$. If, in local coordinates, we have

$$P = \sum_{0\leq|\alpha|\leq k} a_\alpha(x)\frac{\partial^{|\alpha|}}{\partial x^\alpha} \qquad \text{and} \qquad \xi = \sum_{\mu=1}^{m} \xi_\mu dx_\mu \in T^*X_x\,,$$

then

$$\sigma_P(x,\xi) := i^k \sum_{|\alpha|=k} a_\alpha(x)\xi^\alpha : S_x \to S_x\,. \tag{3.1}$$

Now we can formulate:

Definition 3.1. Let X be a Riemannian manifold (with or without boundary) of dimension m and let S be a complex vector bundle over X.

(a) Any first order differential operator $A : C^\infty(X;S) \to C^\infty(X;S)$ with principal symbol of A^2 defining the Riemannian metric, i.e.

$$\sigma_{A^2}(x,\xi) = \sum_{\mu,\nu=1}^{m} g^{\mu\nu}(x)\xi_\mu\xi_\nu\,,$$

shall be called an *operator of Dirac type.* Then the operator A^2 is called the *Dirac Laplacian.*
(b) Let S carry the structure of a (complex) $C\ell(X)$-module bundle over X. We write the left Clifford multiplication as a mapping

$$\mathbf{c} : C^\infty(X;TX \otimes S) \to C^\infty(X;S) .$$

Let $D : C^\infty(X;S) \to C^\infty(X;T^*X \otimes S)$ be a connection on S and let

$$J : C^\infty(X;T^*X \otimes S) \cong C^\infty(X;TX \otimes S)$$

denote the isomorphism of vector and covector fields. The composition

$$A := \mathbf{c} \circ J \circ D$$

defines a canonical first order differential operator $A : C^\infty(X;S) \to C^\infty(X;S)$ called the *(generalized) Dirac operator* of S. Thus, in terms of an orthonormal base $\{e_\nu\}_{\nu=1,\dots,m}$ of $(TX)_x$, we have

$$As(x) = \sum_{\mu=1}^{m} e_\mu \cdot (D_{e_\mu} s)|_x \,, \tag{3.2}$$

where $D_{e_\mu} s|_x$ denotes the covariant derivative of the section $s \in C^\infty(X;S)$, taken at the point $x \in X$ in the direction e_μ, and $\cdot$ denotes the Clifford multiplication.
(c) If the connection D is compatible with the $C\ell(X)$-module structure of S and the extended Levi-Civita connection, we call A a *compatible Dirac operator* or, in short, a *Dirac operator.*

Remarks 3.2. (a) The original Dirac operator is defined for the indefinite relativistic metric and is not elliptic. *Historically* the search for a square root of the Laplacian (for Dirac the Klein-Gordon operator) with its scalar principal symbol was the beginning, see Chapter 5 below. *Systematically* we have a correspondence between operators of Dirac type and Clifford multiplication, since the principal symbol σ_A defines a $C\ell(X)$-module structure on S, if the principal symbol σ_{A^2} has the form of a Riemannian metric. Hence any operator of Dirac type belongs to the subclass of generalized Dirac operators modulo operators of lower order.
(b) An example: We illustrate the construction of the Dirac operator for the exterior algebra $\Lambda(TX)$ of a Riemannian manifold X; see also

Example 1.4 and our comment before Lemma 2.4 above. We identify T^*X with TX and obtain an invariantly defined operator

$$A : C^\infty(\Lambda(T^*X)) \xrightarrow{\nabla} C^\infty(T^*X \otimes \Lambda(T^*X)) \xrightarrow{\mathbf{c}} C^\infty(\Lambda(T^*X)).$$

In terms of an orthonormal basis $\{e_\mu\}$ for TX_x we get

$$A(\alpha)(x) = \sum_\mu (\mathrm{ext}(e_\mu) - \mathrm{int}(e_\mu))\nabla_{e_\mu}(\alpha),$$

since $\mathbf{c} = \mathrm{ext} - \mathrm{int}$. It turns out that A actually coincides with the de Rham complex

$$(d + d^*) : C^\infty(\Lambda(T^*X)) \longrightarrow C^\infty(\Lambda(T^*X)),$$

see e.g. Gilkey [1984; Section 3.1]. The Dirac Laplacian $(d + d^*)^2$ coincides with the Laplace-Beltrami operator Δ, and one derives the *Weitzenböck formula*

$$(d + d^*)^2 = \Delta = -\sum_j \nabla_j \nabla_j + \mathcal{R},$$

where

$$\mathcal{R} = \sum_{\nu\mu\sigma\rho} \tfrac{1}{8} R_{\nu\mu\sigma\rho}\, \mathbf{c}^l_\nu\, \mathbf{c}^l_\mu\, \mathbf{c}^r_\sigma\, \mathbf{c}^r_\rho + \tfrac{1}{4}\kappa$$

with $\mathbf{c}^l = \mathrm{ext}^l - \mathrm{int}^l$ denoting left Clifford multiplication, $\mathbf{c}^r = \mathrm{ext}^r - \mathrm{int}^r$ denoting the corresponding right one, and κ denoting the scalar curvature of X. For details see Gilkey [1991; Lemma 8.1]. In Theorem 4.4 below we shall prove the general Bochner identity as a generalization of the Weitzenböck formula.
(c) We postpone the discussion of the various other classical (natural and geometric) Dirac operators in the Euclidean case and on spin manifolds (Sections 5 and 6 below). Here we shall only notice that some of them, like the operator $d + d^*$ of the de Rham complex over a Riemannian manifold and the Atiyah-Singer operator of the spin complex over a spin manifold, are defined by compatible connections. Others are not, like the operator of the Dolbeault complex

$$\sqrt{2}(\bar\partial + \delta'') = \sum_{\mu=1}^{2n} \mathbf{c}^\mu\, \partial_\mu - r : C^\infty(X; S) \to C^\infty(X; S)$$

$$\text{for } S := \oplus_p \Lambda^{0,p}(X)$$

over an almost complex manifold X, namely when the error term r becomes a non-trivial differential operator of order 0, i.e. a non-trivial local section of $\mathrm{Hom}(S,S)$, cf. Gilkey [1984, Sections 3.5-3.6]. More generally, we have proper inclusions of the three classes of Dirac operators of Definition 3.1:

$$\text{compatible Dirac} \underset{\neq}{\subset} \text{generalized Dirac} \underset{\neq}{\subset} \text{Dirac type.}$$

We shall prove the last inclusion in the following lemma.
(d) Most of the results below for manifolds with boundary hold for arbitrary first order elliptic differential operators. However, the exposition is much simpler for operators of Dirac type because the symbolic calculus can be reinterpreted geometrically. That is also the reason why the unique continuation property holds for operators of Dirac type, see Theorem 8.2 below.

Lemma 3.3. *The principal symbol of a generalized Dirac operator A of a $C\ell(X)$-module S is left Clifford multiplication with $\xi \in T^*X \cong TX$ times the imaginary unit:*

$$\sigma_A(x,\xi) = i\xi\cdot : S_x \to S_x\,. \tag{3.3}$$

For the Dirac Laplacian we obtain the scalar symbol

$$\sigma_{A^2}(x,\xi) = \|\xi\|^2 Id_{S_x} : S_x \to S_x\,. \tag{3.4}$$

In particular, both A and A^2 are elliptic operators.

Proof. Determine the principal symbols of A and A^2: Fix $x \in X$ and an orthonormal basis $\{e_1,\dots,e_m\}$ of TX_x. Choose local coordinates $(x_1,\dots,x_m)$ on X at x such that x corresponds to 0 and e_μ corresponds to $\frac{\partial}{\partial x_\mu}|_0$ for each μ. Under the identification $TX_x \cong T^*X_x$, the tangent vector e_μ also corresponds to the differential $dx_\mu|_0$.

We see that

$$A = \sum e_\mu \cdot \frac{\partial}{\partial x_\mu}|_0 \quad + \quad \text{zero-order terms} \tag{3.5}$$

since the covariant derivative on S takes the form

$$D_{e_\mu} = (\partial/\partial x_\mu)_0 \quad + \quad \text{zero-order terms}$$

for any local trivialization of S near x. Hence, for any cotangent vector $\xi = \sum \xi_\mu (dx_\mu)_0$ at 0 and $s_0 \in S_x$, we get

$$\sigma_A(x,\xi)s_0 = i\sum e_\mu \xi_\mu \cdot s_0 = i\xi \cdot s_0 \,.$$

For the principal symbol of the Dirac Laplacian we find

$$\sigma_{A^2}(x,\xi)s_0 = \sigma_A(x,\xi)\circ\sigma_A(x,\xi)s_0 = -\xi\cdot\xi\cdot s_0 = \|\xi\|^2\,\mathbf{c}(1)s_0 = \|\xi\|^2 s_0 \,.$$

So $\sigma_{A^2}(x,\xi)$ is invertible for all $\xi \neq 0$, hence also $\sigma_A(x,\xi)$, and therefore A is elliptic. □

As permitted by Lemma 2.2, assume that the $C\ell(X)$-module S is endowed with a Hermitian metric which makes Clifford multiplication by unit tangent vectors skew-adjoint. We see at once that the principal symbol of any operator of Dirac type is then self-adjoint:

(3.6)
$$\langle i\xi \cdot s_1; s_2\rangle_x \;=\; i\left\langle \|\xi\|^{-1}\xi \cdot \xi \cdot s_1; \|\xi\|^{-1}\xi \cdot s_2\right\rangle_x \;=\; \langle s_1; i\xi \cdot s_2\rangle_x \,,$$

where $\langle .;.\rangle_x$ denotes the Hermitian metric in the fibre S_x.

If we assume that A is defined by a compatible connection, i.e. A is a Dirac operator, we have much more: not only is the symbol self-adjoint in the fibres, but the operator itself is formally self-adjoint with an explicit boundary integral in Green's formula. That is very important and fundamental to (infinitely) many calculations on manifolds with boundary.

Proposition 3.4. *Let X be a compact Riemannian manifold with or without boundary; let S be a $C\ell(X)$-module, and let A denote its Dirac operator (i.e. built on a compatible connection for S).*
(a) Then A is formally self-adjoint, i.e.

$$(3.7) \qquad \langle As_1; s_2\rangle \;=\; \langle s_1; As_2\rangle \quad \textit{for all } s_1,\, s_2 \in C_0^\infty(X;S)\,,$$

where the inner product on $C^\infty(X;S)$ is induced from the pointwise inner product $\langle ..;..\rangle_x$ by setting

$$(3.8) \qquad \langle s_1; s_2\rangle \;:=\; \int_X \langle s_1(x); s_2(x)\rangle_x \, dvol(x)\,,$$

and $C_0^\infty(X;S)$ denotes the subspace of sections with support in $X\backslash\partial X$.

(b) More generally, the following equality holds for all smooth sections s_1 and s_2:

$$(3.9)\quad \langle As_1; s_2\rangle - \langle s_1; As_2\rangle = -\int_Y \langle G(y)(s_1(y)); s_2(y)\rangle_y \, dvol(y)\,,$$

where Y denotes the boundary of X and $G(y)$ denotes Clifford multiplication by the inward unit normal vector $\mathbf{n} \in TX_y \cong T^*X_y$ for $y \in Y$.

Remark 3.5. We shall refer to (3.9) as a special case of *Green's formula* (for the general case see e.g. Palais [1965a; Theorem XVII.1]), and to G as the *Green's form* of the Dirac operator. From the rule (2.2) for Clifford multiplication by unit vectors in TX we get

$$(3.10)\qquad G^* = -G = G^{-1}\,.$$

Proof of Proposition 3.4. First fix an $x \in X$ and choose an orthonormal tangent frame field $(v_1, \dots, v_m)$ in a neighbourhood of x so that $(D_{v_\nu} v_\mu)_x = 0$ for all μ, ν. This can be achieved by extending a frame at x by parallel translation along geodesic rays emanating from x. Then we have at x:

$$\begin{aligned}
(3.11)\quad \langle As_1; s_2\rangle_x &= \sum_\mu \langle v_\mu \cdot D_{v_\mu} s_1; s_2\rangle_x \\
&\overset{(2.2)}{=} \sum_\mu \langle v_\mu \cdot v_\mu \cdot D_{v_\mu} s_1; v_\mu \cdot s_2\rangle_x = -\sum_\mu \langle D_{v_\mu} s_1; v_\mu \cdot s_2\rangle_x \\
&\overset{(2.4)}{=} \sum_\mu \{\langle s_1; D_{v_\mu}(v_\mu \cdot s_2)\rangle_x - v_\mu \langle s_1; v_\mu \cdot s_2\rangle_x\} \\
&\overset{(2.12)}{=} \sum_\mu \{\langle s_1; (D_{v_\mu} v_\mu) \cdot s_2\rangle_x \\
&\qquad + \langle s_1; v_\mu \cdot (D_{v_\mu} s_2)\rangle_x - v_\mu \langle s_1; v_\mu \cdot s_2\rangle_x\} \\
&= \langle s_1; As_2\rangle_x - \sum_\mu v_\mu \langle s_1; v_\mu \cdot s_2\rangle_x\,.
\end{aligned}$$

To evaluate the second summand, use the following trick: Let r be the vector field defined by the condition that

$$(3.12)\qquad \langle r(x'); w\rangle_{x'} = -\langle s_1(x'); w \cdot s_2(x')\rangle_{x'}$$

for all $w \in TX_{x'}$ and $x' \in X$. Then

$$
\begin{aligned}
-\sum_\mu \{v_\mu \langle s_1; v_\mu \cdot s_2\rangle\}_{|x} &= \sum_\mu \{v_\mu \ \langle r; v_\mu\rangle\}_{|x} \\
(3.13) \qquad &= \sum_\mu \{v_\mu \ \langle r; v_\mu\rangle \ - \ \langle r; D_{v_\mu} v_\mu\rangle\}_{|x} \\
&= \sum_\mu \ \langle D_{v_\mu} r; v_\mu\rangle_x = \operatorname{div}(r)_{|x} \,.
\end{aligned}
$$

From (3.11) and (3.13) we obtain not only

$$
(3.14) \qquad \langle As_1; s_2\rangle_x = \ \langle s_1; As_2\rangle_x + \operatorname{div}(r)_{|x}
$$

at the chosen point x, but in fact

$$
\langle As_1; s_2\rangle_{x'} = \operatorname{div}(r)|_{x'} + \ \langle s_1; As_2\rangle_{x'}
$$

for all $x' \in X$, since all expressions in (3.14) are independent of the chosen frame $\{v_\mu\}$.

Stokes' formula (see e.g. Dubrovin, Fomenko & Novikov [1984; 4.26]) gives

$$
\begin{aligned}
&\int_M \operatorname{div}(r) dvol(x) \\
&\quad = \int_{\partial M} \langle r; -\mathbf{n}\rangle \, dvol(y) \overset{(3.12),\ (2.2)}{=} \int_{\partial M} \langle -\mathbf{n} \cdot s_1; s_2\rangle_y \, dvol(y)
\end{aligned}
$$

which proves (a) and (b). □

4. Dirac Laplacian and Connection Laplacian

We discuss the general Bochner identity which gives an expression of the *Dirac Laplacian* A^2 in terms of the *connection Laplacian* D^*D and certain bundle endomorphisms.

Now we shall express the Dirac Laplacian A^2 in terms of the connection Laplacian D^*D and certain bundle endomorphisms. We begin with the definition of the operator D^*D.

Definition 4.1. Let X be a compact Riemannian manifold and let S be any $C\ell(X)$-module with compatible connection D. The *connection Laplacian*

$$D^*D : C^\infty(X;S) \to C^\infty(X;S)$$

is defined by taking the trace

$$D^*Ds := -\operatorname{tr}(D^2_{\cdot,\cdot}s)\,, \tag{4.1}$$

where $s \in C^\infty(X;S)$ and

$$D^2_{\cdot,\cdot} : C^\infty(X;TX \oplus TX) \to C^\infty(X;S^* \otimes S)$$

denotes the *invariant second derivative* given by

$$D^2_{v,w}s := D_v D_w s - D_{D_v w}s \tag{4.2}$$

for any pair of tangent vector fields v and w.

Remarks 4.2. (a) Clearly the operator $D_{v,\cdot}$ only depends on the value v_x, since this is a general property of the covariant derivative. Also, moreover, $D_{\cdot,w}$ only depends on the value w_x. This follows from the identity

$$D^2_{v,w} - D^2_{w,v} = R_{v,w}\,, \tag{4.3}$$

where R denotes the curvature tensor of S.

(b) In terms of a local orthonormal frame $\{v_\mu\}$ for TX we have

$$D^*Ds = -\sum_{\mu=1}^{m} D^2_{v_\mu, v_\mu} s\,. \tag{4.4}$$

(c) The concept of the connection Laplacian is of great generality in differential geometry and not restricted to the case of $C\ell(X)$-modules solely discussed here.

Proposition 4.3. *The operator $D^*D : C^\infty(X;S) \to C^\infty(X;S)$ has the principal symbol*

$$\sigma_{D^*D}(x,\xi) = \|\xi\|^2 \quad \textit{for } \xi \in T^*X_x\,, \tag{4.5}$$

and so D^*D is elliptic. It is also non-negative and formally self-adjoint. In particular,

$$\int_M \langle D^*Ds_1; s_2\rangle \;=\; \int_M \langle Ds_1; Ds_2\rangle \tag{4.6}$$

for all s_1, $s_2 \in C^\infty(X;S)$, provided that either s_1 or s_2 has support in the interior of X. (As always we assume that X is compact). Here $\langle Ds_1; Ds_2\rangle$ is defined in terms of local frames $\{v_\mu\}$ for TX by the expression

$$\langle Ds_1; Ds_2\rangle \;=\; \sum_\mu \left\langle D_{v_\mu}s_1; D_{v_\mu}s_2\right\rangle\,.$$

Proof. We reproduce the main arguments from the proof of Lemma 3.3 and Proposition 3.4 above, replacing left Clifford multiplication by covariant derivation. Fix an $x \in X$ and choose a local orthonormal frame $\{v_1, \ldots, v_m\}$ for TX with the property that $(D_{v_\mu}v_\nu)_x = 0$ for all μ, ν. Then, at the point x, one has that

$$D^*Ds = -\sum_\mu D_{v_\mu}D_{v_\mu}s \tag{4.7}$$

and so $\sigma_{D^*D}(x,\xi) = \|\xi\|^2$, since $D_{v_\mu} = (\partial/\partial x_\mu)|_0 +$ zero-order terms as observed earlier.

From (4.7) we obtain (all inner products are taken at x) that

$$\begin{aligned}
\langle D^*Ds_1; s_2\rangle \;&=\; -\sum_\mu \left\langle D_{v_\mu}D_{v_\mu}s_1; s_2\right\rangle \\
&= -\sum_\mu \{v_\mu \,\left\langle D_{v_\mu}s_1; s_2\right\rangle \;-\; \left\langle D_{v_\mu}s_1; D_{v_\mu}s_2\right\rangle\} \\
&= -\operatorname{div}(r) + \,\langle Ds_1; Ds_2\rangle\,,
\end{aligned}$$

where r now is the tangent vector field on X defined by the condition $\langle r; w\rangle = \langle D_w s_1; s_2\rangle$ for all $w \in TX$. Then

$$\operatorname{div}(r) = \sum_\mu \langle D_{v_\mu} r; v_\mu\rangle = \sum_\mu v_\mu \langle r; v_\mu\rangle = \sum_\mu v_\mu \langle D_{v_\mu} s_1; s_2\rangle .$$

Now the proposition follows by integration. □

Theorem 4.4. (The general Bochner identity). *Let X be a compact Riemannian manifold (with or without boundary), and let S be a $C\ell(X)$-module with compatible connection. Let A^2 denote the Dirac Laplacian and let D^*D denote the connection Laplacian. Then*

$$A^2 = D^*D + \mathcal{R} .$$

Here $\mathcal{R}$ is a canonical section of Hom(S, S) *defined by the formula*

$$\mathcal{R}(s) := \frac{1}{2} \sum_{\mu,\nu=1}^{m} v_\mu \cdot v_\nu \cdot R_{v_\mu, v_\nu}(s) ,$$

where $\{v_\mu\}$ is any orthonormal tangent frame at the point in question, $R_{v,w}$ is the curvature transformation of S (cf. (4.3) above), and the dot "·" denotes left Clifford multiplication.

Proof. We make the same assumptions about x and $\{v_\mu\}$ as in the preceding proof and find at x that

$$\begin{aligned} A^2 &= \sum_{\mu,\nu} v_\mu \cdot D_{v_\mu}(v_\nu \cdot D_{v_\nu}) = \sum_{\mu,\nu} v_\mu \cdot v_\nu \cdot D_{v_\mu} D_{v_\nu} \\ &\qquad\qquad\qquad\qquad \text{using (2.12) and } D_{v_\mu} v_\nu = 0 \text{ at } x \\ &= \sum_{\mu,\nu} v_\mu \cdot v_\nu D^2_{v_\mu, v_\nu} = -\sum_\mu D^2_{v_\mu} + \sum_{\mu<\nu} v_\mu \cdot v_\nu \cdot (D^2_{v_\mu, v_\nu} - D^2_{v_\nu, v_\mu}) \\ &= D^*D + \mathcal{R} \end{aligned}$$

by (4.3). □

5. Euclidean Examples

Some important examples are discussed: Dirac's search for a square root of the Klein-Gordon operator; the Dirac operator on $\mathbf{R}^2 = \mathbf{C}$ considered as mapping even to odd spinors which is exactly the Cauchy-Riemann operator; the Dirac operator on $\mathbf{H}$-valued functions; and the quaternion analogue for the Cauchy-Riemann operator and its expression by Pauli matrices.

The first class of examples we want to consider are Dirac operators on $\mathbf{R}^m$ or on a compact connected submanifold $X \subset \mathbf{R}^m$ of codimension 0 with smooth boundary. Let $C\ell_m$ denote the Clifford algebra of $\mathbf{R}^m$. It has real dimension 2^m. Let W be a (complex, N-dimensional) $C\ell_m$-module, and let $\mathbf{c} : C\ell_m \to \mathrm{Hom}(W, W)$ denote the related representation, i.e. left Clifford multiplication in W with elements of $C\ell_m$. We study the *Euclidean Dirac operator* $A : C^\infty(X; S) \to C^\infty(X; S)$, where $S := X \times W$ is endowed with the trivial (compatible) connection. It is a constant coefficient operator of the form

$$A = \sum_{\mu=1}^{m} \mathbf{c}_\mu \frac{\partial}{\partial x_\mu},$$

where all $\mathbf{c}_\mu$ are linear maps $\mathbf{c}_\mu : W \to W$ and from their definition as $\mathbf{c}_\mu = \mathbf{c}(e_\mu)$ inherit the relations

$$\mathbf{c}_\mu \circ \mathbf{c}_\nu + \mathbf{c}_\nu \circ \mathbf{c}_\mu = -2\delta_{\mu\nu} Id_W \qquad \text{for all } \mu,\ \nu. \tag{5.1}$$

Here $e_1, \ldots, e_m$ denote the standard basis vectors of $\mathbf{R}^m$. If we choose a basis for W, we can identify $\mathrm{Hom}(W, W)$ with the full matrix algebra $\mathcal{M}(N, \mathbf{C})$, and the $\mathbf{c}_\mu$'s will be represented by elements of the group $GL(N, \mathbf{C})$ because of (5.1) for $\mu = \nu$. Equation (5.1) also implies that $A^2 = \Delta\ Id_W$. In the coordinates for W it is an $N \times N$ diagonal matrix with the (positive) Laplacian $\Delta = -\sum \partial^2/\partial x_\mu^2$ repeated in the diagonal.

Remarks 5.1. (a) The physics literature distinguishes carefully between Euclidean (elliptic) Dirac operators, which are operators of the type discussed here, and Minkowskian (hyperbolic) Dirac operators, which are the operators originally investigated by P.A.M. Dirac in the 1920s. Dirac posed the following question: Can the Klein-Gordon operator $\Box + m^2$, where

$$\Box := \sum g^{\mu\nu} \frac{\partial^2}{\partial x_\mu \partial x_\nu}$$

with $g^{\mu\nu}$ denoting the coefficients of the Minkowski metric, be expressed as a product of two first-order differential operators as follows:

$$\Box + m^2 \stackrel{?}{=} -\left(\sum i\,\mathbf{c}_\mu \frac{\partial}{\partial x_\mu} + m\right)\left(\sum i\,\mathbf{c}_\mu \frac{\partial}{\partial x_\mu} - m\right).$$

Dirac realized that the coefficients $\mathbf{c}_\mu$ must be matrices which generate the algebra $\mathcal{M}(4, \mathbf{C})$ and satisfy the relation (5.1), where the right side is replaced by $2g^{\mu\nu}\delta_{\mu\nu}Id_W$. In physics, the space $C^\infty(\mathbf{R}^4; S)$ with $S = \mathbf{R}^4 \times W$ and $W = \mathbf{C}^4$ is called the *space of (4-component) spinors.* See e.g. Dubrovin, Fomenko & Novikov [1984; Section 40].

(b) If we only consider irreducible $C\ell_m$-modules we can determine the following data for each integer $m \geq 1$: First, find $C\ell_m$. All $C\ell_m$ are explicitly known, they have real dimension 2^m, and there is actually a kind of periodicity with period 8; see Lawson & Michelsohn [1989; Section I.4]. Then determine the irreducible representations $\mathbf{c} : C\ell_m \to \mathrm{Hom}(W, W)$ along the lines of Lemma 1.3. They are uniquely determined (up to isomorphisms) if m is even, i.e. $m = 2n$; then the complex dimension of W is $N = 2^n$; if m is odd, $m = 2n+1$, we have two non-equivalent irreducible representations, each module again of complex dimension $N = 2^n$. Each vector of the canonical basis $\{e_\mu\}$ of $\mathbf{R}^m$ provides an endomorphism $\mathbf{c}_\mu = \mathbf{c}(e_\mu)$ of W which satisfies the relations in (5.1) (as, in the opposite direction, each system of m endomorphisms of W defines a representation of $C\ell_m$, if it satisfies (5.1)). Now choose a basis for W and calculate $\sum \mathbf{c}_\mu\, \partial/\partial x_\mu$ which is an $N \times N$ matrix, the entries of which are linear combinations of the partial derivatives $\partial/\partial x_1, \ldots, \partial/\partial x_m$. This matrix, which is enormous for higher m, has the property that its square is ΔE, where E denotes the $N \times N$ identity matrix. Finally, the principal symbol of the Euclidean Dirac operator restricted to unit cotangent vectors provides a generator $\sigma_A(x, .) : S^{m-1} \to GL(N, \mathbf{C})$ of the homotopy group $\pi_{m-1}(GL(N, \mathbf{C})) \cong \mathbf{Z}$ for m even and $2N \geq m$.

As Lawson & Michelsohn [1989, p. 120] put it: "This explicit form is seldom, if ever, useful. It is always simpler to use the structure of the Clifford module." Nevertheless, below we shall derive the explicit forms in the cases $m = 1, \ldots, 4$ for three reasons:

(1) to illustrate and exercise the basic concepts and constructions of Clifford algebras, Clifford modules, and Dirac operators;
(2) to provide the basic bricks for all geometrically defined operators. This may sound strange, since Euclidean Dirac operators do not immediately extend to elliptic operators over closed manifolds — that is inhibited by the characteristic form of the principal symbols which have non-vanishing local index for even m. However, later in this book, we shall show how to build up arbitrary classical operators by pasting Euclidean Dirac operators together, see Chapters 24 – 26;
(3) and to ease reading the literature, where some calculations are carried out through ingenious manipulations with these matrices, see e.g. Goldstone & Jaffe [1983], Schmidt & Binzer [1987], Schmidt [1987], or Kori [1993].

Let $m = 1$, so that $\mathrm{C}\ell_1 = \mathcal{L}_{\mathbf{R}}\{1, e_1\} = \mathbf{C}$. Then $W = \mathbf{C}$ and $\mathbf{c}(1) = 1$. There are up to equivalence two irreducible representations, depending on whether $\mathbf{c}$ maps e_1 into i or $-i$, and these two representations are not equivalent, since $F : z \mapsto \overline{z}$ is not $\mathbf{C}$-linear. Then the Dirac operator is just

$$A = \pm i \frac{\partial}{\partial x_1} . \tag{5.2}$$

The $\mathbf{Z}_2$-grading $\mathrm{C}\ell_1 = \mathrm{C}\ell_1^+ \oplus \mathrm{C}\ell_1^-$ gives the splitting in real and imaginary parts which the Dirac operator interchanges.

Let $m = 2$, so that $\dim \mathrm{C}\ell_2 = 2^m = 4$ and in fact $\mathrm{C}\ell_2 = \mathcal{L}_{\mathbf{R}}\{1, e_1, e_2, e_1 e_2\} = \mathbf{H}$. The decomposition of $\mathbf{H}$ into $\mathbf{C} \oplus \mathbf{C}$ is natural and corresponds to the $\mathbf{Z}_2$-grading with $\mathrm{C}\ell_2^+ = \mathcal{L}_{\mathbf{R}}\{1, e_1 e_2\}$ and $\mathrm{C}\ell_2^- = \mathcal{L}_{\mathbf{R}}\{e_1, e_2\}$. We are looking for a complex vector space W such that $\mathrm{C}\ell_2 \xrightarrow{\cong} \mathrm{Hom}_{\mathbf{C}}(W, W)$. Then the complex dimension of W has to be $2^{\frac{1}{2}m} = 2$. Actually $W := \mathbf{C} \oplus \mathbf{C}$ is an irreducible $\mathrm{C}\ell_2$-module with Clifford multiplication given by

$$\mathbf{c}(1) := \begin{pmatrix} 1 & 0 \\ 0 & 1 \end{pmatrix}, \qquad \mathbf{c}(e_1) := \begin{pmatrix} 0 & -1 \\ 1 & 0 \end{pmatrix}, \qquad \mathbf{c}(e_2) := \begin{pmatrix} 0 & i \\ i & 0 \end{pmatrix} .$$

With the notation of Example 1.4 we can write down the two basic operators on $\Lambda \mathbf{C}^2 \cong \mathrm{C}\ell_2$:

$$\wp = \mathbf{c}(e_1) + i\,\mathbf{c}(e_2) \qquad \text{and} \qquad \overline{\wp} = \wp_5 = \mathbf{c}(e_1) - i\,\mathbf{c}(e_2)$$

with the relations

$$\begin{aligned}\wp \cdot \wp &= (\mathbf{c}(e_1) + i\,\mathbf{c}(e_2))(\mathbf{c}(e_1) + i\,\mathbf{c}(e_2)) \\ &= -\,\mathbf{c}(1) + i\,\mathbf{c}(e_1)\,\mathbf{c}(e_2) + i\,\mathbf{c}(e_2)\,\mathbf{c}(e_1) + i^2\,\mathbf{c}(e_2)^2 = 0\end{aligned}$$

and

$$\begin{aligned}\overline{\wp} \cdot \overline{\wp} &= 0 \\ \wp \cdot \overline{\wp} = \cdots &= -2(\mathbf{c}(1) + i\,\mathbf{c}(e_1)\,\mathbf{c}(e_2)) \\ \overline{\wp} \cdot \wp = \cdots &= -2(\mathbf{c}(1) - i\,\mathbf{c}(e_1)\,\mathbf{c}(e_2)) \quad \text{so} \\ \wp\overline{\wp} + \overline{\wp}\wp &= -4\end{aligned}$$

as required in Example 1.4d. Then

$$\wp \cdot \overline{\wp} = -2(\mathbf{c}(1) + i\,\mathbf{c}(e_1)\,\mathbf{c}(e_2)) \quad \text{and} \quad 1 \cdot \overline{\wp} = \overline{\wp} = (\mathbf{c}(e_1) - i\,\mathbf{c}(e_2))$$

form a basis for W, and Clifford multiplication with $\wp$ and $\overline{\wp}$ does not lead out of the subspace W of $C\ell_2$. Indeed, we have $\wp \cdot W \subset W$, since

$$\wp \cdot (\wp \cdot \overline{\wp}) = 0 \quad \text{and trivially} \quad \wp \cdot (1 \cdot \overline{\wp}) = (\wp \cdot \overline{\wp}),$$

and $\overline{\wp} \cdot W \subset W$, since

$$\overline{\wp} \cdot (\wp \cdot \overline{\wp}) = (-4 - \wp\overline{\wp})\overline{\wp} = -4\overline{\wp} \quad \text{and} \quad \overline{\wp} \cdot (1 \cdot \overline{\wp}) = 0.$$

To get an invariant description of W^+ and W^-, determine the ± 1-eigenspaces of the orientation form

$$\Gamma_5 = i\,\mathbf{c}(e_1)\,\mathbf{c}(e_2).$$

For the ease of presentation we omit the $\mathbf{c}$ in $\mathbf{c}(e_j)$ and write just e_j for Clifford multiplication with e_j. Then we get

$$\Gamma_5(\wp\overline{\wp}) = -2ie_1e_2(1 + ie_1e_2) = \cdots = -2(1 + ie_1e_2) = \wp\overline{\wp},$$

hence $\wp\overline{\wp}$ generates W^+; and

$$\Gamma_5(\overline{\wp}) = ie_1e_2(e_1 - ie_2) = \cdots = -(e_1 - ie_2) = -\overline{\wp},$$

hence $\overline{\wp}$ generates W^-.

Correspondingly the Dirac operator $A = e_1 \cdot (\partial/\partial x_1) + e_2 \cdot (\partial/\partial x_2)$ has the form

$$A = \begin{pmatrix} 0 & -\frac{\partial}{\partial z} \\ \frac{\partial}{\partial \bar{z}} & 0 \end{pmatrix}, \tag{5.3}$$

where $\partial/\partial z := \partial/\partial x_1 - i\partial/\partial x_2 = 2\partial$, and $\partial/\partial \bar{z} := \partial/\partial x_1 + i\partial/\partial x_2 = 2\bar{\partial}$ is twice the standard Cauchy-Riemann operator. Thus the Dirac operator on $\mathbf{R}^2 = \mathbf{C}$, considered as mapping even to odd spinors, is exactly the (double) Cauchy-Riemann operator.

Note. Our choice of $\mathbf{c}(e_1)$ and $\mathbf{c}(e_2)$ is by no means canonical. There are actually three matrices, the Euclidean analogues of the classical *Pauli matrices*

$$\sigma_1 := \begin{pmatrix} i & 0 \\ 0 & -i \end{pmatrix}, \quad \sigma_2 := \mathbf{c}(e_1), \quad \text{and } \sigma_3 := \mathbf{c}(e_2),$$

which generate the full matrix algebra $\mathcal{M}(2, \mathbf{C})$ and satisfy the relations (5.1). Every two of them together generate an irreducible representation of $C\ell_2$ in complex form. By Lemma 1.3 these representations are equivalent. This is easily checked, e.g. for the two representations $\mathbf{c}_j : C\ell_2 \to \mathcal{M}(2, \mathbf{C})$ defined by

$$\mathbf{c}_1 := \mathbf{c} \qquad \text{and} \qquad \mathbf{c}_2(e_\mu) := \sigma_\mu \quad \text{for } \mu = 1, 2\,.$$

Then $\mathbf{c}_1(e_\mu) = F^{-1}\,\mathbf{c}_2(e_\mu)F$ for $\mu = 1, 2$ if $F = \begin{pmatrix} a & ia \\ a & -ia \end{pmatrix}$ for arbitrary $a \neq 0$.

Let $m = 3$. One finds $C\ell_3 = \mathbf{H} \oplus \mathbf{H}$ and $W = \mathbf{C} \oplus \mathbf{C} = \mathbf{H}$; hence $W = C\ell_3$. Clearly we have two non-equivalent representations of $\mathbf{H}$ on $\mathbf{H}$, namely, left quaternion multiplication and right quaternion multiplication. We begin with left quaternion multiplication which gives the following expression for the Dirac operator on $\mathbf{H}$-valued functions on $\mathbf{R}^3$:

$$A_{\text{left}} = i\frac{\partial}{\partial x_1} + j\frac{\partial}{\partial x_2} + k\frac{\partial}{\partial x_3}\,. \tag{5.4}$$

If we write $q \in \mathbf{H} = \mathbf{R}^4$ in the form $q = a + bi + cj + dk$ with real coefficients a, b, c, d and the canonical basis $\{1, i, j, k\}$, we can write left quaternion multiplication as 4×4 real matrices (corresponding to

$L(i)q = i \cdot q = -b + ai - dj + ck$, $L(j)q = \ldots$, and $L(k)q = \ldots$) and obtain

$$\text{(5.5)} \qquad A_{\text{left}} = \begin{pmatrix} 0 & -\partial_1 & -\partial_2 & -\partial_3 \\ \partial_1 & 0 & -\partial_3 & \partial_2 \\ \partial_2 & \partial_3 & 0 & -\partial_1 \\ \partial_3 & -\partial_2 & \partial_1 & 0 \end{pmatrix},$$

where $\partial_\mu := \partial/\partial x_\mu$.

If we choose right quaternion multiplication, we obtain a different Dirac operator

$$\text{(5.6)} \qquad A_{\text{right}} = R(i)\frac{\partial}{\partial x_1} + R(j)\frac{\partial}{\partial x_2} + R(k)\frac{\partial}{\partial x_3}$$

which can be written in real coordinates (corresponding to $R(i)q = q \cdot i = -b + ai - dj + ck$, $R(j)q = \ldots$, and $R(k)q = \ldots$) as

$$\text{(5.7)} \qquad A_{\text{right}} = \begin{pmatrix} 0 & -\partial_1 & -\partial_2 & -\partial_3 \\ \partial_1 & 0 & \partial_3 & -\partial_2 \\ \partial_2 & -\partial_3 & 0 & \partial_1 \\ \partial_3 & \partial_2 & -\partial_1 & 0 \end{pmatrix}.$$

Now decompose $\mathbf{H} = \mathbf{C} \oplus \mathbf{C}$ and rewrite

$$q = a + bi + cj + dk = ((a + bi), (c + di)).$$

Then right quaternion multiplication may be described by complex 2×2 matrices. Check that $R(i) = \sigma_1$, $R(j) = \sigma_2$, and $R(k) = \sigma_3$, where σ_μ are the Pauli matrices defined above.

Let $m = 4$. Then $C\ell_4 = \mathcal{M}(2, \mathbf{H})$ and $W = \mathbf{H} \oplus \mathbf{H} \cong \mathbf{C}^4$. The representation $\mathbf{c} : C\ell_4 \to \mathrm{Hom}_{\mathbf{C}}(W, W)$ is uniquely determined up to isomorphisms. So we can just play around with the Pauli matrices until we get complex 4×4 matrices $\mathbf{c}_\mu = \mathbf{c}(e_\mu)$ for $\mu = 0, 1, 2, 3$ which satisfy the Clifford relations (5.1). Clearly the following matrices will do:

$$\text{(5.8)} \qquad \mathbf{c}_0 = \begin{pmatrix} 0 & -1 \\ 1 & 0 \end{pmatrix} \quad \text{and} \quad \mathbf{c}_\mu = \begin{pmatrix} 0 & \sigma_\mu \\ \sigma_\mu & 0 \end{pmatrix} \qquad \text{for } \mu = 1, 2, 3\,.$$

Then the Dirac operator can be expressed with respect to the splitting $\mathbf{H} \oplus \mathbf{H}$ as

$$\text{(5.9)} \qquad A = \begin{pmatrix} 0 & -\frac{\partial}{\partial q} \\ \frac{\partial}{\partial \bar{q}} & 0 \end{pmatrix},$$

where

$$\frac{\partial}{\partial \overline{q}} := \frac{\partial}{\partial x_0} + i\frac{\partial}{\partial x_1} + j\frac{\partial}{\partial x_2} + k\frac{\partial}{\partial x_3},$$

and

$$\frac{\partial}{\partial q} := \frac{\partial}{\partial x_0} - i\frac{\partial}{\partial x_1} - j\frac{\partial}{\partial x_2} - k\frac{\partial}{\partial x_3}.$$

Note the analogy with dimension two. For a thorough exploitation of the analogy, see e.g. Delanghe, Sommen & Souček [1992], where quaternionic analysis and other higher dimensional analogies of complex analysis are described. An analogy of holomorphic differential forms in higher dimensions builds on formal similarities between the Cauchy-Riemann operator and the other Euclidean Dirac operators.

6. The Classical Dirac (Atiyah-Singer) Operators on Spin Manifolds

We consider a spin manifold with a spin structure on its tangent bundle, and a spinor bundle endowed with its canonical connection. We formulate the Lichnerowicz vanishing theorem.

One might argue that this chapter is merely of historical interest now that the fact is established that, for any bundle S of complex left modules over $C\ell(X)$, there always exists a connection which is compatible with Clifford multiplication and extends the Levi-Civita connection on X to S; see Chapter 2. Since that is true for any bundle of Clifford modules over a Riemannian manifold X, this somewhat removes the prominence of spin manifolds in earlier work in index theory and the exclusiveness of spin manifolds in admitting non-Euclidean Dirac operators.

Nevertheless, for various calculations, one can take great advantage of the symmetry of spin manifolds, of the associated spinor bundles, and of the related Dirac (Atiyah-Singer) operators. Introductions to the subject can nowadays be found in many places. We especially refer to Gilkey [1984; 3.2 and 3.3] or Lawson & Michelsohn [1989; II.2, II.5, and II.8]. Here it suffices to explain the relation to our previous introduction of isomorphism classes of irreducible Clifford modules. We shall begin with an intuitive argument.

Let X be a compact oriented Riemannian manifold of even dimension $m = 2n$. For every point $x \in X$, we have three different objects:

(1) the tangent space $(TX)_x$ of real dimension m,
(2) the Clifford algebra $C\ell(TX_x)$ of real dimension 2^m which contains TX_x,
(3) and, by Lemma 1.3 (up to isomorphism), the uniquely determined irreducible complex $C\ell(TX_x)$-module Δ of complex dimension 2^n, which we shall denote by S_x.

Whereas the families $\{TX_x\}_{x\in X}$ and $\{C\ell(TX_x)\}_{x\in X}$ carry natural topological and differential structures which make them bundles, namely, TX and $C\ell(X)$, the parametrized family $\{S_x\}_{x\in X}$ admits the structure of a continuous and smooth bundle only if X is a *spin man-*

ifold. In other words, a principal bundle $P_{\text{Spin}}(TX)$ over X is given which has Spin(m) as structural group and is (roughly speaking) a double covering of the principal bundle $P_{\text{SO}}(TX)$ of orthogonal oriented frames, or (more precisely) accompanied by an oriented vector bundle isomorphism

$$TM \approx P_{\text{Spin}}(TX) \times_\rho \mathbf{R}^m ,$$

where $\rho : P_{\text{Spin}} \to P_{\text{SO}}$ denotes the double cover.

Given such a spin structure we make, of the parametrized family $\{S_x\}_{x\in M}$, the *associated bundle* $\mathcal{S}_{\mathbf{C}}$ by regarding S_x as Spin(TX_x)-modules and identifying

$$(x, g_{\iota\kappa}(x)s, \iota) \sim (x, s, \kappa),\ x \in U_\iota \cap U_\kappa ,$$

where $g_{\iota\kappa} : U_\iota \cap U_\kappa \to \text{Spin}(TX_x)$ are suitable coordinate transformations for the spin manifold X. Then the associated bundle carries a natural Hermitian structure satisfying (2.2) and is endowed with a natural compatible connection induced by the Levi-Civita connection for TX.

Definition 6.1. (a) For spin manifolds of even dimension, one has a (unique) irreducible complex spinor bundle for any choice of a spin structure. We shall denote this bundle by $\mathcal{S}_{\mathbf{C}}$. The Dirac operator in this case will be written $\not{D}$.
(b) More generally, a (complex) *spinor bundle*, associated with TX, is a bundle S of modules over $C\ell(X)$ of the form

$$S = P_{\text{Spin}}(TX) \times_\rho \mathcal{M} ,$$

where $\mathcal{M}$ is a complex left module for the complexified Clifford algebra $C\ell(\mathbf{R}^m) \otimes \mathbf{C}$. Then, once again, the Hermitian structure of S is compatible with Clifford multiplication (in the sense of equation (2.2)) and S carries a natural compatible connection as explained, for example, in Lawson & Michelsohn [1989; II.4]. The Dirac operator will in this case be called (following Lawson & Michelsohn) the *Atiyah-Singer operator*. We write A or A^S.

The classical necessary and sufficient topological conditions for a Riemannian manifold to admit a spin structure is derived and discussed, for example, in Lawson & Michelsohn, namely, the vanishing of the second Stiefel-Whitney class. The most elementary examples are:

- All 2-connected manifolds (like homotopy spheres and simply-connected Lie groups) carry a unique spin structure.
- Any manifold, whose tangent bundle is stably parallelizable (i.e. trivial after adding a suitable trivial bundle), is spin. This includes any Lie group and any orientable manifold of dimension ≤ 3.
- A complex manifold is spin, if and only if its first Chern class is even.

We have:

Theorem 6.2. (A. Lichnerowicz). *Let X be a spin manifold and let S be a spinor bundle associated with TX with spin connection D. Then*

$$A^2 = D^*D + \tfrac{1}{4}\kappa . \tag{6.1}$$

Here A^2 denotes the Dirac Laplacian, D^*D denotes the connection Laplacian (see Definition 4.1), and κ denotes the scalar curvature, expressed by the formula

$$\kappa = - \sum_{\mu,\nu=1}^{m} \langle R_{v_\mu, v_\nu}(v_\mu), v_\nu \rangle , \tag{6.2}$$

where $\{v_\mu\}$ is any local orthonormal tangent frame at the point in question and $R_{v,w}$ the curvature transformation of S; see equation (4.3).

Corollary 6.3. *We have*

$$\ker A^S = 0 \tag{6.3}$$

for any spinor bundle associated with TX, if the underlying manifold is spin and closed (compact and without boundary) and if the scalar curvature is ≥ 0 and > 0 at some point.

Theorem 6.2 is an immediate consequence of the general Bochner identity (Theorem 4.4), and the corollary follows by integration of (6.1). For details, cf. Lawson & Michelsohn [1989; p. 161].

Examples 6.4. (a) We can see $C\ell(X)$ as a bundle of left modules over itself by left Clifford multiplication and endow $S := C\ell(X)$ with its canonical Riemannian connection. The Dirac operator in this case is a square root of the classical Hodge Laplacian.
(b) By tensoring we can generate ever new examples from the basic examples. More precisely: Let S be a (complex) $C\ell(X)$-module over X (not necessarily spin) with compatible connection D^S, and let E be an arbitrary Hermitian vector bundle with connection D^E over X. Then the tensor product $S \otimes E$ is again a $C\ell(X)$-module and is equipped with a compatible connection.

7. Dirac Operators and Chirality

We emphasize the decomposition of a $C\ell(X)$-bundle $S = S^+ \oplus S^-$ and the related splitting of Dirac operators. It is illuminating to treat the signature operator and other geometrically defined operators in this context.

Now we are back to our general situation, where X is a Riemannian oriented (not necessarily spin) manifold of dimension m and S a $C\ell(X)$-module with compatible connection. We want to globalize the $\mathbf{Z}_2$-grading explained in Remark 1.1e and decompose the Dirac operator. We do that in a series of lemmata.

Let ω be a section of $C\ell(X)$ given at any point $x \in X$ by

$$\omega := e_1 \cdot \ldots \cdot e_m \,, \tag{7.1}$$

where $\{e_\mu\}$ is a positively oriented orthonormal basis of TX_x. This provides a well-defined global section of $C\ell(X)$. Since ω is independent of the choice of basis, for any x , we may choose a local frame $\{e_\mu\}$ such that $D_{e_\mu} e_{\nu|_x} = 0$ for all μ, ν. Hence we have the following lemma:

Lemma 7.1. *On any oriented Riemannian X of dimension m, we have a canonically defined section ω of $C\ell(X)$ which satisfies the following relations:*

$$\omega^2 = (-1)^{\frac{1}{2}m(m+1)} \,, \tag{7.2}$$

$$\omega v = (-1)^{m-1} v\omega \quad \textit{for any section } v \textit{ of } TX \,, \tag{7.3}$$

$$D\omega = 0 \,. \tag{7.4}$$

Now define a bundle automorphism $\lambda_\omega : S \to S$ by left Clifford multiplication

$$\lambda_\omega(s) := \omega \cdot s \,. \tag{7.5}$$

It follows from Lemma 7.1 that, for even-dimensional X, the bundle S splits:

$$S = S^+ \oplus S^- \,, \tag{7.6}$$

with $v \cdot S^+ \subset S^-$ and $v \cdot S^- \subset S^+$, where $S^\pm$ are the subbundles spanned by the eigensections of λ_ω corresponding to the eigenvalue $\pm i$, if $m \equiv 2 \mod 4$, or ± 1 if $m \equiv 0 \mod 4$.

Lemma 7.2. *For the (total) Dirac operator* $A : C^\infty(X; S) \to C^\infty(X; S)$ *we have*

$$A\lambda_\omega = -\lambda_\omega A .$$

From this lemma it follows that the Dirac operator splits in even dimensions. The components

$$A^\pm := A|_{C^\infty(X;S^\pm)} : C^\infty(X; S^\pm) \to C^\infty(X; S^\mp) \tag{7.7}$$

are well-defined elliptic operators. That construction provides interesting indices, because in difference to A the operators $A^\pm$ are not formally self-adjoint.

We collect our results:

Theorem 7.3. *Let X be a Riemannian, oriented manifold with or without boundary and let S be a* $\mathrm{C}\ell(X)$*-module with compatible connection. If* $\dim X$ *is even, the bundle S and the total Dirac operator A decompose to*

$$S = S^+ \oplus S^- \qquad \textit{and} \qquad A = \begin{pmatrix} 0 & A^- \\ A^+ & 0 \end{pmatrix}$$

and the operator A^- is formally adjoint to the operator A^+ .

Examples 7.4. (a) One of the most prominent Dirac operators is the signature operator. Let X be a compact oriented, Riemannian manifold (with or without boundary) of dimension $4k$. Usually one defines the signature operator via differential forms: Recall that the operator $A := d + d^*$ acts on the space $\Omega^* := C^\infty(X; \Lambda(T^*X))$ of all differential forms. Denoting by $\Omega^\pm$ the ± 1-eigenspaces of the involution $\tau : \Omega^* \to \Omega^*$, defined by $\Omega^p \ni s \mapsto \sqrt{-1}^{\,p(p-1)+2k} * s \in \Omega^{4k-p}$, it follows that A interchanges Ω^+ and Ω^- and hence by restriction defines an operator

$$A^+ : \Omega^+ \to \Omega^-$$

which is called the *signature operator.* This corresponds to the preceding definition given in (7.7) when we considered the basic case of $\mathrm{C}\ell(X)$-module bundles, namely,

$$S := \mathrm{C}\ell(X) = \mathrm{C}\ell^+(X) \oplus \mathrm{C}\ell^-(X) =: S^+ \oplus S^- .$$

Here $\mathrm{C}\ell(X)$ is identified with $\Lambda(T^*X)$. Moreover, the $\mathbf{Z}_2$-grading is given by the volume element ω once again.

(b) To get the required form

$$A^+ = G(\partial_u + B),$$

close to the boundary Y of X, assume that X is isometric to a product near the boundary. Identifying the restriction of Ω^+ to Y with the space $\Omega^*(Y) = C^\infty(\Lambda(T^*Y))$ of all differential forms on Y, one finds that

$$Bs = (-1)^{k+p+1}(\pm 1 * d - d *)s,$$

where the sign in the formula depends on the parity p of the differential form; see Atiyah, Patodi & Singer [1975; (4.6)] and Gilkey [1984; pp. 261 ff]. One sees that B is a self-adjoint operator on Y which preserves the parity of forms on Y and commutes with $s \mapsto (-1)^p * s$, so that B splits into $B = B^{\text{ev}} \oplus B^{\text{odd}}$ and B^{ev} is isomorphic to B^{odd}. We get

$$B^{\text{ev}} s := \sum_p (-1)^{k+p+1}(*d - d*)s_{2p},$$

where $s = \sum_p s_{2p} \in C^\infty(X; V)$ and $V := \oplus_p \Lambda^{2p}(T^*X)$; see also Branson & Gilkey [1992a, Example 2].

(c) For Dirac operators coupled to vector potentials equal to Dirac operators with coefficients in an auxiliary bundle, we refer to Example 6.4b and Atiyah & Singer [1984].

(d) A simple construction of Dirac operators on S^2 with coefficients in line bundles is given in Chapter 26.

Remarks 7.5. (a) Instead of our choice of the orientation section ω in (7.1) we could have taken a normalized orientation τ like in our note to Lemma 1.3. This makes only one difference, namely, $\tau^2 = 1$ instead of the more complicated formula (7.2) for ω^2. Then our subbundles $S^\pm$ become the ± 1 eigenspaces of τ.

(b) From Lemma 1.3 we obtain that the smallest fibre dimension for any total Dirac operator over an even-dimensional manifold X is the dimension of the irreducible module, i.e. $N = 2^{m/2}$ for $\dim X = m$. Therefore we obtain

$$N = 2^{\frac{m}{2}-1}$$

as minimal fibre dimension for the split operators. In fact there is no first-order elliptic differential operator over an open subset of $\mathbf{R}^m$ defined on a smaller number of complex valued functions; see also Atiyah [1970].

8. Unique Continuation Property for Dirac Operators

We give a direct proof of the unique continuation property of a Dirac operator by exploiting its simple product decomposition.

Definition 8.1. An operator A over a smooth manifold X (with or without boundary) has the *unique continuation property*, if any solution s of $As = 0$, which vanishes on an open subset of X also vanishes on the whole connected component of the manifold.

All classical (Euclidean) Dirac operators *eo ipso* satisfy the unique continuation property when they are not coupled to a vector potential (that is, not tensored with an auxiliary coefficient bundle; see Chapter 5 and Example 6.4b above). This follows immediately from the Cauchy-Kovalevskaya theory of elliptic equations with analytic coefficients, since the Euclidean Dirac operators are analytic and can even be written as operators with constant coefficients.

It is a remarkable property of all other operators of Dirac type that they also obey the unique continuation principle. In recent years several authors have exploited this property; see, for example, Roze [1970], Jaffe & Taubes [1980], Kalf [1981], Simon [1982], Freed & Uhlenbeck [1984], Donaldson and Kronheimer [1990], and Taubes [1990]. Usually reference is given to the Aronszajn-Cordes uniqueness theorem for second order differential operators $A = \sum_{|\alpha|\leq 2} a_\alpha(x)\partial^{|\alpha|}/\partial x^\alpha$ with elliptic scalar real valued principal symbol.

We want to keep our treatment of Dirac operators self-contained and show that the unique continuation principle for Dirac operators is a consequence of their simple product form. For unique continuation, the decisive property of Dirac operators is that even in the interior of the manifold does the locally well-defined product $A = G(y,u)(\partial_u + B_u)$ (resp. to a chosen Riemannian metric) have locally a tangential part B_u with *elliptic* self-adjoint part $\frac{1}{2}(B_u + B_u^*)$.

Theorem 8.2. *Let X be a compact Riemannian manifold (with or without boundary) and S a $\mathrm{C}\ell(X)$-module with $\mathrm{C}\ell(X)$-compatible connection D. Then the unique continuation principle is valid for the corresponding Dirac operator $A : C^\infty(X;S) \to C^\infty(X;S)$.*

We shall mostly apply the unique continuation principle in the following form which, via Green's formula, is an easy consequence of the preceding theorem (for details, cf. Lemma 9.2).

Corollary 8.3. *Let $X = X_+ \cup X_-$ be a connected partitioned manifold with $X_+ \cap X_- = \partial X_\pm = Y$. Let $s \in C^\infty(X;S)$ satisfy $As = 0$ and $s|_Y = 0$. Then $s = 0$ on X.*

Remark 8.4. In most parts of this book we assume that the Riemannian structure of X and the Hermitian structure of S are products near Y. This implies a product form $A = G(y)(\partial_u + B)$ with unitary G and self-adjoint B close to Y, where G and B do not depend on the normal variable u. Then, close to Y, the unique continuation property of A follows immediately from elementary harmonic analysis when we expand a solution of $As = 0$ near the boundary in the form

$$s(u, y) = \sum_\lambda f_\lambda(u)\varphi_\lambda(y)\,,$$

where $\{\lambda; \varphi_\lambda\}$ is a spectral resolution of $L^2(Y; S|_Y)$ generated by B (cf. the discussions in Chapter 22; e.g. the proof of Proposition 22.20). The crucial point of Theorem 8.2 is that one obtains the unique continuation property also in the interior, where no product structure of the metrics can be assumed *a priori*. In the interior we may of course also introduce a product structure locally for the ease of calculation. But then we can no longer assume that a given Dirac operator will split there in a product with the parts G and B independent of a local "normal" parameter u.

Proof of Theorem 8.2. Let $s \in C^\infty(X;S)$ be a solution of $As = 0$, and assume that $s = 0$ in an open set $V \underset{\neq}{\subset} X$. Let $x_0 \in \partial V$. We shall work locally in the neighbourhood of x_0. By fixing a local chart for X we get spherical coordinates locally. The situation is described in Figure 8.1.

First choose a positive real r sufficiently small and a point $p \in V$ at a distance r from x_0 such that the ball with center at p and radius r is contained in V. We are going to show that for some real $T > 0$, the section s is equal 0 in the ball with center at p and radius $r + \frac{T}{2}$. This proves Theorem 8.2 since X is connected.

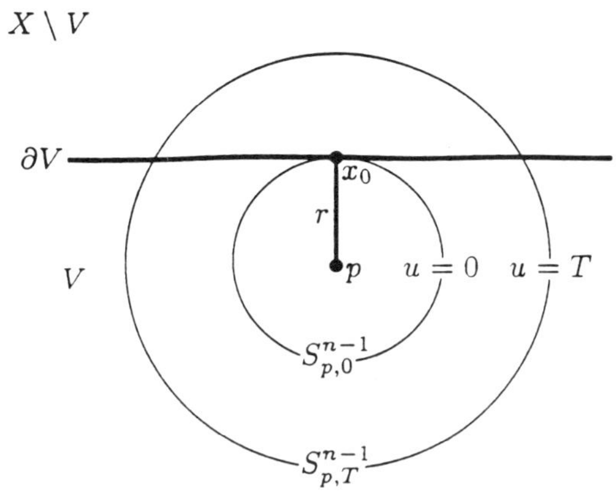

Fig. 8.1 Local specification for the Carleman estimate

The result is a consequence of the following estimate of Carleman type:

$$
\begin{aligned}
(8.1)\quad R\int_{u=0}^{T}\int_{S^{n-1}_{p,u}} & e^{R(T-u)^2}\,\|v(u,y)\|^2\,dy\,du \\
&\le C\int_{u=0}^{T}\int_{S^{n-1}_{p,u}} e^{R(T-u)^2}\,\|Av(u,y)\|^2\,dy\,du\,,
\end{aligned}
$$

where R is a sufficiently large positive number and

$$
v(u,y) := \varphi(u)s(u,y)
$$

with a smooth cut-off function φ chosen in such a way that $\varphi(u) = 1$ for $u \le \frac{8T}{10}$ and $\varphi(u) = 0$ for $u > \frac{9T}{10}$. It follows (see Figure 8.1) that

$$
(8.2)\quad \operatorname{supp} v \subset [0, \tfrac{8T}{10}] \times S^{n-1}_{p,0} \quad \text{and} \quad \operatorname{supp} Av \subset [\tfrac{8T}{10}, \tfrac{9T}{10}] \times S^{n-1}_{p,0}\,.
$$

The proof of Theorem 8.2 is split into two lemmas. First we show that (8.1) ensures that s also vanishes in the annular region $u < \frac{T}{2}$. Next the inequality (8.1) is proven. □

Lemma 8.5. *If (8.1) holds for any $R > 0$ sufficiently large, then s is equal to 0 for $u < \frac{T}{2}$.*

Proof. We have

$$
\begin{aligned}
e^{RT^2/4} \int_0^{\frac{T}{2}} \int_{S^{n-1}_{p,u}} \|s(u,y)\|^2 \, dy \, du
&= \int_0^{\frac{T}{2}} \int_{S^{n-1}_{p,u}} e^{RT^2/4} \, \|s(u,y)\|^2 \, dy \, du \\
&\le c \int_0^{T} \int_{S^{n-1}_{p,u}} e^{R(T-u)^2} \, \|v(u,y)\|^2 \, dy \, du \\
&\overset{(8.1)}{\le} \tfrac{C}{R} \int_0^{T} \int_{S^{n-1}_{p,u}} e^{R(T-u)^2} \, \|Av(u,y)\|^2 \, dy \, du \\
&\overset{(8.2)}{\le} \tfrac{C}{R} e^{RT^2/25} \int_0^{T} \int_{S^{n-1}_{p,u}} \|Av(u,y)\|^2 \, dy \, du \,;
\end{aligned}
$$

hence we have

$$
\begin{aligned}
\text{(8.3)} \quad & \int_0^{\frac{T}{2}} \int_{S^{n-1}_{p,u}} \|s(u,y)\|^2 \, dy \, du \\
& \qquad \le \tfrac{C}{R} e^{-21RT^2/100} \int_0^{T} \int_{S^{n-1}_{p,u}} \|Av(u,y)\|^2 \, dy \, du \,,
\end{aligned}
$$

which gives the result as $R \to \infty$. □

Now (8.1) must be established.

Lemma 8.6. *For T sufficiently small there exists a constant C such that the inequality (8.1) holds for any positive R.*

Proof. First consider a few technical points. The Dirac operator A has the form $G(u)(\partial_u + B_u)$ on the annular region $[0,T] \times S^{n-1}_{p,0}$, and it is obvious that we may consider the operator $\partial_u + B_u$ instead of A. The operator B_u is a self-adjoint elliptic differential operator on $S^{n-1}_{p,u}$. Note that the metric structures depend on the normal variable u.

To simplify the computations, deform the Riemannian and the Hermitian structure in a neighbourhood of $S^{n-1}_{p,u}$ such that they do not depend on u; but keep the operator A fixed. Still it is enough to prove (8.1) for the new metric. Then, however, the tangential part B_u is not a self-adjoint operator with respect to the new structures, but its self-adjoint parts

$$B_+ = (B_u)_+ := \tfrac{1}{2}(B_u + B_u^*)$$

are elliptic differential operators (at least when T is sufficiently small). Now make the substitution

$$v =: e^{-R(T-u)^2/2} v_0 \tag{8.4}$$

which replaces (8.1) by

$$\begin{aligned} R \int_0^T \int_{S_{p,u}^{n-1}} \|v_0(u,y)\|^2 \, dy \, du \\ \leq C \int_0^T \int_{S_{p,u}^{n-1}} \left\| \tfrac{\partial v_0}{\partial u} + B v_0 + R(T-u) v_0 \right\|^2 dy \, du \, . \end{aligned} \tag{8.5}$$

Now we prove (8.5). Decompose $\frac{\partial}{\partial u} + B + R(T-u)$ into its symmetric part $B_+ + R(T-u)$ and its anti-symmetric part $\partial_u + B_-$, where

$$B_- := \tfrac{1}{2}(B - B^*) \, .$$

This gives

$$\begin{aligned} &\int\int \left\| \frac{\partial v_0}{\partial u} + B v_0 + R(T-u) v_0 t \right\|^2 dy \, du \\ &= \int\int \left\| \frac{\partial v_0}{\partial u} + B_- v_0 \right\|^2 dy \, du \\ &+ \int\int \|(B_+ + R(T-u)) \, v_0\|^2 \, dy \, du \\ &+ 2\Re \int\int \left\langle \frac{\partial v_0}{\partial u} + B_- v_0 ; B_+ v_0 + R(T-u) v_0 \right\rangle dy \, du \, . \end{aligned} \tag{8.6}$$

Integrate by parts and use the identity for the real part

$$\Re\langle f; Pf \rangle = \frac{1}{2} \langle f; (P + P^*) f \rangle$$

in order to investigate the last term on the right side of (8.6). This yields

$$
\begin{aligned}
2\Re\int\!\!\int\left\langle \frac{\partial v_0}{\partial u} + B_- v_0; B_+ v_0 + R(T-u)v_0\right\rangle dy\,du
&= 2\Re\int\!\!\int\left\langle \frac{\partial v_0}{\partial u}; B_+ v_0 + R(T-u)v_0\right\rangle dy\,du \\
&\quad + 2\Re\int\!\!\int\langle B_- v_0; B_+ v_0\rangle\, dy\,du \\
&= -2\Re\int\!\!\int\left\langle v_0; \left\{\frac{\partial}{\partial u}(B_+ + R(T-u))\right\} v_0\right\rangle dy\,du \\
&\quad - 2\Re\int\!\!\int\langle v_0; B_- B_+ v_0\rangle\, dy\,du \\
&= \int\!\!\int\left\langle v_0; -\frac{\partial B_+}{\partial u} v_0 + R v_0\right\rangle dy\,du \\
&\quad + \int\!\!\int\langle v_0; [B_+; B_-] v_0\rangle\, dy\,du \\
&= R\int_0^T \|v_0\|_0^2\, du \\
&\quad + \int\!\!\int\left\langle v_0; -\frac{\partial B_+}{\partial u} v_0 + [B_+; B_-] v_0\right\rangle dy\,du\,,
\end{aligned}
\tag{8.7}
$$

where $\|\cdot\|_m$ denotes the m-th Sobolev norm on $S|_{S^{n-1}_{p,u}}$. It follows from (8.6) and (8.7) that the proof of (8.5) will be completed when it is shown that

$$
\begin{aligned}
&\left|\int\!\!\int\left\langle v_0; -\frac{\partial B_+}{\partial u} v_0 + [B_+; B_-] v_0\right\rangle dy\,du\right| \\
&\qquad \le k\left(R\int_0^T \|v_0\|_0^2\, du + \int_0^T \|(B + R(T-u))v_0\|^2\, du\right)
\end{aligned}
\tag{8.8}
$$

for some constant $0 < k < 1$. The operators $B_+ = (B_u)_+$ are elliptic of first order, hence

$$\|f\|_1 \le c\,(\|f\|_0 + \|B_+ f\|)$$

for any section f of S on $S^{n-1}_{p,u}$ (and $0 \le u \le T$). Then

(8.9)
$$\left|\int\int \left\langle v_0; -\frac{\partial B_+}{\partial u} v_0 + [B_+;B_-]v_0 \right\rangle dy\, du\right|$$
$$\le \int_0^T \|v_0\|_0 \left\| -\frac{\partial B_+}{\partial u} v_0 + [B_+;B_-]v_0 \right\|_0 \le c_1 \int_0^T \|v_0\|_0 \, \|v_0\|_1 \, du$$
$$\le c_1 c \int_0^T \|v_0\|_0 \left(\|Bv_0\|_0 + \|v_0\|_0\right) du$$
$$\le c_1 c \int_0^T \|v_0\|_0 \left\{\|(B + R(T-u))v_0\|_0 + (R(T-u)+1)\|v_0\|_0\right\} du$$
$$\le c_1 c(RT+1) \int_0^T \|v_0\|_0^2 \, du$$
$$+ c_1 c \int_0^T \|(B_+ + R(T-u))v_0\|_0 \, \|v_0\|_0 \, du\,.$$

The integrand of the second summand is equal to

$$(8.10) \quad \frac{\|(B_+ + R(T-u))v_0\|_0}{\sqrt{c_1 c}} \left(\sqrt{c_1 c}\, \|v_0\|_0\right)$$
$$\le \tfrac{1}{2}\left\{\frac{1}{c_1 c}\|(B_+ + R(T-u))v_0\|_0^2 + c_1 c \|v_0\|_0^2\right\}$$

with the inequality due to the estimate $ab < \frac{1}{2}(a^2+b^2)$. By inserting (8.10) in (8.9) we obtain

$$\left|\int\int \langle v_0; \dots\rangle \, dy\, du\right| \le \frac{1}{2}\int_0^T \|(B_+ + R(T-u))v_0\|_0^2 \, du$$
$$+ c_1 c R\left(T + \frac{c_1 c + 2}{R}\right)\int_0^T \|v_0\|_0^2 \, du\,.$$

So the desired result holds for T and $\frac{1}{R}$ sufficiently small. □

Remark 8.7. Clearly the split Dirac operators introduced in Chapter 7 inherit the unique continuation property of the full Dirac operator, since, by decomposition of A, any solution u of $A^+u = 0$ provides a solution $A\begin{pmatrix} u \\ 0 \end{pmatrix} = 0$.

9. Invertible Doubles

We discuss an important *extension* result for Dirac operators on manifolds with boundary which provides an invertible differential operator on the closed double.

Next we derive an important *extension* result for operators of Dirac type on manifolds with boundary. Once again it holds for any elliptic operator of first order, but the restriction simplifies the notation. We recall: Our main object on even-dimensional manifolds is the operator

$$A^+ : C^\infty(X; S^+) \to C^\infty(X; S^-) .$$

In a collar neighbourhood of the boundary Y of X it takes the product form

$$A^+ = G(\partial_u + B_u) ,$$

where G is a unitary morphism $G : S^+ \longrightarrow S^-$ and B_u a family of Dirac operators over Y. When investigating the η-invariant, the focus is on *odd* dimensions and on $B = B_0$. (Here the η-function of B will appear only a few times.) If the odd-dimensional compact manifold Y is not closed, but has a boundary Z, then, in a collar neighbourhood of Z, the odd-dimensional Dirac operator B takes the form

$$B = \Gamma(\partial_u + C) = \begin{pmatrix} i & 0 \\ 0 & -i \end{pmatrix} \cdot \left(\partial_u + \begin{pmatrix} 0 & C^- \\ C^+ & 0 \end{pmatrix} \right)$$

with even-dimensional Dirac operators C, C^+, C^- over Z. They play the role of A, A^+, A^-. This time, however, we have $\Gamma^2 = -\operatorname{Id}$.

Since we are mainly interested in the even-dimensional case, we concentrate on the extension of the operator A^+. Also the treatment becomes more transparent due to the chiral notation. However, all arguments can easily be transferred to the odd-dimensional case. Transcriptions are left to the reader as an exercise.

Theorem 9.1. *Let $A^+ : C^\infty(X; S^+) \to C^\infty(X; S^-)$ be a Dirac operator on an even-dimensional manifold X with boundary Y. There* *exists a closed manifold* $\widetilde{X}$*, which contains* X*, and a Dirac operator* $\widetilde{A^+}$ *on* $\widetilde{X}$ *such that*

$$\widetilde{A^+}|_X = A^+ \quad \textit{and} \quad \widetilde{A^+} \text{ is an invertible operator.} \tag{9.1}$$

More precisely,

$$\ker \widetilde{A^+} = \{0\} = \operatorname{coker} \widetilde{A^+} \tag{9.2}$$

and $(\widetilde{A^+})^{-1}$ *is an elliptic pseudo-differential operator of order* -1.

Note. The existence of an invertible extension of A^+ (and of A^- as well) is a nice property of Dirac operators which is not found in the literature. We shall use it in Chapter 12 in order to construct the Calderón projection onto the Cauchy data of A^+.

Proof. It is standard that A^+ takes the form

$$A^+ = G(u,y)(\partial_u + B_u) \qquad \text{in a collar neighbourhood } N \text{ of } Y \text{ in } X. \tag{9.3}$$

Here u denotes the inward oriented normal coordinate which gives a parametrization of N as $I \times Y$; $G(u)$ is Clifford multiplication by the unit inward vector, hence $G(u,y)$ is a unitary bundle isomorphism; B_u denotes the corresponding (total, self-adjoint) Dirac operator on the manifold $Y \times \{u\}$ which may be identified with Y. In general, G and B depend on the u-coordinate. Only in the cylindrical case can we obtain that the B_u equal a fixed Dirac operator $B : C^\infty(Y; S^+|_Y) \to C^\infty(Y; S^+|_Y)$ independent of u, and that Clifford multiplication with the unit inward tangent vector in a distance u from a boundary point y yields the same as Clifford multiplication at the boundary point under identification of the fibres $S^+_{u,y}$ and $S^+_{0,y} = S^+_y$ by local orthonormal frames.

To eliminate the u-dependence and obtain a true product form of A^+, we therefore follow the usual procedure and attach a cylinder $(-1,0] \times Y$ to X and extend the operator A^+ in such a way that, on a small collar neighbourhood of $\{-\frac{1}{3}\} \times Y$, it takes the form

$$A^+ = G(y)(\partial_u + B) \qquad \text{close to } \{u = -\tfrac{1}{3}\}, \tag{9.4}$$

where G and B do not depend on the normal coordinate u. This can always be done and corresponds to a deformation of the Riemannian structure on X and of the Hermitian structure on S into a *product* in a collar neighbourhood of the boundary. Therefore, A^+ extends to an operator A_1^+ on the manifold $X_1 := X \cup ((-1,0] \times Y)$ such that A_1^+ is still of Dirac type and that A_1^+ restricted to $(-1,-\frac{1}{2}] \times Y$ is of the form (9.4).

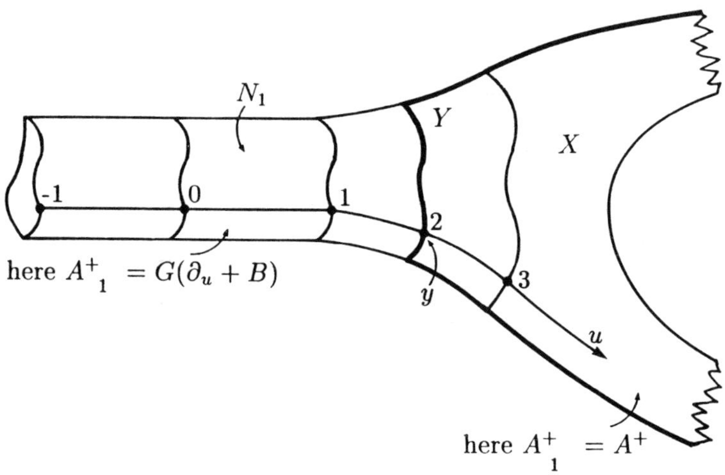

Fig. 9.1 The extension of A^+ into product form

To simplify the notation we change the parametrization in the normal direction in such a way that we obtain a collar neighbourhood $N_1 := (-1,3) \times Y$ with the following properties:

(9.5)

$$X = X_1 \setminus ((-1,2] \times Y),$$

$$A_1^+|_X = A^+ \quad \text{and} \quad A_1^+ = G(\partial_u + B) \quad \text{on } (-1,1) \times Y, \quad \text{and}$$

$$A_1^+ : C^\infty(X_1; S_1^+) \to C^\infty(X_1; S_1^-) \quad \text{of Dirac type},$$

(see Figure 9.1). Of course the point is that the extensions $S_1^\pm$ of $S^\pm$ are again spinor bundles of positive/negative *chirality*, namely, as splittings of the extension of the $C\ell(X)$-module S with compatible connection D to a $C\ell(X_1)$-module S_1 with compatible connection D_1. That means that A_1^+ is still a true Dirac operator and does not lose essential properties, like the unique continuation property, under the deformations and extensions. To avoid excessive use of subscripts, denote the extended bundles $S_1^\pm$ again by $S^\pm$ and the extended Dirac operator A_1^+ again by A^+.

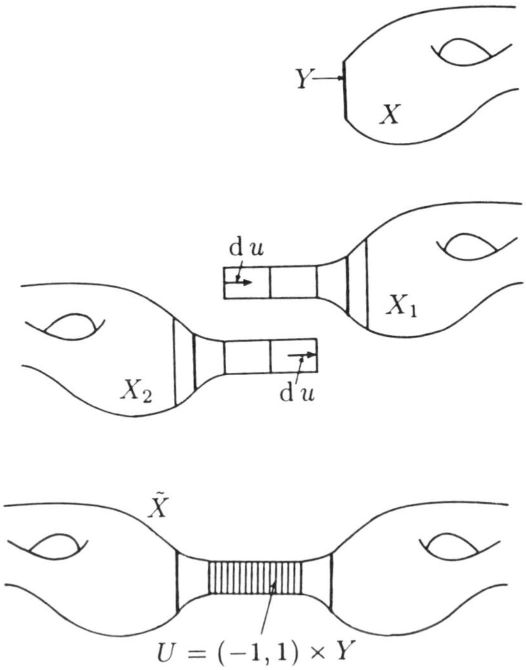

Fig. 9.2 The making of $\widetilde{X}$

Now take $X_2 := (-X_1)$, a copy of X_1 with reversed orientation. Hence its corresponding collar neighbourhood of Y is of the form $N_2 = [-3,1) \times Y$. Glue these two copies together to obtain the (extended) double $\widetilde{X}$ with $U := X_1 \cap X_2 = (-1,1) \times Y$ in $\widetilde{X}$; see Figure 9.2. To introduce the bundles $\widetilde{S}^+$ and $\widetilde{S}^-$, take the bundles S^+ over X_1 and S^- over X_2 and identify any element $s \in S^+_{u,y}$ with the element $G(y)s \in S^-_{u,y}$ for $(u,y) \in U$. Notice that Clifford multiplication with a tangent vector by definition toggles between S^+ and S^-; see (1.5) and (7.5)/(7.6). However, the orientation on X_2 has been reversed so that the *negative* spinors become the *positive* ones. Therefore denote the bundle obtained in this way by

$$\widetilde{S^+} := S^+ \cup_G S^-. \tag{9.6}$$

In the same way define the bundle of spinors of negative chirality on $\widetilde{X}$:

$$\widetilde{S^-} := S^- \cup_{G^{-1}} S^+. \tag{9.7}$$

A smooth section of $\widetilde{S^+}$ is a couple (s_1, s_2) such that s_1 is a smooth section of S^+ over X_1, s_2 is a smooth section of S^- over X_2, and on U the following coupling condition is satisfied (see Figure 9.3):

$$s_2(u,y) = G(y)s_1(u,y) \quad \text{for } y \in Y \text{ and } u \in (-1,1)\,. \tag{9.8}$$

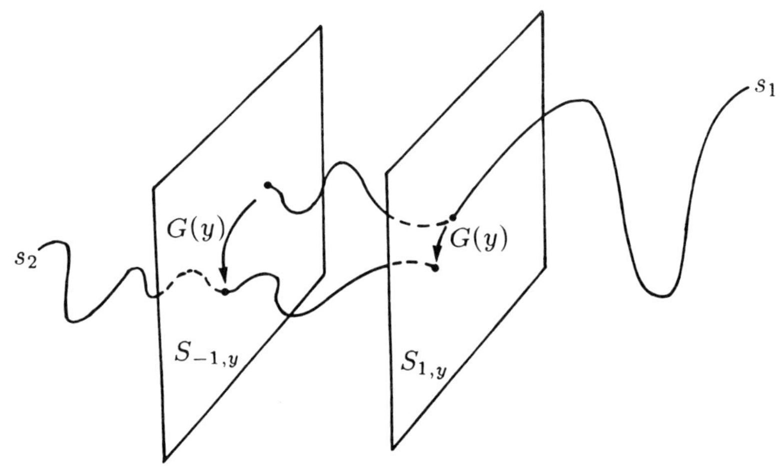

Fig. 9.3 Cut through a global section

Now we shall show that $\widetilde{A^+}$ defined by the formula

$$\widetilde{A^+}(s_1, s_2) := (A^+s_1, A^-s_2)$$

(notice $A^- = (A^+)^*$ by Theorem 7.3) provides a well-defined elliptic differential operator of first order $\widetilde{A^+} : C^\infty(\widetilde{X}; \widetilde{S^+}) \to C^\infty(\widetilde{X}; \widetilde{S^-})$. In fact, by (9.6) and (9.7), the bundles $\widetilde{S}^\pm$ can be considered as the *positive* and *negative* chirality subbundles of a well-defined $C\ell(\widetilde{X})$-module $\widetilde{S}$ with compatible connection; and $\widetilde{A^+}$ appears just as the Dirac operator associated with the bundle $\widetilde{S^+}$. To prove this we only have to check the pasting condition on U for the couple (A^+s_1, A^-s_2). We need to obtain a section of $\widetilde{S^-}$. In order to check the compatibility of A^+s_1 with A^-s_2 in the sense of (9.7), recall that s_1 and s_2 satisfy (9.8) and that the operator A^- is formally adjoint to A^+. This implies that A^- is of the following form on U:

$$A^- = (\partial_u + B)G^{-1} = G^{-1}(\partial_u + GBG^{-1}) \quad \text{for } u \in (-1,1)\,. \tag{9.9}$$

The choice of signs in (9.9) needs an explanation: Formally, the formal adjoint to an operator of the form (9.4) is given by $(-\partial_u+B)G^{-1}$. But this is the formula on X_1, where u is the inward normal coordinate. On X_2 the coordinate u is the outward normal coordinate. Therefore we have to change the sign in front of ∂_u. Now it is easy to check the pasting condition (9.7). We have the following equalities on U:

$$\begin{aligned} A^- s_2 &= (\partial_u + B)G^{-1}s_2 = (\partial_u + B)G^{-1}(Gs_1) = (\partial_u + B)s_1 \\ &= G^{-1}G(\partial_u + B)s_1 = G^{-1}A^+ s_1 \,. \end{aligned} \tag{9.10}$$

From the construction it follows that

$$(A^+ \cup A^-)^* \;=\; A^- \cup A^+ \,; \tag{9.11}$$

hence index $\widetilde{A^+} = 0$. Now we want to show that $\widetilde{A^+}$ is a truly invertible operator. We postpone the proof that $\widetilde{A^+}$ is invertible until after the next lemma, and the proof that hence $(\widetilde{A^+})^{-1}$ is an elliptic pseudo-differential operator of order -1 to Proposition 9.5. □

Lemma 9.2. *Let X be a closed partitioned Riemannian manifold of even dimension, $X = X_1 \cup X_2$ with $X_1 \cap X_2 = Y = \partial X_1 = \partial X_2$, and let $A^+ : C^\infty(X; S^+) \to C^\infty(X; S^-)$ be a Dirac type operator, where $S^\pm$ are the $\pm$-chirality components of a $C\ell(X)$-module S with compatible connection. Then any $s_1 \in C^\infty(X_1; S^+|_{X_1})$ with $A^+ s_1 = 0$ and $s_1|_Y = 0$ can be continued to a smooth solution for the operator A^+ over the whole manifold X by setting*

$$\widetilde{s} := \begin{cases} s_1 & \text{on } X_1 \\ 0 & \text{on } X \setminus X_1 \,. \end{cases}$$

Proof. We show that the extension of s_1 by zero is a weak solution for A^+ over the whole of X. Apply Green's formula (3.9):

$$\begin{aligned} \langle \widetilde{s}; A^- v\rangle &= \int_M \langle \widetilde{s}(x); A^- v(x)\rangle \, dvol(x) \\ &= \int_{X_1} \langle s_1(x); A^- v(x)\rangle \, dvol(x) + \int_{X_2} 0 \\ &= \int_{X_1} \langle A^+ s_1(x); v(x)\rangle \, dvol(x) \\ &\quad + \int_Y \langle G(0,y)s_1(y); v(y)\rangle \, dvol(y) = 0 \end{aligned}$$

for any $v \in C^\infty(X; S^-)$. By the regularity of the solutions of elliptic equations over closed manifolds it follows that $\widetilde{s} \in C^\infty(X; S^+)$ and $\operatorname{supp} \widetilde{s} \subset X_1$. □

Proposition 9.3. *The operator $\widetilde{A^+}$ defined above is an invertible operator (i.e. has trivial kernel and cokernel).*

Proof. For any $s \in \ker \widetilde{A^+}$ we show that $s|_Y = 0$. Recall from the definition of $\widetilde{A^+}$ that s can be written as a couple $(s_1, s_2) \in C^\infty(\widetilde{X}; \widetilde{S^+})$ subject to the coupling condition (9.8). Then $s \in \ker \widetilde{A^+}$ means $A^+ s_1 = 0$ and $A^- s_2 = 0$; hence

$$
\begin{aligned}
0 &= \langle A^+ s_1; s_2 \rangle - \langle s_1; A^- s_2 \rangle \\
&\overset{\text{Green's formula}}{=} - \int_Y \langle G(s_1|_Y); s_2|_Y \rangle \, dy = \|s_2|_Y\|^2 .
\end{aligned}
$$

It follows from Lemma 9.2 and the unique continuation property (Theorem 8.2 and Remark 8.7) that $s_1 \equiv 0$ and $s_2 \equiv 0$, hence $\ker(\widetilde{A^+}) = \{0\}$. We apply the same argument to the operator $\widetilde{A^-} = A^- \cup A^+$ to show that

$$\operatorname{coker} \widetilde{A^+} \cong \ker(\widetilde{A^+})^* = \ker \widetilde{A^-} = \{0\} .$$ □

To complete the proof of Theorem 9.1, we must show that the inverse $(\widetilde{A^+})^{-1}$ is an elliptic *pseudo-differential* operator of order -1. This follows from standard pseudo-differential theory. We begin with a proposition from elementary functional analysis which we shall also apply later on.

Proposition 9.4. *Let $T : H_1 \to H_2$ be a bounded operator acting between (complex separable) Hilbert spaces H_1, H_2. Then* range(T) *is closed, if and only if the inverse operator of $T|_{(\ker T)^\perp}$ is bounded.*

Proof. Let $\widetilde{T}$ denote the operator $T|_{(\ker T)^\perp}$. Assume that range$(T)=$ dom$(\widetilde{T})^{-1}$ is closed. Then the operator $(\widetilde{T})^{-1}$ is bounded by the closed graph theorem.

For the opposite direction we give an elementary proof. Assume that there exists $c > 0$ such that

$$\|(\widetilde{T})^{-1} v\| < c\|v\| \qquad \text{for any } v \in \operatorname{range}(T) \tag{9.12}$$

and that range(T) is not closed. Hence we have a sequence

$$\{x_n\} \subset (\ker T)^{\perp} \qquad \text{with } Tx_n \to y \neq \text{range}(T)\,.$$

Next notice that $\{x_n\}$ is an unbounded sequence. If $\|x_n\| < k$ for some $k > 0$, then $\{x_n = (\widetilde{T})^{-1}(Tx_n)\}$ would be a Cauchy sequence (because $(\widetilde{T})^{-1}$ is bounded) and we would have

$$x := \lim_{n\to\infty} x_n \in (\ker T)^{\perp} \qquad \text{and} \qquad y = Tx\,.$$

Finally, by eventually passing to a subsequence, we may assume that

$$\|x_n\| > n \text{ and } \|Tx_n\| < \|y\| + 1\,.$$

This yields

$$\|(\widetilde{T})^{-1}(Tx_n)\| = \|x_n\| > n\,,$$

hence $(\widetilde{T})^{-1}$ is not bounded. □

Proposition 9.5. *Let $A : C^{\infty}(M;E) \to C^{\infty}(M : F)$ be a bijective elliptic pseudo-differential operator of order m over a closed manifold M. Then A^{-1} is a pseudo-differential operator of order $-m$.*

Proof. Since $A^{-1} = (A^*A)^{-1}A^*$, we can assume that A is a self-adjoint and non-negative operator on $C^{\infty}(M;E)$. Now A is a Fredholm operator; hence for any k the extension

$$A^{(k)} : H^k(M;E) \to H^{k-m}(M;E)$$

of A has a closed range. By the preceding Proposition 9.4, this means that the inverse operator of $A^{(k)}$ is a bounded operator

$$(A^{(k)})^{-1} : H^{k-m}(M;E) \to H^k(M;E)\,.$$

Let Q denote a parametrix of A

$$QA = \mathrm{Id} + K\,, \tag{9.13}$$

where K is a smoothing operator. We multiply (9.13) by A^{-1} from the right which yields the following formula for A^{-1}:

$$A^{-1} = Q - KA^{-1}\,. \tag{9.14}$$

The operator KA^{-1} is a smoothing operator, hence it has a smooth kernel and A^{-1} is a pseudo-differential operator of order $-m$. □

Remarks 9.6. (a) The operator $\widetilde{A^+}$ is not a self-adjoint operator though it is isomorphic to $\widetilde{A^-} = (\widetilde{A^+})^*$ by (9.11). To be precise, let $J : \widetilde{X} \to \widetilde{X}$ denote the isometry which interchanges the two factors X_1 and X_2 of $\widetilde{X}$ by mapping x of the one copy into $J(x) := x$ of the other copy. The pullback $J^*(\widetilde{S^+})$ is equal to $\widetilde{S^-}$ and we can extend J to a bundle automorphism of $\widetilde{S^+}$ which on sections is given by the formula

$$J(s_1, s_2) := (s_2, s_1) .$$

We have the following formula:

$$J\widetilde{A^+} = \widetilde{A^-}J$$

which gives the required isomorphism.
(b) The construction of operators of the form $A \cup A^*$ was introduced in Wojciechowski [1985b], where the analytical realization of the relative cycles for the K-homology group $K_0(X, Y)$ is discussed. See Douglas & Wojciechowski [1989] for a detailed exposition of this aspect of the theory. There the odd-dimensional case is also treated and operators of the form $B \cup (-B)$ are constructed, where B is a formally self-adjoint Dirac operator. Our construction is also applied in some further work, where we discuss spectral invariants of self-adjoint operators and boundary problems; see Booß & Wojciechowski [1989], Douglas & Wojciechowski [1991], Klimek & Wojciechowski [1992], [1993], and Wojciechowski [1993].

We can apply the glueing construction in a more general context as well. We only need

(1) a partitioning of a closed manifold $M = U \cup V$ into two submanifolds with common boundary Y , and
(2) Dirac operators A_1 on U and A_2 on V which are equal on Y, i.e. with common Green form G and boundary Dirac operator B.

Then we can always construct the operator $A_1 \cup A_2^*$ on M. The details are left to the reader; see also Wojciechowski [1985b].

10. Glueing Constructions. Relative Index Theorem

We explain the relative index theorem of Cheeger, Gromov & Lawson and offer an analytical interpretation which also works in the odd-dimensional case. We also give a simple proof of a special case of the excision principle for indices.

Here we want to give an analytical interpretation of the relative index theorem of Cheeger-Gromov-Lawson (see Gromov & Lawson [1983] and Lawson & Michelsohn [1987]). We set aside some definitions and technical details, since there are quite a few excellent expositions of the theorem including also various K-theoretical interpretations (see also Borisov, Müller & Schrader [1988], Julg [1988], and Bunke [1992a], [1992b]). We only want to offer a simple analytical interpretation that works in the odd-dimensional case as well.

The situation is as follows: We have E_1 and E_2 bundles of Clifford modules over X_1 and X_1 respectively, where X_1 and X_2 are two complete Riemannian manifolds (not necessarily compact). Assume that there exist K_j compact subsets of X_j, such that there exists $\Phi : E_1|_{X_1 \setminus K_1} \to E_2|_{X_2 \setminus K_2}$ a unitary isomorphism of Clifford bundles. Let f denote an isometry of $X_1 \setminus K_1$ onto $X_2 \setminus K_2$ covered by Φ. This assumption implies that the corresponding Dirac operators D_j are equal outside of K_j, $j = 1, 2$. Now assume that $\dim X_j$ is even. Then the following fact is straightforward.

Proposition 10.1. *Let $\mathcal{E}_j(t; x, y)$ denote the kernel of the operator $e^{-tD_j^2}$. Then*

(10.1)

$$\operatorname{index}(D_1, D_2) := \lim_{t \to 0} \left[\int_{X_1} \operatorname{str} \mathcal{E}_1(t; x, x) dvol_1 - \int_{X_2} \operatorname{str} \mathcal{E}_2(t; x, x) dvol_2 \right]$$

is a well-defined integer which we call the relative index of the operators D_1 and D_2. Here the symmetric trace is defined by

$$\operatorname{str} e^{-tD_j^2} = \operatorname{tr} e^{-tD_j^- D_j^+} - \operatorname{tr} e^{-tD_j^+ D_j^-}$$

as usual.

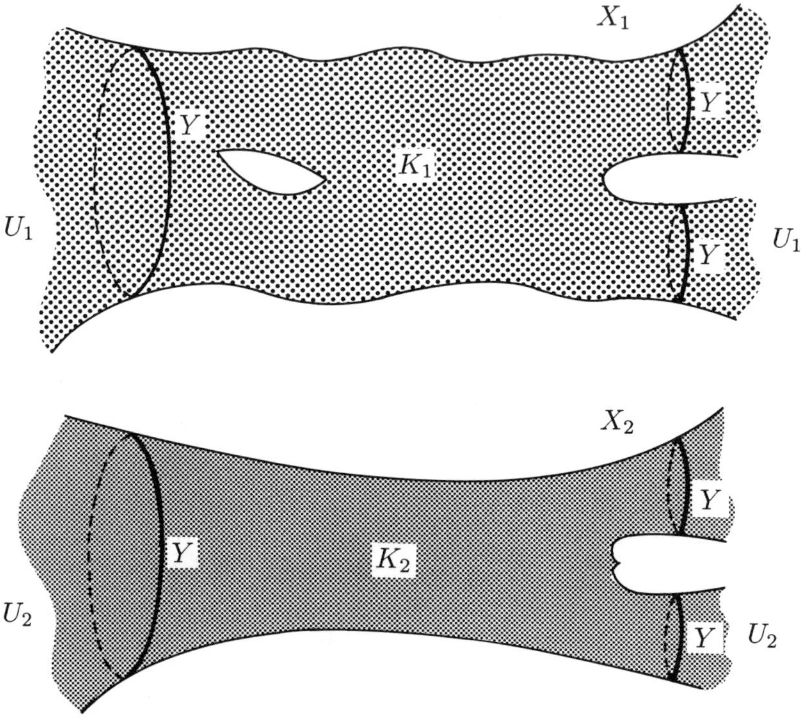

Fig. 10.1 The D_j live on X_j

Proof. Recall that $\alpha_j(x) = \lim_{t\to 0} \operatorname{str} \mathcal{E}_j(t;x,x)$ is the index density of the operator D_j and is given by a local expression which only depends on the coefficients of the operator D_j at the point x (see Gilkey [1984], Berline, Getzler & Vergne [1992]). Therefore, we have to integrate only over the compact sets K_j. □

We assume that $U_j := X_j \setminus K_j$ is an open submanifold such that ∂U_j is equal to a fixed Riemannian manifold Y. We can eventually deform the metrics in a bicollar neighbourhood $N \approx I \times Y$ of Y in X_j in such a way that we obtain a product metric in N. As a result we obtain an operator D'_j which is the perturbed operator D_j. Now we can construct an operator $A = A(N, \Phi)$ on the manifold $K_1 \cup_Y K_2$ (see also Figure 10.1 and Figure 10.2):

$$A = A(N, \Phi) := D'_1|_{K_1} \cup_Y (D'_2)^*|_{K_2} \,. \tag{10.2}$$

Theorem 10.2. *Under the assumptions above the following formula holds:*

$$\operatorname{index}(D_1, D_2) = \operatorname{index} A\,. \tag{10.3}$$

In particular there always exists a Dirac operator A on the closed manifold $K_1 \cup_Y K_2$ such that it realizes the relative index of the operators D_1 and D_2. The index of the operator $A(N, \Phi)$ does not depend on the choice of the product structure on N.

The proof of this theorem follows from the construction of A and the local index theorem.

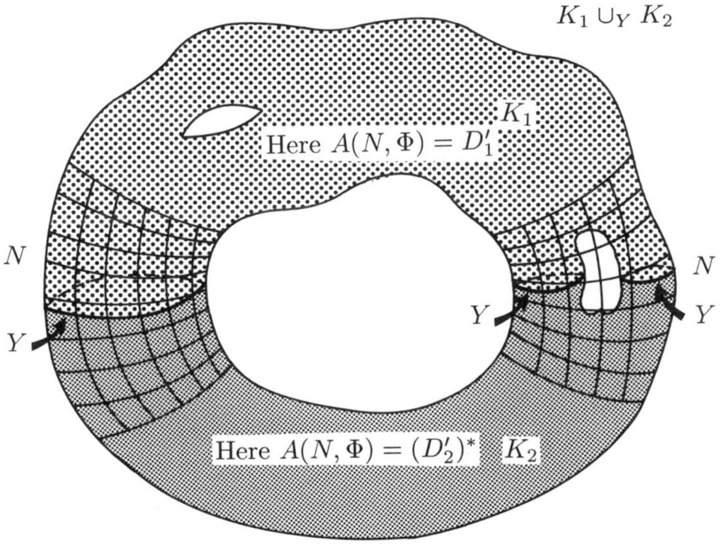

Fig. 10.2 $A(N,\Phi)$ lives on $K_1 \cup_Y K_2$

Remark 10.3. Following the presentation of Douglas & Wojciechowski [1991], we can formulate an *odd* version of the relative index theorem. Once again we can show that the difference between the integrals of the corresponding *local* η-densities is well-defined and in fact equal to the η-invariant of a suitable Dirac operator on $K_1 \cup_Y K_2$. Since the technical details are quite complicated and assume the application of the Duhamel principle, we shall not discuss the *odd* case in this presentation.

We can also apply our glueing construction to prove the excision principle for indices. We obtain an explicit operator expression for the differences. Following Donaldson & Kronheimer [1990; Section 7.1] (see also Seeley [1965, Appendix]) we consider two sets of data

$$(M;\ U,\ V,\ D_1,\ D_2) \quad \text{and} \quad (N;\ W,\ Z,\ A_1,\ A_2),$$

where M is a closed Riemannian manifold (compact and without boundary) which is the union of two open submanifolds $M = U \cup V$. Let

$$D_i : C^\infty(M; E_i) \longrightarrow C^\infty(M; F_i),\ i = 1, 2$$

denote a pair of elliptic differential operators over M. Assume that there exist bundle isomorphisms

$$\Phi : E_1|_V \longrightarrow E_2|_V \quad \text{and} \quad \Psi : F_1|_V \longrightarrow F_2|_V$$

such that

$$D_2 \quad = \quad \Psi D_1 \Phi^{-1} \qquad \text{over } V.$$

We make corresponding assumptions about the second set of data, the second quintuple, i.e. roughly speaking: $N = W \cup Z$ and A_1 is isomorphic to A_2 over Z. (See Figure 10.3.)

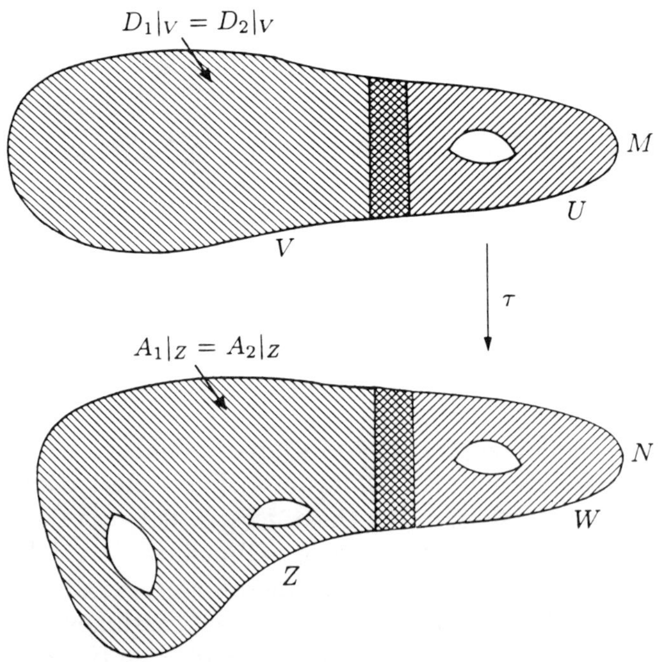

Fig. 10.3 The excision data: two quintuples

In addition to Seeley and Donaldson & Kronheimer, we make two further assumptions. First consider only (compatible) Dirac operators. Second, assume that $U \setminus V$ has a smooth boundary Y and that all structures are products in a bicollar neighbourhood of Y. This gives the following variant of the relative index theorem:

Proposition 10.4. *Let a quintuple $(M; U, V, D_1, D_2)$ be given, i.e. a closed Riemannian manifold $M = U \cup V$ given as the union of two open submanifolds and two Dirac operators D_1, D_2 over M which coincide under suitable bundle isomorphisms over V. We assume that $U \setminus V$ has a smooth boundary Y and that all structures are products in a bicollar neighbourhood of Y. Then we have*

$$\operatorname{index} D_1 - \operatorname{index} D_2 = \operatorname{index} D_{1,U} \cup D_{2,U}^* ,$$

where $D_{j,U}$ denotes the restriction

$$D_{j,U} := D_j|_{U \setminus V}, \qquad j = 1, 2 ,$$

and $D_{1,U} \cup D_{2,U}^*$ denotes the Dirac operator obtained over the closed double of $U \setminus V$ by glueing the pieces together along Y.

Now assume that we have two quintuples as above:

$$(M;\ U,\ V,\ D_1,\ D_2) \quad \text{and} \quad (N;\ W,\ Z,\ A_1,\ A_2) .$$

Assume also that there exists a diffeomorphism $\tau : U \to V$ covered by suitable bundle isomorphisms such that

$$D_1|_U \cong A_1|_W \qquad \text{and} \qquad D_2|_U \cong A_2|_W .$$

Theorem 10.5. *Under the preceding assumptions we get the excision property for indices*

$$\operatorname{index} A_1 - \operatorname{index} A_2 = \operatorname{index} D_1 - \operatorname{index} D_2 .$$

Proof.

$$\begin{aligned} \operatorname{index} A_1 - \operatorname{index} A_2 &\overset{\text{Prop. } 10.4}{=} \operatorname{index} A_{1,W} \cup A_{2,W}^* \\ &\overset{\tau}{=} \operatorname{index} D_{1,U} \cup D_{2,U}^* \overset{\text{Prop. } 10.4}{=} \operatorname{index} D_1 - \operatorname{index} D_2 . \end{aligned}$$

□

Remark 10.6. The explicit description of the index differences is due to two circumstances, namely that we deal with Dirac operators and not arbitrary elliptic differential operators, and that we restrict ourselves to the smooth, relatively compact case.

Part II
Analytical and Topological Tools

11. Sobolev Spaces on Manifolds with Boundary

We investigate the continuity properties of taking traces in Sobolev spaces over manifolds with boundary.

Let X be a smooth compact n-dimensional manifold with boundary Y and fix a Riemannian structure on X. Use the Riemannian structure to construct a collar neighbourhood $N = Y \times I$ of the boundary Y and denote the (inward) normal coordinate by $t \in I = [0,1]$. Actually, we may assume that X is a submanifold of a closed (= compact, without boundary) smooth manifold M, e.g. $M = X \cup_Y X$, the closed double of X. We extend the identification $Y \times [0,1] = N$ to an imbedding $\Psi : Y \times [-1,1] \hookrightarrow M$.

Definition 11.1. The following chains of *Sobolev spaces* are naturally associated with our data ($s \in \mathbf{R}$, $s \geq 0$):
(a) The space $H^s(M)$ consists of all complex valued L^2-functions u over M which yield elements of $H^s(\mathbf{R}^n)$ in local coordinates. More precisely: Let $\{U_\iota, \kappa_\iota\}$ with U_ι open in M and $\kappa_\iota : U_\iota \to \mathbf{R}^n$ be a smooth atlas for M. Then, for any smooth function ϕ over M with compact support contained completely in U_ι, there exists an element $u_\iota \in H^s(\mathbf{R}^n)$ such that $\phi \cdot u|_{U_\iota} = u_\iota \circ \kappa_\iota$.
(b) The space $H^s(Y)$ is defined in the same way.
(c) The space $H^s(X)$ consists of the restrictions $\{r^+u \mid u \in H^s(M)\}$, where $r^+ : L^2(M) \to L^2(X)$ denotes the restriction operator $u \mapsto u|_X$.

Note. The preceding concepts extend to $s < 0$ by duality.

For convenience we recall the definition of

$$H^s(\mathbf{R}^n) := \{v \mid v \in L^2(\mathbf{R}^n) \text{ and } \xi \mapsto (1+|\xi|^2)^{s/2}\hat{v}(\xi) \in L^2(\mathbf{R}^n)\},$$

where the Fourier transform is defined by

$$\hat{v}(\xi) := \int_{\mathbf{R}^n} e^{-ix\cdot\xi} v(x)dx\,. \tag{11.1}$$

Then the Fourier inversion formula yields

$$v(x) = (2\pi)^{-n} \int_{\mathbf{R}^n} e^{ix\cdot\xi}\hat{v}(\xi)d\xi\,. \tag{11.2}$$

Note that

$$\|v\|_s := (2\pi)^{-n/2}\|(1+|\xi|^2)^{s/2}\hat{v}(\xi)\|_{L^2(\mathbf{R}^n)} \tag{11.3}$$

provides a norm for $H^s(\mathbf{R}^n)$ which can be regarded as $L^2(\mathbf{R}^n;(1+|\xi|^2)^s)\,d\xi$, the L^2-space with respect to the measure $(1+|\xi|^2)^s$. From this we get norms for the spaces $H^s(M)$ and $H^s(X)$ which depend on the choice of a smooth atlas $\{U_\iota,\kappa_\iota\}$ and a corresponding C^∞-partition $\{\phi_\iota\}$ of the unity with $\phi_\iota \geq 0$, $\operatorname{supp}\phi_\iota \subset U_\iota$, and $\sum_\iota \phi_\iota = 1$:

$$\|u\|_s := \sum \|(\phi_\iota \cdot u)\circ \kappa_\iota^{-1}\|_s \tag{11.4}$$

for $u \in H^s(M)$ and similarly for $u \in H^s(X)$.

It is remarkable that all these norms for $H^s(M)$ are equivalent and define the same topology:

Proposition 11.2. *Let Φ be a diffeomorphism of an open set $U_1 \subset \mathbf{R}^n$ onto another open set $U_2 \subset \mathbf{R}^n$. Let U_2' be an open, relatively compact subset of U_2. Then we have for all s*

$$\|v\circ\Phi\|_s \leq C\|v\|_s \quad \text{for all } v \in C_0^\infty(U_2'),$$

where C is independent of v.

Proof. Accounted for e.g. in Gilkey [1984; p. 28] or Booß & Bleecker [1985; p. 177]. □

Corollary 11.3. *Let U_1 be an open subset of a closed manifold M_1 (or of $\mathbf{R}^n$) and Φ a diffeomorphism of U_1 onto an open subset U_2 of another closed manifold M_2 (or of $\mathbf{R}^n$). Let K be a compact subset of U_1 and let $H^s(K)$ denote the space of elements of $H^s(M_1)$ with support in K. Then $(\Phi^{-1})^* : v \mapsto v\circ\Phi^{-1}$ maps $H^s(K)$ continuously into $H^s(\Phi(K))$.*

The aim of this chapter is to prove the following trace theorem.

Theorem 11.4. *For $u \in C^\infty(M)$ we define $u_t(y) := u(\Psi(y,t))$, where $\Psi : Y \times [-1,1] \hookrightarrow M$ is a fixed parametrization of a collar neighbourhood of Y. Then for $t \in (-1,1)$, the mapping $u \mapsto u_t$ from $C^\infty(M)$ to $C^\infty(Y)$ can be extended to the whole space $H^s(M)$ for $s > 1/2$ by continuity. It provides a continuous linear map $\gamma_t : H^s(M) \to H^{s-1/2}(Y)$ which is uniformly bounded for t belonging to any compact subinterval of $(-1,1)$. For given $u \in H^s(M)$, the* *mapping* $t \mapsto \gamma_t(u) = u_t$ *is a continuous mapping from the interval* $(-1,1)$ *into* $H^{s-1/2}(Y)$.

The delicacy of the trace theorem is due to the fact that we cannot in general apply the bounded Sobolev embedding $H^s(\mathbf{R}^n) \subset C^0(\mathbf{R}^n)$ which is valid only for $s > n/2$. A special feature of Theorem 11.4 is the emphasis on the dependence and independence of all constants of the normal parameter t.

The main ingredient of the proof is the following lemma:

Lemma 11.5. *For each* $t \in \mathbf{R}$, $s \in \mathbf{R}$, *and* $s > 1/2$ *we have*

$$(11.5)\quad \|f_t\|^2_{s-1/2} \leq (2\pi)^{-2}\|f\|^2_s \int_{-\infty}^{\infty} (1+\tau^2)^{-s} d\tau \quad \textit{for } f \in C_0^\infty(\mathbf{R}^n),$$

where f_t *is the restriction of* f *on* $\mathbf{R}^{n-1} \times \{t\}$.

Proof. First assume $t = 0$. Note that $\hat{f}_0(y) = \frac{1}{2\pi}\int_{-\infty}^{\infty} \hat{f}(y,t)dt$ for all $y \in \mathbf{R}^{n-1}$. In fact

$$(11.6)\qquad \hat{f}_0(y) = \int_{\mathbf{R}^{n-1}} e^{-iy\eta} f(\eta,0) d\eta$$
$$= \frac{1}{2\pi}\int_{\mathbf{R}^{n-1}}\int_{-\infty}^{\infty}\int_{-\infty}^{\infty} e^{-iy\cdot\eta}e^{i0\tau}e^{-it\tau} f(\eta,t)\, dt\, d\tau\, d\eta$$

because of the Fourier inversion formula for $f(\eta,0)$,

$$= \frac{1}{2\pi}\int_{-\infty}^{\infty}\int_{\mathbf{R}^n} e^{-iy\cdot\eta-it\tau} f(\eta,\tau)\, d\eta\, d\tau\, dt\,.$$

(Since f has compact support, the multiple integrals converge absolutely and the order of integration can be inverted.) Therefore,

$$|\hat{f}_0(y)|^2 = (2\pi)^{-2}\left|\int_{-\infty}^{\infty} \hat{f}(y,t)dt\right|^2$$
$$\leq (2\pi)^{-2}\left(\int_{-\infty}^{\infty} |\hat{f}(y,t)|(1+y^2+t^2)^{s/2}(1+y^2+t^2)^{-s/2}dt\right)^2$$
$$\leq (2\pi)^{-2}\int_{-\infty}^{\infty} |\hat{f}(y,t)|^2(1+y^2+t^2)^s dt \int_{-\infty}^{\infty}(1+y^2+t^2)^{-s}dt\,.$$

Since $s > 1/2$, the second integral exists and we can apply the Hölder-Schwarz inequality. Now if we make the substitution $t = (1+y^2)^{1/2}\tau$ we get

$$\int_{-\infty}^{\infty}(1+y^2+t^2)^{-s}dt = (1+y^2)^{-s+1/2}\int_{-\infty}^{\infty}(1+\tau^2)^{-s}d\tau\,.$$

Therefore,

$$(1+y^2)^{s-1/2}|\hat{f}_0(y)|^2$$
$$\leq (2\pi)^{-2}\int_{-\infty}^{\infty}(1+\tau^2)^{-s}d\tau\int_{-\infty}^{\infty}|\hat{f}(y,t)|^2(1+y^2+t^2)dt\,,$$

and integrating with respect to y gives the desired result for the restriction $f\mapsto f_t$ for $t=0$.

For general t we make the dilatation $g(y,s):=f(y,s+t)$ and receive

$$\begin{aligned}\|f_t\|_{s-1/2}=\|g_0\|_{s-1/2}&\leq\frac{1}{2\pi}\|g\|_s\Big(\int_{-\infty}^{\infty}(1+\tau^2)^{-s}d\tau\Big)^{1/2}\\&=\frac{1}{2\pi}\|f\|_s\Big(\int_{-\infty}^{\infty}(1+\tau^2)^{-s}d\tau\Big)^{1/2},\end{aligned}$$

because the dilatation of the variable yields only a *phase shift* of the Fourier transform:

$$\begin{aligned}\hat{g}(\xi)&=\int_{\mathbf{R}^n}e^{-iy\eta}e^{-is\tau}f(y,s+t)\,ds\,dy\\&=e^{it\tau}\int_{\mathbf{R}^n}e^{-iy\eta}e^{-i\sigma\tau}f(y,\sigma)\,d\sigma\,dy\\&=e^{it\tau}\hat{f}(\xi)\end{aligned}$$

for $\xi=(\eta,\tau)\in\mathbf{R}^n$, hence $|\hat{g}(\xi)|=|\hat{f}(\xi)|$ and $\|g\|_s=\|f\|_s$. □

Proof of Theorem 11.4. We suppose that $H^s(M)$ and $H^{s-1/2}(Y)$ are equipped with fixed norms $\|.\|_{H^s(M)}$ and $\|.\|_{H^{s-1/2}(Y)}$. We choose an open covering $\{U_\iota\}$ of Y, an atlas of C^∞ charts κ_ι of U_ι onto open subsets of $\mathbf{R}^{n-1}$, and a corresponding C^∞ partition $\{\phi_\iota\}$ of unity. Let t belong to a compact subinterval J of $(-1,1)$. We choose a real C^∞ function χ which is equal 1 on J and has support in $(-1,1)$.

Then for any $s>1/2$ and $u\in C^\infty(M)$ we have

$$\|u_t\|_{H^{s-1/2}(Y)}\leq c_1\sum\|\phi_\iota u_t\|_{H^{s-1/2}(Y)}\,,\tag{11.7}$$

where c_1 does not depend on u. Also

$$\|\phi_\iota u_t\|_{H^{s-1/2}(Y)}=c_2|\chi(t)^{-1}|\cdot\|(v_\iota)_t\|_{H^{s-1/2}(\mathbf{R}^{n-1})}\,,\tag{11.8}$$

where $v_\iota(\kappa_\iota(y),t) := \chi(t)\kappa_\iota^{*-1}(\phi_\iota u_t)(y)$.

Now, using Lemma 11.5, note that

$$\|(v_\iota)_t\|_{H^{s-1/2}(\mathbf{R}^{n-1})} \leq c_3 \|v_\iota\|_{H^s(\mathbf{R}^n)}\,. \tag{11.9}$$

Corollary 11.3 here says

$$\|v_\iota\|_{H^s(\mathbf{R}^n)} \leq c_4 \|u\|_{H^s(M)}\,. \tag{11.10}$$

All the inequalities together imply $\|u_t\|_{H^{s-1/2}(Y)} \leq c\|u\|_{H^s(M)}$ with c independent of t and u. Therefore the mapping $u \mapsto u_t$ can be extended to all of $H^s(M)$, $s > \frac{1}{2}$, by continuity, and yields, for each t, a continuous map of $H^s(M)$ into $H^{s-1/2}(Y)$ which is uniformly bounded for t varying in any compact subinterval of $(-1,1)$.

To finish our proof we shall show that the parametrization $t \mapsto u_t$ is a continuous mapping from $(-1,1)$ into $H^{s-1/2}(Y)$ for any given $u \in H^s(M)$. This follows from the fact that the function $(y,t) \mapsto u_t(y)$ is an infinitely differentiable function in $Y \times (-1,1)$ for $u \in C_0^\infty(M)$ and from the earlier derived continuity properties. □

We shall now give various reformulations and extensions of the trace theorem.

Corollary 11.6. *Theorem 11.4 remains true, if we replace the closed manifold M by the manifold X with boundary. More precisely: For $s > 1/2$ and a given parametrization $\Psi : Y \times [0,1] \hookrightarrow X$ of the collar of Y in X there exists a natural separately continuous mapping $\gamma : H^s(X) \times [0,1) \to H^{s-1/2}(Y)$ which is uniformly bounded on any compact subinterval of* $[0,1)$.

For the proof of this corollary and for later use, we recall various restrictions, extensions, and reflections in function spaces over manifolds with boundary. Not all of the mappings are natural; some depend on the choice of smoothing functions and local coordinates.

Definition 11.7. We distinguish the following natural mappings (for simplicity $s \geq 0$):

(a) The *restriction* operator

$$r^+ : H^s(M) \to H^s(X) \tag{11.11}$$

is given by $u \mapsto u|_X$.

(b) The *extension by zero* operator $e^+ : L^2(X) \to L^2(M)$ is defined by

$$e^+ u := \begin{cases} u & \text{on } X \\ 0 & \text{on } M \setminus X. \end{cases} \tag{11.12}$$

Then the composition $e^+ r^+$ yields the *cut-off operator.*
(c) The *reflection* operator

$$J : u(y,t) \mapsto u(y,-t) \tag{11.13}$$

is only well-defined for functions with support close to Y. It can be defined for the whole function spaces by multiplication with a smooth function χ which is equal to 1 near Y and vanishes outside a neighbourhood of Y.
(d) A *continuous extension* operator of codimension 0

$$\ell_k^+ : H^k(X) \to H^k(M) \quad \text{for } k \in \mathbf{N} \cup \{0\}$$

is given by graded globalized reflection

$$\ell_k^+ u(y,t) := \begin{cases} u(y,t) & \text{if } t \geq 0 \\ \chi(t) \sum_{j=1}^{k+1} a_j u(y,-jt) & \text{if } t < 0\,, \end{cases} \tag{11.14}$$

where χ is a smoothing function equal to 1 close to Y and vanishing outside a neighbourhood of Y and the a_j will be picked to yield

$$\sum_{j=1}^{k+1} (-j)^m a_j = 1 \quad \text{for } m = 0, 1, ..., k\,.$$

The operator ℓ_k^+ extends to continuous operators $\ell^+ : H^s(X) \to H^s(M)$ for $s \leq k$ with the property $r^+ \ell^+ = \mathrm{id}_{H^s(X)}$.
(e) A *continuous extension* operator of codimension 1

$$\mathcal{E} : H^{s-\frac{1}{2}}(Y) \to H^s(M) \quad \text{for } s > \frac{1}{2}$$

is given close to Y in local coordinates by

$$\mathcal{E}_\phi(g)(y,t) := \chi(t) \int_{\mathbf{R}^{n-1}} \phi(\eta) g(y - t\eta)\, d\eta\,, \tag{11.15}$$

where χ is a smoothing function equal to 1 close to Y and vanishing outside a neighbourhood of Y, and ϕ is C^∞ with compact support on a coordinate patch of Y. Note that $\gamma_\epsilon \mathcal{E}_\phi(g) \to (\int \phi) g$ as $\epsilon \to 0$.

The claimed properties of the *natural maps* follow immediately from the definition. Then Theorem 11.4 implies Corollary 11.6 by the continuous extension $\ell^+ : H^s(X) \to H^s(M)$.

Note that the definitions and statements of this section remain valid, if we replace the function spaces by spaces of sections in arbitrary smooth Hermitian vector bundles. By local trivializations we obtain:

Corollary 11.8. *Given a bundle E over X we can extend E over M. Near Y we can represent it as $E|_Y \times (-1,1)$ so that sections u of E can be written in the form $u_t(y) = u(y,t)$ with $y \in Y$, $t \in (-1,1)$ and $u_t(y) \in E|_Y = E|_{Y\times\{0\}} \cong E|_{Y\times\{t\}}$. Then, for $s > 1/2$, the trace map*

$$\gamma : H^s(E) \times [0,1) \to H^{s-1/2}(E|_Y)$$

is well-defined, separately continuous and uniformly bounded on any compact subinterval of $[0,1)$.

We end this chapter with a brief discussion of a few examples.

Examples 11.9. Let $\mathbf{T}^n$ denote the n-dimensional torus $\mathbf{R}^n/2\pi\mathbf{Z}^n$.
(a) The space $L^2(\mathbf{T}^n)$ is generated by the complete orthonormal system of C^∞ functions $\{e_\nu\}_{\nu\in\mathbf{Z}^n}$ with $e_\nu(x) := (2\pi)^{-n/2}e^{i\nu\cdot x}$.
(b) For arbitrary s, the space $H^s(\mathbf{T}^n)$ consists of the completion of $C^\infty(\mathbf{T}^n)$ under the norm

$$\|u\|_s^2 := \sum_{\nu\in\mathbf{Z}^n} |\hat{u}(\nu)|^2(1+|\nu|^2)^s\,,$$

where the ν-th Fourier coefficient is given by

$$\hat{u}(\nu) := (2\pi)^{-n/2}\int_{\mathbf{T}^n} e^{-i\nu\cdot x}u(x)dx\,.$$

In other words, any $u \in L^2(\mathbf{T}^n)$ belongs to $H^s(\mathbf{T}^n)$ exactly when the sequence $\sum \hat{u}(\nu)e_\nu$ converges absolutely to u in the $\|\cdot\|_s$ norm.
(c) Let $y \mapsto (y,0)$ be the natural embedding of $\mathbf{T}^{n-1}$ in $\mathbf{T}^n \approx \mathbf{T}^{n-1}\times S^1$. Then for any $u \in H^s(\mathbf{T}^n)$ and arbitrary s, r, the restriction $u|_{\mathbf{T}^{n-1}}$ is well-defined and belongs to $H^r(\mathbf{T}^{n-1})$ exactly when

$$\|u|_{\mathbf{T}^{n-1}}\|_r^2 \;:=\; \frac{1}{2\pi}\sum_{\lambda\in\mathbf{Z}^{n-1}}\left|\sum_{\mu\in\mathbf{Z}}\hat{u}(\lambda,\mu)\right|^2(1+|\lambda|^2)^r \;<\; \infty\,.$$

Comments on the example. (a) is well-known from the theory of Fourier series. (b) may be deduced from our Definition 11.1 by straightforward calculations or, more smoothly, by a characterization of $H^s(M)$ as the domain of the differential operator $\Delta^{s/2}$ in $L^2(M)$. This characterization is valid for arbitrary Riemannian M. We obtain (c) from noticing $u(y,\theta) = \sum \hat{u}(\lambda,\mu)e_\lambda(y)e_\mu(\theta)$, where the sum is over all $(\lambda,\mu) \in \mathbf{Z}^{n-1} \times \mathbf{Z} = \mathbf{Z}^n$, and the convergence on $\mathbf{T}^n$ is uniform. Since $e_\mu(0) = (2\pi)^{-1/2}$, it follows that

$$u(y,0) = \sum_{\lambda \in \mathbf{Z}^{n-1}} e_\lambda(y) \sum_{\mu \in \mathbf{Z}} \hat{u}(\lambda,\mu)\,,$$

where the series converges uniformly on $\mathbf{T}^n$. Thus we have

$$(\widehat{u|_{\mathbf{T}^{n-1}}})(\lambda) = (2\pi)^{-1/2} \sum_{\mu \in \mathbf{Z}} \hat{u}(\lambda,\mu)\,. \qquad \square$$

Remarks 11.10. (a) From Example 11.9c we gain insight into the possible loss of summability under restrictions of Sobolev spaces (trace theorem) in contrast to the gain of differentiability in C^k theory. Take e.g. the Fourier series over $\mathbf{T}^n$

$$\hat{u}(\lambda,\mu) := \frac{1}{\lambda^\alpha \mu^\beta}\,.$$

Then the series $\sum_\lambda \sum_\mu |\hat{u}(\lambda,\mu)|^2(1+\lambda^2+\mu^2)^s$ converges for $s=0$ and $\alpha,\ \beta > \frac{1}{2}$, and

$$u(y,0) := \frac{1}{2\pi} \sum_{\lambda=1}^{\infty} \sum_{\mu=1}^{\infty} \frac{1}{\lambda^\alpha \mu^\beta} e^{i(\lambda y + \mu 0)}$$

defines an element of $H^0(\mathbf{T}^2) = L^2(\mathbf{T}^2)$, whereas the series

$$\frac{1}{2\pi} \sum_\lambda \left| \sum_\mu \hat{u}(\lambda,\mu) \right|^2 (1+\lambda^2)^r$$

does not converge for $r=0$ and $\beta \leq 1$. Hence $u|_{S^1}$ is not well-defined in $H^0(S^1) = L^2(S^1)$.

(b) One can also easily construct examples of functions which lie in $H^s(\mathbf{R}^n)$, but with trace not belonging to $H^s(\mathbf{R}^{n-1})$. For example does the function $u(x) := |x|^\alpha \chi(x)$, where χ is a C^∞ function on $\mathbf{R}^n$ with compact support and equal to 1 on the unit ball, belong to $H^1(\mathbf{R}^n)$ exactly when $2\alpha > 2-n$. So $u(x) := |x|^{-1/4}\chi(x)$ is an element of $H^1(\mathbf{R}^3)$, but neither does the first trace $u(.,.,0)$ belong to $H^1(\mathbf{R}^2)$, nor the second trace $u(.,0,0)$ to $H^1(\mathbf{R})$. The example furthermore shows that there is no continuous linear map $H^s(\mathbf{R}^n) \to H^s(\mathbf{R}^{n-1})$ which extends the trace map $C_0^\infty(\mathbf{R}^n) \to C_0^\infty(\mathbf{R}^{n-1})$.

12. Calderón Projector for Dirac Operators

For any Dirac operator over a compact manifold with smooth boundary, the orthogonal projection of sections over the boundary onto the space of Cauchy data is a pseudo-differential operator of order zero with principal symbol given by the related spectral projection.

We need to fix and explain the notation. As in the preceding chapter we assume that M is a closed smooth n-dimensional manifold which is partitioned into two manifolds X_+ and X_- with $\partial X_+ = \partial X_- = Y$ by a smooth closed submanifold Y of codimension 1. A Riemannian structure is fixed on M which defines a parametrization $Y \times [-1, 1]$ of a bicollar neighbourhood of Y in M.

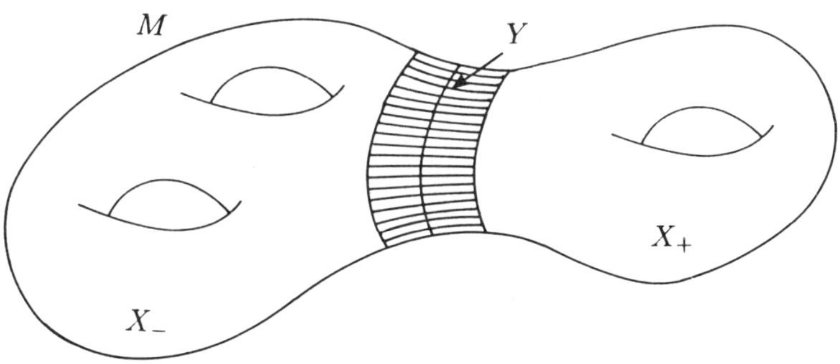

Fig. 12.1 The partition of $X = X_- \cup X_+$

Following the fundamental tenets of numerical analysis, we build our analysis on the concept of the spaces of Cauchy data. The point is that the following concepts and results are well-defined and true for the total Dirac operator A and the component A^+ (and for even and odd dimensions); hence in the following definition A may denote any operator of Dirac type.

Definition 12.1. For any Dirac operator

$$A : C^\infty(M; E) \to C^\infty(M; F)$$

over a partitioned manifold $M = X_- \cup X_+$ with $X_- \cap X_+ = Y$ we define
(a) the spaces of *Cauchy data* (along Y) by

$$(12.1) \qquad H_\pm(A) := \{u|_Y \mid u \in C^\infty(E) \text{ and } Au = 0 \text{ in } X_\pm\},$$

and we denote the closure of these spaces in $H^{s-\frac{1}{2}}(E|_Y)$ by $H_\pm(A,s)$ for any real s;
(b) for any real s we define the *solution spaces (null spaces)* by

$$(12.2) \qquad \ker_\pm(A,s) := \{u \in H^s(E|_{X_\pm}) \mid Au = 0 \text{ in } X_\pm \setminus Y\}.$$

Remarks 12.2. (a) Since Dirac operators satisfy the unique continuation property (see Chapter 8), there are no non-trivial sections in $\ker A$ or in $\ker_+(A,s)$ with support contained in $X_+ \setminus Y$. Notice that the unique continuation property also excludes non-trivial solutions which vanish identically on the boundary. This is standard: Let $u \in C^\infty(E|_{X_+})$ and $Au = 0$ in X_+ and $u|_Y = 0$. Then *the extension by zero* e^+u is a weak solution for A over the whole of M because

$$\begin{aligned}
\langle e^+u; A^*v\rangle_E &= \int_M (e^+u; A^*v) = \int_{X_+} (u; A^*v|_{X_+}) + \int_{X_-} 0 \\
&= \int_{X_+} (Au; v|_{X_+}) + \int_Y (G(y,0)u(y); v(y)) \\
&\qquad \text{by decomposition of Green's formula (Proposition 3.4)} \\
&= 0 \qquad \text{for any } v \in C^\infty(F).
\end{aligned}$$

By the regularity of the solutions of elliptic equations over closed manifolds, it follows that $w := e^+u \in C^\infty(M;E)$ as desired; hence $w \equiv 0$ by unique continuation.
(b) As in Chapter 9, we shall in the present chapter restrict our presentation to the case of even-dimensional manifolds. The only reason is to take advantage of the easy to follow chiral notation in the even-dimensional split case. However, the results remain true and the arguments remain applicable also in the odd-dimensional case. We leave the necessary transcriptions to the reader.

To avoid an inflation of subscripts and superscripts, we shall use the notations X and X_+, S^+ and S, and A^+ and A interchangeably throughout this chapter, namely, for the objects on the primary (*plus*) side of the closed double $\widetilde{X}$. This ambiguity seems tolerable, since all results are true for the total Dirac operator and the chiral Dirac operator as well. Note that, corresponding to the ambiguity in the notation of S and A (whether they refer to the formally self-adjoint total Dirac operator *or* its non-self-adjoint chiral component), we have an ambiguity in the notation S^- and A^-. The ambiguity is whether they designate the same bundle and operator as on the plus side, but copied to the *secondary* (minus) side of $\widetilde{X}$, *or* the image bundle and the formally adjoint operator, which are the same for the total Dirac operator, but different bundles and different operators in the chiral split case. Both of these ambiguities disappear once it is specified whether one argues about the total or the chiral split Dirac operator.

Let $\widetilde{A}$ (instead of $\widetilde{A^+}$) denote the invertible extension $A^+ \cup A^-$ of A^+ on the closed double constructed in Chapter 9. We begin our analysis of the behaviour of Dirac operators close to the boundary by presenting the following fact.

Lemma 12.3. *Let $A^+ : C^\infty(X; S^+) \to C^\infty(X; S^-)$ be a Dirac operator over an even-dimensional manifold X with boundary Y and let $\widetilde{A}$ denote its extension over the closed double $\widetilde{X}$ constructed in Chapter 9. Then the spaces H_+, H_- of Cauchy data of $\widetilde{A}$ intersect only in the 0 section.*

Proof. Let $u \in H_+(\widetilde{A}) \cap H_-(\widetilde{A})$. Then there exists $s_+ \in C^\infty(X; S^+)$ and $s_- \in C^\infty(X; S^-)$ such that $A^+ s_+ = 0$ and $A^- s_- = 0$ and

$$s_+|_Y = u = s_-|_Y \,. \tag{12.3}$$

Define a continuous section s of $\widetilde{S^+}$ on $\widetilde{X}$ by

$$s = \begin{cases} s_+ & \text{on } X_+ = X \\ s_- & \text{on } X_- \,. \end{cases} \tag{12.4}$$

We show that s is a weak solution of $\widetilde{A}s = 0$. Let $w \in C^\infty(\widetilde{S}^-)$. Then we have over the closed manifold $\widetilde{X}$

(12.5)

$$\langle \widetilde{A}s; w \rangle \;=\; \langle s, (\widetilde{A})^* w \rangle \;=\; \langle s_+; ((\widetilde{A})^* w)|_{X_+} \rangle + \langle s_-; ((\widetilde{A})^* w)|_{X_-} \rangle \,.$$

Recall from Chapter 9 that $\widetilde{A}(s_+, s_-) = (A^+(s_+), A^-(s_-))$, where $\widetilde{A}$ takes the form $G(\partial_t + B)$ on X_+ close to Y, and the form $G(-(-\partial_t) + B) = (-G)(-\partial_t - B)$ on X_- close to Y, with the normal parameter

t inward (and therefore opposite) oriented in both cases. Hence, by Green's formula (Proposition 3.4),

$$\langle s_+ ; ((\widetilde{A})^* w)|_{X_+}\rangle \;=\; \langle A^+ s_+ ; w|_{X_+}\rangle + \langle G\cdot(s_+|_Y); w|_Y\rangle$$

and

$$\langle s_- ; ((\widetilde{A})^* w)|_{X_-}\rangle \;=\; \langle A^- s_- ; w|_{X_-}\rangle + \langle (-G)\cdot(s_-|_Y); w|_Y\rangle\,.$$

Since A^+s_+ and A^-s_- vanish, we finally obtain

$$\text{(12.6)}\qquad \langle \widetilde{A}s; w\rangle \;=\; \langle G\cdot(s_+|_Y - s_-|_Y); w|_Y\rangle = 0\,.$$

By the regularity theorem for elliptic operators over closed manifolds this implies that $s \in C^\infty$, hence $s = 0$ for the injectivity of $\widetilde{A}$ (see Chapter 9) and therefore $u = s|_Y = 0$. □

The main result of this chapter is the following construction of a *precise* and *pseudo-differential* projection onto the space of Cauchy data of a Dirac operator.

Theorem 12.4. (A. P. Calderón, R. T. Seeley). *Let X be a smooth compact even-dimensional Riemannian manifold with boundary Y; let S be a bundle of $C\ell(X)$-modules with compatible metric, and let the corresponding Dirac operator be denoted by $A^+ : C^\infty(X;S^+) \to C^\infty(X;S^-)$. We assume that the Riemannian structure of X is a product near Y. (Hence the operator A extends to an invertible operator $\widetilde{A}$ over the closed double $\widetilde{X}$, and near Y it takes the form $G(y)(\partial_t + B)$ as explained above in Chapter 9. In the following, the tangential part B does not have to be self-adjoint although our models, the Dirac operators, have self-adjoint B). Let r^+ denote the restriction operator from sections over $\widetilde{X}$ to sections over $X_+ = X$ and let $\gamma_\epsilon : H^s(\widetilde{X};\widetilde{S}^+) \to H^{s-\frac{1}{2}}(Y;S^+|_Y)$ be the trace map which was introduced in Chapter 11 and is well-defined and continuous for $s > \frac{1}{2}$. Then:*

(a) *For any real s, the mapping*

$$\text{(12.7)}\qquad K_+ := r^+\widetilde{A}^{-1}\gamma_0^* G : C^\infty(S^+|_Y) \to C^\infty(S^+|_{X_+\setminus Y})$$

extends to a continuous map $K_+ : H^{s-\frac{1}{2}}(S^+|_Y) \to \ker_+(A,s)$ which is surjective, and is bijective if restricted to the Cauchy data space $H_+(A,s)$. Here γ_0^* denotes the adjoint of

$$\text{(12.8)}\qquad \gamma_0^- : H^s(\widetilde{X};\widetilde{S}) \to H^{s-\frac{1}{2}}(Y;S^-|_Y)\,.$$

It is continuous from $H^{r-\frac{1}{2}}(Y; S^-|_Y) \to H^{r-1}(\widetilde{X}; \widetilde{S})$ *for* $r < \frac{1}{2}$.
(b) *For all* $g \in C^\infty(S^+|_Y)$, *the sections* $\gamma_\varepsilon \widetilde{A}^{-1}\gamma_0^* G g$ *converge uniformly to limits* $\mathcal{P}_+ g$ *as* $\varepsilon \to 0_+$, *and to limits* $-\mathcal{P}_- g$ *as* $\varepsilon \to 0_-$; *the operators* $\mathcal{P}_\pm$ *are projections onto the spaces of Cauchy data* $H_\pm$ *along* $H_\mp$ *and are pseudo-differential operators of order 0. Let* b *denote the principal symbol of the tangential operator* B. *Then the principal symbol* p_+ *of* $\mathcal{P}_+$ *at* (y, η), $y \in Y$, *and* $\eta \in T^*Y \setminus 0$ *is the projection onto the subspace of* S^+ *spanned by the eigenvectors corresponding to the eigenvalues of* $b(y, \eta) : S_y^+ \to S_y^+$ *with positive real part.*

Note. The operator $\mathcal{P}_+$ is called the *Calderón projector* of A. By a *projection* here, we mean an operator P with $P^2 = P$, but not necessarily $P^* = P$ (see also Lemma 12.8 below). In other words, we show that $\mathcal{P}_+$ is idempotent.

The operator $(\widetilde{A})^{-1} \sim (A^+ \cup A^-)^{-1}$ is sometimes called the *volume potential* of A; it plays the role of the *Newtonian potential* in classical potential theory. Recall that the Newtonian potential $V(x) := c \log |x|$ provides the first stage in finding the solution of the Dirichlet problem

$$\Delta u = v \text{ in } X \quad \text{and} \quad u|_Y = g \tag{12.9}$$

for a given compact domain $X \subset \mathbf{R}^2$ with smooth boundary Y and given functions v on X and g on Y: Let v be extended in any convenient way to the whole $\mathbf{R}^2$. Then $v'(x) := \int V(|x - \xi|)v(\xi)d\xi$ is a solution of $\Delta v' = v$, but in general with $g_1 := v'|_Y \neq g$. To get the boundary values right, the next step is to solve the problem

$$\Delta w = 0 \quad \text{and} \quad w|_Y = g - g_1 =: g_2 \,. \tag{12.10}$$

This is done for the disc $\{x \mid |x| < R\}$ by the classical *Poisson formula*

$$w(r\cos\varphi, r\sin\varphi) := \frac{1}{2\pi R}\int_0^{2\pi} \frac{g_2(R\cos\theta, R\sin\theta)(R^2 - r^2)}{\sqrt{R^2 + r^2 - 2Rr\cos\theta}}\, d\theta$$

and for general, not necessarily circular, Y by the classical method of *double layer potentials* by writing w in the form

$$w(x) := \int_Y \frac{\partial V(x - y)}{\partial \mathbf{n}} f(y) dy \,,$$

where $\mathbf{n}$ denotes the normal outward derivative and f has to be determined by the boundary integral equation

$$cf(y) + \int_Y L(y,\zeta)f(\zeta)d\zeta = g_2(y) \quad \text{for } y \in Y$$

with a certain pseudo-differential kernel L. Then $u := v' + w$ solves the original problem.

The operator K_+ may be called the *surface potential* or the *Poisson operator* of A; it plays the role of the mentioned *double layer potential.* The use of these three operators $\mathcal{P}_+$, $\widetilde{A}^{-1}$, and K_+ in solving elliptic boundary value problems for Dirac operators is explained below; see especially Chapters 18 and 20.

Proof. The essential part of the statement and of the proof of all parts of this theorem is the uniform convergence in part (b) with its delicate estimates. However, for a better demonstration of what is going on, we prefer to begin with the proof of part (a) (assuming that some of the results of (b) have already been proved).

(a) Let g belong to $C^\infty(S^+|_Y)$. The distribution $\gamma_0^* g$ is defined by

$$\langle v; \gamma_0^* g\rangle := \int_Y (v|_Y; g) \tag{12.11}$$

for $v \in C^\infty(\widetilde{X}; \widetilde{S})$. So $\gamma_0^* g$ is nothing but the distribution $\delta_Y \otimes g$. When we apply the Green form G we obtain a distribution $w = G(\delta_Y \otimes g)$ on the whole of $\widetilde{X}$, which in fact belongs to $H^{-\frac{1}{2}}$, and $\widetilde{A}^{-1}w$ is C^∞ out of Y simply because of the fundamental property of pseudo-differential operators, namely, the pseudo-locality

$$\operatorname{sing\,supp} \widetilde{A}^{-1} w \subset \operatorname{sing\,supp} w\,.$$

We shall in (b) below show that $\gamma_\epsilon \widetilde{A}^{-1}\gamma_0^* Gg$ and all its derivatives converge uniformly as $\epsilon \to 0_+$. This means that $Kg := \widetilde{A}^{-1}\gamma_0^* Gg \in L^2(\widetilde{X}; \widetilde{S})$. Then $\widetilde{A}Kg = \gamma_0^* Gg$ is supported (as a distribution) in Y; hence $\widetilde{A}Kg = 0$ in $\widetilde{X} \setminus Y$ and $K_+ g = r^+ Kg \in \ker_+(A, 0)$.

Now we show that for any $g \in C^\infty(S^+|_Y)$ and any real s

$$\|K_+ g\|_s = \|r^+ \widetilde{A}^{-1}\gamma_0^* Gg\|_s \le C\|g\|_{s-\frac{1}{2}} \tag{12.12}$$

with a constant C independent of g. We shall do this for s equal to an integer $k \ge 1$. (For $s < \frac{1}{2}$ the estimate follows immediately from bounds on $\widetilde{A}^{-1}$ and γ_0^*.) Set $u := K_+g = r^+\widetilde{A}^{-1}\gamma_0^* Gg$ and $g_0 := \lim_{\epsilon\to 0+}\gamma_\epsilon u$. By part (b) of this theorem, the limit exists uniformly and $\|g_0\|_{k-\frac{1}{2}} \le C_1\|g\|_{k-\frac{1}{2}}$. Now choose a continuous extension operator $\mathcal{E} : H^{k-\frac{1}{2}}(Y) \to H^k(\widetilde{X})$ as in Definition 11.7e and obtain a section $v = \mathcal{E}g_0 \in H^k(\widetilde{X})$ with v smooth over $\widetilde{X}\setminus Y$, $\gamma_0 v = g_0$, and $\|v\|_k \le C_2\|g_0\|_{k-\frac{1}{2}}$. Next assemble a section $w \in L^2(\widetilde{X})$ by

$$w = \begin{cases} u & \text{for } t \ge 0 \\ r^-(v) & \text{for } t < 0\,, \end{cases} \tag{12.13}$$

where r^- denotes the restriction to $X_-\setminus Y = \{t<0\}$. Note that u and v coincide on $Y = \{t = 0\}$.

It must now be shown that w and hence u belong to H^k. To do that we show that $\widetilde{A}w$ has support in X_-: In fact for any smooth section ϕ we have

$$\begin{aligned}\langle \widetilde{A}w;\phi\rangle &= \int_{\widetilde{X}} (w;(\widetilde{A})^*\phi) = \lim_{\epsilon\to 0+}\int_{|t|>\epsilon}(w;(\widetilde{A})^*\phi)\\ &= \lim\int_{|t|>\epsilon}(\widetilde{A}w;\phi) + \lim\int_Y (G(y,\epsilon)\gamma_\epsilon w;\gamma_\epsilon\phi)\\ &\qquad\qquad - \lim\int_Y (G(y,-\epsilon)\gamma_{-\epsilon}w;\gamma_{-\epsilon}\phi)\,.\end{aligned}$$

Here the two integrals over Y cancel as $\epsilon \to 0$, since $r^+w = u = K_+g$ and $r^-w = r^-v$ have the same trace over Y; recall also that $Au = AK_+g$ vanishes in the interior of X. This yields

$$\langle \widetilde{A}w;\phi\rangle = \int_{t\le 0}(\widetilde{A}w;\phi) = \int_{X_-}(r^-(\widetilde{A}v);r^-(\phi))\,.$$

But $\widetilde{A}$ is an operator of first order, so $\widetilde{A}v \in H^{k-1}(\widetilde{X})$ with $\|\widetilde{A}v\|_{k-1} \le C_3\|v\|_k$ and also $\widetilde{A}w \in H^{k-1}(\widetilde{X})$ with the same norm because the support of $\widetilde{A}w$ is completely contained in X_-. Since $\widetilde{A}^{-1}$ is an operator of order -1, we finally obtain $w \in H^k(\widetilde{X})$ and

$$\begin{aligned}\|u\|_k \le \|w\|_k = \|\widetilde{A}^{-1}\widetilde{A}w\|_k &\le C_1\|\widetilde{A}w\|_{k-1} = C_1\|Av\|_{k-1}\\ &\le C_2\|v\|_k = C_2\|\mathcal{E}g_0\|_k \le C_3\|g\|_{k-\frac{1}{2}}\end{aligned}$$

which proves (12.12).

Thus the potential operator K_+ extends to a continuous map $H^{s-\frac{1}{2}}(S^+|_Y) \to \ker(A,s)$ for real $s < \frac{1}{2}$ and for integer $s \geq 1$ and by interpolation for all real s. Also K_+ is surjective:

Let u belong to $\ker(A,s)$, i.e. $u \in H^s(X;S^+)$ and $Au = 0$ in $X \setminus Y$; $s > \frac{1}{2}$. We set $g := \gamma_0 u$ and $v := K_+ g$. Then $K_+ g$ also belongs to $\ker(A,s)$ and it follows that $A(u-v) = 0$ in $X \setminus Y$ and $(u-v)|_Y = 0$, hence $u - v = 0$ because the Dirac operators fulfill the unique continuation property. This ends the proof of statement (a) of Theorem 12.4 for $s > \frac{1}{2}$. The general case needs additional arguments and will be discussed in the next chapter.

(b.1) First we want to show that the limits $\mathcal{P}_+ g = \lim_{\epsilon \to 0_+} \gamma_\epsilon \widetilde{A}^{-1} \gamma_0^* G g$ exist, are uniformly approached, and in fact provide a pseudo-differential operator $\mathcal{P}_+$ of order 0 with the claimed principal symbol. It suffices to consider the local situation. Since the Green form G is invertible, it may be neglected. Thus we consider operators of the form $M_{\phi_1} \widetilde{A}^{-1} M_{\phi_2} \gamma_0^*$ with $M_{\phi_j}(u)(x) := \phi_j(x) u(x)$, where the ϕ_j are locally supported real valued C^∞-functions on $\widetilde{X}$ of a C^∞-partition of unity suited to a finite covering of $\widetilde{X}$ with open sets.

Distinguish three situations. (i) When the support of ϕ_2 does not intersect Y, then all is trivial since $M_{\phi_2} \gamma_0^* g = 0$. (ii) When ϕ_1 and ϕ_2 have disjoint support, $M_{\phi_1} \widetilde{A}^{-1} M_{\phi_2}$ has a smooth kernel as integral operator and $M_{\phi_1} \widetilde{A}^{-1} M_{\phi_2} \gamma_0^* : H^s(Y) \to C^\infty(M)$ is continuous for any $s < 0$ and therefore for all s. From the well-known Sobolev lemma (or from the stronger Theorem 11.4) we then have that $\gamma_\epsilon M_{\phi_1} \widetilde{A}^{-1} M_{\phi_2} \gamma_0^* g$ converges uniformly as $\varepsilon \to 0_+$ and the limit yields an operator of order $-\infty$.
(iii) Suppose now that ϕ_1 and ϕ_2 have support in an open set U which is contained in the strip $\{|t| < 1\}$. We choose coordinates $\chi : U \to \mathbf{R}^n$ and assume that $S^+|_U$ is trivial. Unfortunately, no direct approach is available for investigating $\lim_{\varepsilon \to 0_+} \gamma_\varepsilon M_{\varphi_1} \widetilde{A}^{-1} M_{\varphi_2} \gamma_0^*$. But a slight modification of $M_{\varphi_1} \widetilde{A}^{-1} M_{\varphi_2}$ will lead to an operator T with well-defined traces $\gamma_0 T \gamma_0^* g$. Then the investigation of the *correcting* terms will yield the claimed convergence, pseudo-differential and symbol properties.

More precisely, choose a C^∞-function φ with $\operatorname{supp} \phi \subset U$ and $\phi \equiv 1$ on $\operatorname{supp}(|\phi_1| + |\phi_2|) =: V$. Then $M_\phi \widetilde{A}^{-1} M_\phi$ is a pseudo-differential operator with the (total) symbol given explicitly by $\sum_{k=1}^\infty c_{-k}$ with $c_{-1} = (a_1)^{-1}$ on V. Here $a_1 = \sigma_1(\widetilde{A})$ denotes the principal symbol of $\widetilde{A}$. We denote by $c_k(x,D)$ the homogeneous pseudo-differential operator of order k over U given there in the local coordinates χ by

$$(12.14)\qquad c_k(x,D)u(x) = (2\pi)^{-n}\int e^{i\langle x,\xi\rangle} c_k(x,\xi)\hat{u}(\xi)d\xi\,.$$

Set $C_k := M_{\phi_1} c_k(x,D) M_{\phi_2}$ for $k=-1,-2,\dots$. Then for any $k_0\in\mathbf{N}$

$$(12.15)\qquad T := M_{\phi_1}\widetilde{A}^{-1}M_{\phi_2} - \sum_{k<k_0} C_{-k}$$

is an operator of order $-k_0$. By Theorem 11.4 we obtain a continuous mapping

$$(12.16)\qquad T\gamma_0^* : H^{-\frac12}(S^+|_Y) \xrightarrow{\gamma_0^*} H^{-1}(\widetilde{S}) \to H^{k_0-1}(\widetilde{S})$$

which, for sufficiently high k_0, by the Sobolev embedding theorem gives a continuous map $H^{-\frac12}(S^+|_Y)\to C^0(\widetilde{S})$. Hence $T\gamma_0^* g$ has a well-defined trace on Y and it remains to prove that the *errors* $\gamma_\epsilon C_{-k}\gamma_0^* g$ converge uniformly as $\epsilon\to 0_+$. This is not a trivial result, since even for $g\in C^\infty$ we have, *a priori*, only $C_{-k}\gamma_0^* g\in H^{k-\frac12}$ which, for example, for $k=1$ does not necessarily have a well-defined trace. To overcome that difficulty it is crucial to know certain details about the symbol of $C_{-k}\gamma_0^*$:

Let $\alpha\in C^\infty(\mathbf{R})$ with $\operatorname{supp}\alpha\subset(-1,-\frac12)$ and $\int\alpha = 1$. We set $\alpha_m(t) := m\alpha(mt)$. Then the sequence $\{\alpha_m\otimes g\}_{m=1,2,\dots}$ converges weakly to $\gamma_0^* g$ in $H^{-s}(\widetilde{X})$ for each $s>\frac12$. In fact

$$\langle \alpha_m\otimes g; v\rangle = \int\!\!\int (\alpha_m(t)g(y); v(y,t))\,dydt \longrightarrow \int g(y)v(y,0)dy$$

for $v\in H^s(\widetilde{X})$ as $m\to\infty$. Now recall

$$\begin{aligned}\widehat{\alpha_m g}(\tau,\eta) &= \int e^{-i\tau t}e^{-i\langle\eta,y\rangle}\alpha_m(t)g(y)dtdy\\ &= \int e^{-i\tau t} m\alpha(mt)dt\int e^{-i\langle\eta,y\rangle}g(y)dy\\ &= \int e^{-i\frac{\tau}{m}mt}\alpha(mt)d(mt)\hat{g}(\eta) = \hat{\alpha}(\tau/m)\hat{g}(\eta)\,.\end{aligned}$$

Then we obtain

$$\begin{aligned}&(12.17)\quad C_{-k}\gamma_0^* g(y,t)\\ &= \lim_{m\to\infty}(2\pi)^{-n}\int_{\mathbf{R}^{n-1}}\int_{-\infty}^{\infty} e^{i\langle y,\eta\rangle + it\tau} c_{-k}(y,t;\eta,\tau)\widehat{\alpha_m g}(\tau,\eta)d\tau d\eta\\ &= \lim_{m\to\infty}(2\pi)^{-n}\int_{\mathbf{R}^{n-1}} e^{i\langle y,\eta\rangle}\hat{g}(\eta)\Big(\int_{-\infty}^{\infty} e^{it\tau}\hat{\alpha}(\tau/m)c_{-k}(y,t;\eta,\tau)d\tau\Big)\,d\eta\,.\end{aligned}$$

Now we replace the integral $\int_{-\infty}^{\infty}$ for any $\eta \in \mathbf{R}^{n-1}$ by an integral over a finite path $\Gamma(\eta)$ in $\{\Im(\tau) \geq 0\}$.

In a certain sense this replacement is the decisive step in establishing the Calderón projection, since it provides the uniform boundedness of the integrands which permits passing $\lim_{m\to\infty}$ inside the integrals. Then for (12.17), this yields the expression

(12.18)

$$C_{-k}\gamma_0^* g(y,t) = (2\pi)^{-n} \int_{\mathbf{R}^{n-1}} e^{i\langle y,\eta\rangle} \left\{ \int_{\Gamma(\eta)} e^{it\tau} c_{-k}(y,t;\eta,\tau) d\tau \right\} \hat{g}(\eta) d\eta\,,$$

since $\hat{\alpha}(\tau/m)$ converges boundedly to $\hat{\alpha}(0) = \int \alpha = 1$ as $m \to \infty$.

Before drawing conclusions from (12.18), return to the replacement of $\int_{-\infty}^{\infty}$. Let $q(\tau)$, $\tau \in \mathbf{R}$ denote the integrand of $\int_{-\infty}^{\infty}$ in (12.17)

$$q(\tau) := e^{it\tau}\hat{\alpha}(\tau/m)c_{-k}(y,t;\eta,\tau)\,.$$

We are going to show that q

- is continuous on $\mathbf{R}$ (for $\eta \neq 0$),
- extends to a holomorphic function in $\Omega_R := \{\tau \in \mathbf{C} \mid \Im(\tau) \geq 0, \quad |\tau| \geq R\}$ for some R,
- and suffices $q(\tau) = o(\tau^N)$ for some N, when $\tau \to \infty$ in Ω_R.

Then one obtains

$$\int_{-\infty}^{\infty} q(\tau)d\tau = \int_{-R}^{R} q(\tau)d\tau - \int_{\pi}^{0} q(Re^{i\theta})Rie^{i\theta}d\theta$$

which is sometimes emphasized by denoting both sides of the equation by the expression $\int^+ q(\tau)d\tau$. We shall do likewise.

First find the necessary holomorphic extension of $c_{-k}(y,t;\eta,\tau)$ in an appropriate region of $\{\Im(\tau) \geq 0\}$. All the c_k are rational functions of τ, hence meromorphic. In fact, their poles are the roots of a_1. More generally we can argue as follows:

From the definition of $\{c_{-k}\}_{k=1,\dots,\infty}$ and the symbol composition formula for pseudo-differential operators, we obtain the following relations for $(y,t) \in \operatorname{supp}(|\phi_1| + |\phi_2|)$:

$$c_{-1}a_1 = \mathrm{Id}$$

$$c_{-1}a_0 + c_{-2}a_1 - i\frac{\partial c_{-1}}{\partial \tau}\frac{\partial a_1}{\partial t} - i\sum_{j=2}^{n}\frac{\partial c_{-1}}{\partial \eta_j}\frac{\partial a_1}{\partial y_j} = 0$$

$$\dots\dots\dots\dots\dots$$

$$\dots\dots\dots\dots\dots$$

$$c_{-k}a_1 + c_{-k+1}a_0 + \dots c_{-1}a_{-k+2} - i\frac{\partial c_{-k+1}}{\partial \tau}\frac{\partial a_1}{\partial t} - \dots$$
$$+ i\frac{\partial^2 c_{-k+2}}{\partial \tau^2}\frac{\partial^2 a_1}{\partial t^2} + \dots - i\dots + i\dots\dots = 0\,,$$

i.e. c_{-k} can be written as

(12.19)
$$a_1^{-1} \times (\text{sums and products of derivatives of } a_1, a_0, \dots, a_{-k+2}$$
$$\text{and of } c_{-1}, c_{-2}, \dots, c_{-k+1})\,.$$

Here $\sum_{k\le 1} a_k$ denotes the asymptotic expansion of the (total) symbol of $\widetilde{A}$.

Now $a_1(y,t;\eta,\tau)$ is linear in (η,τ) and non-singular for $|\eta|^2+\tau^2>0$, $\eta\in\mathbf{R}^{n-1}$ and $\tau\in\mathbf{R}$. Allowing τ to be complex, there is a compact set $\mathcal{Z}$ in the complex plane $\mathbf{C}$, not intersecting the real axis, such that for $|\eta|=1$ and $(y,t)\in\operatorname{supp}\phi$, the complex values of τ, for which a_1 is not invertible, lie in $\mathcal{Z}$. Thus $c_{-1}=a_1^{-1}$ has a holomorphic extension to $\mathbf{C}\setminus\mathcal{Z}$ for such η and (y,t). Then by induction it follows from (12.19) that all c_{-k} and their derivatives have such extensions, each of which is bounded by some polynomial in $|\tau|$. See also Figure 12.2.

Let $\mathcal{Z}_+$ denote the part of $\mathcal{Z}$ which belongs to $\{\Im(\tau)>0\}$. We choose a positive real R such that $\mathcal{Z}_+\subset\{|\tau|\le R\}$.

For any $\eta\in\mathbf{R}^{n-1}$ define a closed path

$$\Gamma(\eta) := \partial\left(\{\tau \mid |\tau| \le \max(1;R|\eta|)\} \cap \{\Im(\tau)\ge 0\}\right)$$

with counter clockwise orientation and a closed path $\Gamma_0(\eta)$ as indicated in Figure 12.3.

Our choice of R ensures that the integrand $e^{it\tau}\hat{\alpha}(\tau/m)c_{-k}(y,t;\eta,\tau)$ of our real integral $\int_{-\infty}^{\infty}$ in (12.17) is holomorphic for τ in the region which is enclosed by $\Gamma_0(\eta)$ for all natural m and k. Let us denote

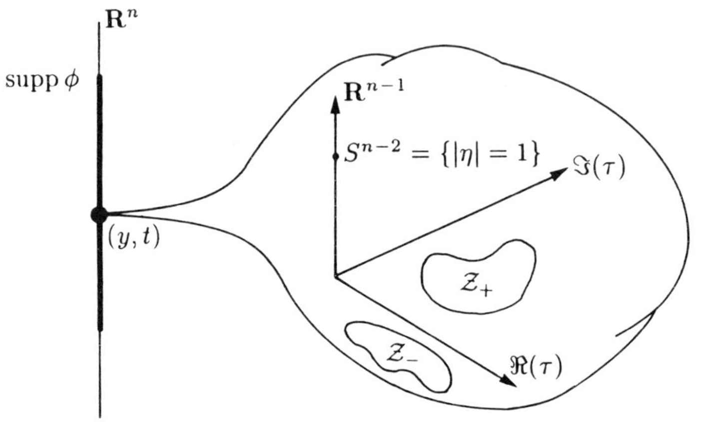

Fig. 12.2 The complex values of τ for which $a_1(y,\ t;\ \eta,\ \tau)$ is not invertible

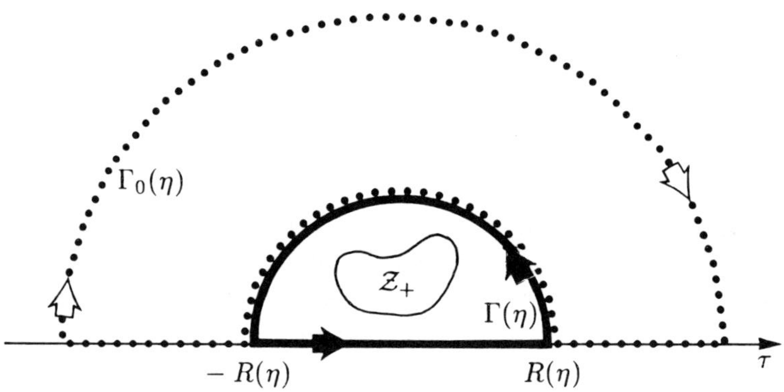

Fig. 12.3 The closed paths $\Gamma(\eta)$ and $\Gamma_0(\eta)$

the joint arc of the contours $\Gamma(\eta)$ and $\Gamma_0(\eta)$ above $\mathcal{Z}_+$ by Γ_1, and the segment of $\Gamma_0(\eta)$ close to ∞ by Γ_2. Note that $e^{it\tau}$ is bounded for $t \geq 0$ and $\Im(\tau) \geq 0$ and that

$$\hat{\alpha}(\tau) = \int_{-1}^{-\frac{1}{2}} e^{-is\tau}\alpha(s)ds$$

decays exponentially as $|\tau| \to \infty$ in $\{\Im(\tau) \geq 0\}$. From all this we get

$$\int_{-\infty}^{\infty} \cdots = \int_{-\infty}^{\infty} \cdots + \int_{\Gamma_0} \cdots = \int_{-R|\eta|}^{R|\eta|} \cdots + \int_{\Gamma_1} \cdots + \int_{\Gamma_2} \cdots$$
$$= \int_{\Gamma(\eta)} \cdots + \int_{\Gamma_2} \cdots = \int_{\Gamma(\eta)} \cdots .$$

since we can move the segment Γ_2 arbitrarily close to ∞ and thereby let the suppressed integrals go to zero without changing $R(\eta)$ and the contour $\Gamma(\eta)$.

Having replaced $\int_{-\infty}^{\infty}$ by the contour integral over the finite path $\Gamma(\eta)$ we obtain a dominated convergence for the integrands (bounded for all m by a common polynomial in $|\tau|$). We may therefore pass $\lim_{m\to\infty}$ inside the integrals. This proves (12.18).

It is clear that $C_{-k}\gamma_0^* g(y,t)$ in (12.18) is C^∞ for $t > 0$, and all its derivatives in y and t extend continuously to $\{t \geq 0\}$. This gives the uniform convergence of $\gamma_\varepsilon C_{-k}\gamma_0^* g$ in $L^2(Y; S^+|_Y)$ for $\varepsilon \to 0_+$.

From (12.18) it also follows that the hereby well-defined operator $\lim_{\varepsilon\to 0_+} \gamma_\varepsilon C_{-k}\gamma_0^*$ is a pseudo-differential operator of order $1-k$, since its symbol

$$p_{-k}(y,\eta) := \frac{1}{2\pi} \int_{\Gamma(\eta)} c_{-k}(y,0;\eta,\tau)d\tau$$

in the integral representation

$$(\lim_{\varepsilon\to 0_+} \gamma_\varepsilon C_{-k}\gamma_0^* g)(y) = (2\pi)^{-(n-1)} \int_{\mathbf{R}^{n-1}} p_{-k}(y,\eta)e^{i\langle y,\eta\rangle}\hat{g}(\eta)d\eta$$

is homogeneous of degree $1-k$ in η for large $|\eta|$ and hence belongs to the standard symbol class S^{1-k}. More precisely, we have for real $s > 1$,

$$2\pi p_{-k}(y,s\eta) = \int_{\Gamma(s\eta)} c_{-k}(0,y;\tau,s\eta)d\tau = \int_{\Gamma(s\eta)} s^{-k}c_{-k}(0,y;\tau/s,\eta)d\tau,$$

since c_{-k} remains homogeneous in its holomorphic extension,

$$= s^{-k+1} \int_{\Gamma(s\eta)} c_{-k}(0,y;\tau/s,\eta)s^{-1}d\tau = s^{-k+1} \int_{\Gamma'(s\eta)} c_{-k}(0,y;\tau',\eta)d\tau'$$

with $\tau' = \tau/s$ and $\Gamma' = s^{-1}\cdot\Gamma$

$$= s^{-k+1} \int_{\Gamma(\eta)} c_{-k}(0,y;\tau',\eta)d\tau',$$ since $\Gamma(\eta)$ is the boundary of $\{|\tau| \leq R|\eta|\} \cap \{\Im(\tau) \geq 0\}$ and therefore equal $s^{-1}\Gamma(s\eta)$,

$$= s^{1-k}\, 2\pi\, p_{-k}(y,\eta).$$

(It is not crucial to the argument that $s^{-1}\Gamma(s\eta) = \Gamma(\eta)$, but only that the two paths are 0-homologic — form a boundary — in the region where the integrand is analytic.)

In our investigation of $\lim_{\varepsilon\to 0_+} \gamma_\varepsilon M_{\varphi_1}\widetilde{A}^{-1}M_{\varphi_2}\gamma_0^*$, let us see how far we have gone. *First* we saw that the *corrected* operator $T = M_{\phi_1}\widetilde{A}^{-1}M_{\phi_2} - \sum_{k<k_0} C_{-k}$ of (12.15) is an operator of order $-k_0$ and that $\gamma_0 T\gamma_0^* g \in C^0(Y; S^+|_Y)$ is well-defined for $g \in C^\infty(Y; S^+|_Y)$. Of course $\gamma_0 T\gamma_0^*$ also extends on all Sobolev spaces and is pseudo-differential. This follows by the same argument, as in deriving (12.16), replacing T by DT, where D is any differential operator of order $\leq k$, $k \in \mathbf{N}$. Then for sufficiently large k_0, namely, $k_0 > 1 + k + n/2$, once again Sobolev's theorem and the continuity of

$$\gamma_0^* : H^{-\frac{1}{2}}(S^+|_Y) \to H^{-1}(\widetilde{S})$$

show that also $DT\gamma_0^* G$ is continuous from $H^{-\frac{1}{2}}(S^+|_Y)$ to $C^0(\widetilde{S})$. Hence $\gamma_\varepsilon DT\gamma^* Gg$ converge uniformly as $\varepsilon \to 0$. Hence $T\gamma_0^*$ is well-defined on all Sobolev spaces. *Then* we proved that for any k

$$\mathcal{P}_{-k} = \lim_{\varepsilon\to 0_+} \gamma_\varepsilon C_{-k}\gamma_0^*$$

is a well-defined pseudo-differential operator over Y of order $-k+1$. This shows that the sections $\gamma_\varepsilon DM_{\phi_1}\widetilde{A}^{-1}M_{\phi_2}\gamma_0^* g$ converge uniformly to limits

$$\gamma_0 DT\gamma_0^* g + \sum_{k<k_0} \lim_{\varepsilon\to 0_+} \gamma_\varepsilon DC_{-k}\gamma_0^* g\,.$$

Repeating the argument we get

$$\|\gamma_\varepsilon D\widetilde{A}^{-1}\gamma_0^* Gg\|_s \leq \|g\|_{s+k} \tag{12.20}$$

for g in C^∞, with c independent of ε. This establishes the existence of $\mathcal{P}_+$, which by definition is a pseudo-differential operator of order 0.

We determine the principal symbol of $\mathcal{P}_+$. Without further suppression of the Green form $G : S^+|_Y \to S^-|_Y$ we compute

$$\begin{aligned}\sigma^L(\mathcal{P}_{-1})(y,\eta) &= \sigma(\mathcal{P}_{-1})(y,\eta) = \sigma(\lim \gamma_\varepsilon C_{-1}\gamma_0^* G)(y,\eta)\\ &= \frac{1}{2\pi}\int_{\Gamma(\eta)} a_1^{-1}(0,y;\tau,\eta)d\tau \cdot G(y)\\ &= \frac{1}{2\pi}\{G(y)(i\tau + b(y,\eta))\}^{-1}G(y)d\tau\\ &= \frac{1}{2\pi}\int_{\Gamma(\eta)} \{i\tau + b(y,\eta)\}^{-1}G^{-1}(y)G(y)d\tau\\ &= \frac{1}{2\pi i}\int_{\Gamma(\eta)} \{\tau - ib(y,\eta)\}^{-1}d\tau\\ &= \frac{-1}{2\pi i}\int_{\Gamma(\eta)} \{ib(y,\eta) - \tau\}^{-1}d\tau\,,\end{aligned}$$

which is the desired spectral projection.

(b.2) Now define the operator $\mathcal{P}_-$ in an analogous way. Then only two steps remain for the proof of (b), namely, to show (i) $\mathcal{P}_+ + \mathcal{P}_- = \mathrm{Id}$, and (ii) range $\mathcal{P}_\pm \subset H_\pm$. Since it was already shown in Lemma 12.3 that $H_+ \cap H_- = 0$, it follows that $\mathcal{P}_+$ and $\mathcal{P}_-$ are complementary projections onto $H_\pm$ along $H_\mp$.

First we show $\mathcal{P}_+ + \mathcal{P}_- = \mathrm{Id}$. Let $g \in C^\infty(Y; S^+|_Y)$. Then for any $v \in C^\infty(\widetilde{X}; \widetilde{S})$ one has

$$\int_Y (G\cdot g; v|_Y) = \langle G\cdot g; \gamma_0 v\rangle_{L^2(Y)} = \gamma_0^* Gg[v]\,, \tag{12.21}$$

where the symbol $[v]$ means that the distribution $\gamma_0^* Gg$ is evaluated on v. Set $Kg := \widetilde{A}^{-1}\gamma_0^* Gg$. Then (12.21) may be continued:

$$\begin{aligned}&= (\widetilde{A}\widetilde{A}^{-1}\gamma_0^* Gg)[v] = (\widetilde{A}Kg)[v] = (Kg)[(\widetilde{A})^* v]\\ &\qquad = \int_{\widetilde{X}} (Kg; (\widetilde{A})^* v) = \lim_{\varepsilon\to 0_+} \int_{|t|>\varepsilon} (Kg; (\widetilde{A})^* v)\,.\end{aligned}$$

Here we used the observation from the beginning of our proof of part (a), namely, that Kg belongs to $L^2(\widetilde{X}; \widetilde{S})$. This permits the introduction of $\lim_{\varepsilon\to 0_+}$ by bounded convergence. From Green's formula (Proposition 3.4), we obtain

$$\begin{aligned}&\int_{|t|>\varepsilon} (Kg; (\widetilde{A})^* v)\\ &= \int_{|t|>\varepsilon} (\widetilde{A}Kg; v) + \int_{Y_\varepsilon} (G\cdot(Kg)|_{Y_\varepsilon}; v|_{Y_\varepsilon}) - \int_{Y_{-\varepsilon}} (G\cdot(Kg)|_{Y_{-\varepsilon}}; v|_{Y_{-\varepsilon}})\,,\end{aligned}$$

where $Y_{\pm\varepsilon} = \{t = \pm\varepsilon\}$ denotes the two submanifolds of $\widetilde{X}$ of codimension 1 defined by the normal coordinate $t = \pm\varepsilon$. Since $\widetilde{A}Kg$ vanishes outside of Y, the two integrals $\int_{t>\varepsilon}$ and $\int_{t<-\varepsilon}$ vanish and we obtain

$$(12.22) \quad \int_Y (G \cdot g; v|_Y)$$

$$= \lim_{\varepsilon \to 0_+} \int_{Y_\varepsilon} (G \cdot (\gamma_\varepsilon(\widetilde{A}^{-1}\gamma_0^* Gg)); \gamma_\varepsilon v)$$

$$- \lim_{\varepsilon \to 0_+} \int_{Y_{-\varepsilon}} (G \cdot (\gamma_{-\varepsilon}(\widetilde{A}^{-1}\gamma_0^* Gg)); \gamma_{-\varepsilon} v)$$

$$= \int_Y (G \cdot (\mathcal{P}_+ g + \mathcal{P}_- g); v|_Y) \,.$$

One could pass $\lim_{\varepsilon \to 0_+}$ inside the integrals because all sections are C^∞, the domain Y is compact, and all $Y_{\pm\varepsilon}$ may be identified with Y. Since G is invertible and $\{v|_Y \mid v \in C^\infty(\widetilde{X})\} = C^\infty(Y)$, equation (12.22) proves $\mathcal{P}_+ + \mathcal{P}_- = \mathrm{Id}$.

No new arguments are needed to show range $\mathcal{P}_+ \subset H_+$. We proved already that $K_+ g \in C^\infty(S^+|_{X_+ \setminus Y})$ and that $AK_+ g = \gamma_0^* Gg$ is supported in Y and therefore $AK_+ g = 0$ in the interior of X_+; in fact, $K_+ g \in \ker(A, s)$ for all $s \geq 0$ if $g \in C^\infty(S^+|_Y)$. □

Remarks 12.5. (a) One observes that the principal symbol p_+ of the Calderón projector $\mathcal{P}_+$ has locally (over the connected components of Y) constant rank for all non-zero covectors (y, η): Consider the subspace $M^>_{y,\eta}$ of S^+_y spanned by the eigenvectors corresponding to the eigenvalues of $b(y, \eta) : S^+_y \to S^+_y$ with positive real part. Since A takes the form $G(\partial_t + B)$ close to Y, we get

$$a(t, y; \tau, \eta) = G(y)(i\tau + b(y, \eta))$$

for the principal symbol of A close to Y as noticed above. The ellipticity of A therefore implies that $b(y, \eta)$ has no purely imaginary eigenvalues. Hence no eigenvalues can pass from the one complex halfplane into the other, and the dimension of $M^>_{y,\eta}$ cannot change locally.

(b) Now consider the space

(12.23)

$$\mathcal{M}^>_{y,\eta} := \{w : \mathbf{R} \to S^+_y \mid a(0, y; \frac{d}{dt}, \eta) w = 0 \text{ and } w(t) \to 0 \text{ as } t \to \infty\} \,.$$

The space consists of those solutions of the given system of ordinary differential equations which are exponentially decreasing on $\mathbf{R}_+$. If we identify the solutions with their initial values, we obtain the space $M^{>}_{y,\eta}$.

Applying the preceding theorem, we shall give a simple geometric proof of the following result, observed in Booß & Wojciechowski [1985]. It answers the question of how to describe the kernel of the Poisson operator

$$K_+ : H^{s-\frac{1}{2}}(Y; S^+|_Y) \longrightarrow \ker_+(A,s)\,. \tag{12.24}$$

By Theorem 12.4a , $K_+|_{H_+(A,s)} : H_+(A,s) \to \ker_+(A,s)$ is bijective, and by Theorem 12.4b the closures of $H_+(\widetilde{A})$ and $H_-(\widetilde{A})$ are complementary. This complementarity relies on our construction of the invertible double $\widetilde{A}$ and remains valid for all invertible Dirac operators over a partitioned manifold, but clearly breaks down when a given Dirac operator over a partitioned manifold is not invertible. On the contrary, the following corollary expresses the kernel of (12.24) intrinsically in terms of the given manifold with boundary and provides a short exact sequence for elliptic equations on manifolds with boundary, namely,
(12.25)

$$0 \longrightarrow G^{-1}(H_+(A^*,s)) \hookrightarrow H^{s-\frac{1}{2}}(Y; S^+|_Y) \xrightarrow{K_+} \ker_+(A,s) \longrightarrow 0\,.$$

Corollary 12.6. *Under the assumptions of the preceding theorem, the subspaces $H_+(A)$ and $G^{-1}(H_+(A^*))$ of $C^\infty(Y; S^+|_Y)$ are orthogonal and complementary.*

Proof. First extend the Dirac operator $A = A^+ : C^\infty(X; S^+) \to C^\infty(X; S^-)$ to the operator $\widetilde{A} = A^+ \cup A^-$ over the closed double. Denote by $\mathcal{P}(A)$ and $\mathcal{P}(A^*)$ the Calderón projections of $L^2(Y; S^+|_Y)$ onto the closure of $H_+(A) = H_+(\widetilde{A})$ and of $H_-(\widetilde{A}) = G^{-1}(H_+(A^-)) = G^{-1}(H_+(A^*))$. Then

$$\mathcal{P}(A) = \mathcal{P}_+(\widetilde{A}) \quad \text{and} \quad \mathcal{P}(A^*) = \mathcal{P}_-(\widetilde{A})$$

and the corollary follows from Theorem 12.4b. □

Another interesting consequence of Theorem 12.4 is that it provides with the operator $r^+(\widetilde{A})^{-1}e^+$ a left-parametrix for A. The point is

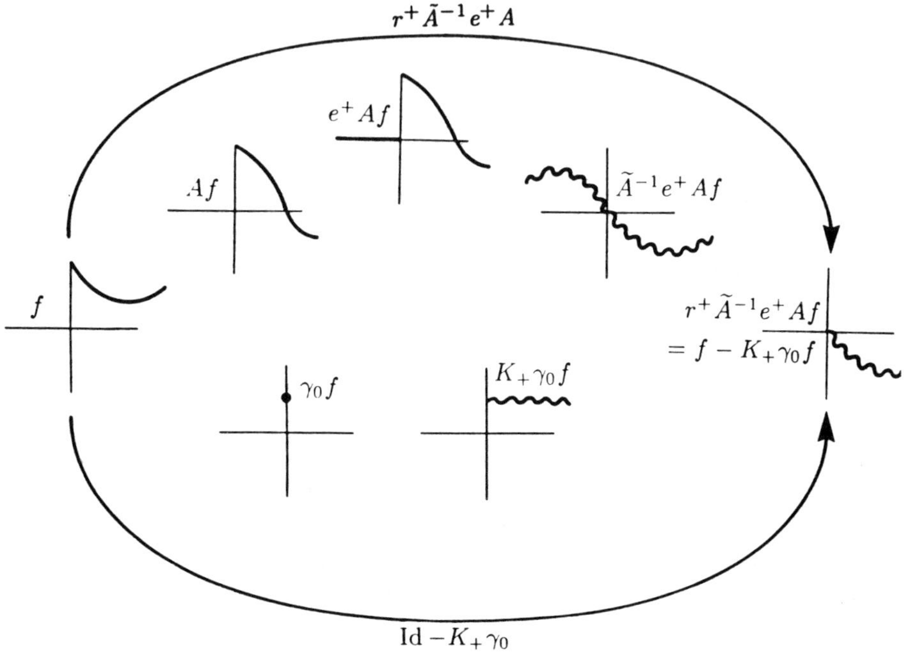

Fig. 12.4 Sketch of $r^+\tilde{A}^{-1}e^+Af = f - K_+\gamma_0 f$

that of course $\widetilde{A}^{-1}\widetilde{A} = \mathrm{Id}$ over the closed manifold $\widetilde{X}$, but there appears a correction term $K_+\gamma_0$ coming from the boundary data when we restrict the operation to the manifold X with boundary Y. For a rough visualization see Figure 12.4.

Lemma 12.7. *For any Dirac operator $A : C^\infty(X;S^+) \to C^\infty(X;S^-)$ over a compact manifold X with smooth boundary Y, the operator*

$$r^+(\widetilde{A})^{-1}e^+A : C^\infty(X;S^+) \longrightarrow C^\infty(X;S^+) \tag{12.26}$$

is well-defined and we have

$$r^+\widetilde{A}^{-1}e^+A = \mathrm{Id} - K_+\gamma_0\,. \tag{12.27}$$

Proof. Consider the operator of (12.26) on the distributional level.

Let $f \in H^s(X;S^+)$ and $h \in C^\infty(X;S^+)$. We have

$$
\begin{aligned}
(12.28)\ &\langle r^+\widetilde{A}^{-1}e^+(Af);h\rangle_X \\
&= \langle \widetilde{A}^{-1}e^+(Af);e^+h\rangle_{\widetilde{X}} = \int_X \langle Af; r^+(\widetilde{A}^{-1})^* e^+h\rangle dx \\
&= \int_X \langle f; A^* r^+(\widetilde{A}^{-1})^* e^+h\rangle dx - \int_Y \langle G(f|_Y); \left((\widetilde{A}^{-1})^* e^+h\right)|_Y\rangle dy \\
&= \langle f; r^+h\rangle_X - \langle G\gamma_0 f; \left(\gamma_0(\widetilde{A}^{-1})^*\right)(e^+h)\rangle_Y \\
&= \langle f; r^+h\rangle_X - \langle r^+\widetilde{A}^{-1}\gamma_0^* G(\gamma_0 f); h\rangle_X \\
&= \langle (\mathrm{Id} - K_+\gamma_0)f; h\rangle_X\,. \qquad \square
\end{aligned}
$$

It was pointed out that $\mathcal{P}(A)$ is just a projection, i.e. a not necessarily orthogonal idempotent. Actually one can always replace $\mathcal{P}(A)$ by the corresponding orthogonal projection in the case of product structure near the boundary and when A is an operator of Dirac type (namely with unitary G and self-adjoint tangential part B). The orthogonal projection obtained is a pseudo-differential operator with principal symbol equal to the principal symbol of $\mathcal{P}(A)$.

Lemma 12.8. (a) *Let P be a projection in a separable Hilbert space. Then*

$$P_{\mathrm{ort}} := PP^*\left(PP^* + (\mathrm{Id} - P^*)(\mathrm{Id} - P)\right)^{-1}$$

is an orthogonal projection onto the range of P.
(b) *If P is a pseudo-differential projection with orthogonal principal symbol p, we have*

$$\sigma_0(P_{\mathrm{ort}}) \;=\; \sigma_0(P) \;=\; p\,,$$

hence $P_{\mathrm{ort}} - P$ is a compact operator (operator of order -1).

Note. In the terminology to be introduced in Chapter 15, this means that P and P_{ort} belong to the same Grassmannian Gr_p. They also belong to the same connected component of Gr_p. The reason is that $P_{\mathrm{ort}}|_{\mathrm{range}\,P} = P|_{\mathrm{range}\,P} = \mathrm{Id}$, so

$$\mathrm{index}\{PP_{\mathrm{ort}} : \mathrm{range}\,P_{\mathrm{ort}} \to \mathrm{range}\,P\} = 0\,,$$

cf. Theorem 15.12 below.

Proof. *(a):* First $\{PP^* + (\mathrm{Id} - P^*)(\mathrm{Id} - P)\}$ is invertible. In fact,

$$PP^* + (\mathrm{Id} - P^*)(\mathrm{Id} - P) > 0\,,$$

since

$$\langle \{PP^* + (\mathrm{Id} - P^*)(\mathrm{Id} - P)\}\, u; u\rangle = \|P^*u\|^2 + \|(\mathrm{Id} - P)u\|^2 \geq 0$$

with equality, if and only if $Pu = u$ and $P^*u = 0$, hence

$$0 = \langle P^*u; u\rangle = \langle P^*Pu; u\rangle = \|Pu\|^2 = \|u\|^2$$

and so $u = 0$.

Next observe the vanishing of the commutator

$$[PP^*; PP^* + (\mathrm{Id} - P^*)(\mathrm{Id} - P)]$$

and hence of the commutator $[PP^*; \{PP^* + (\mathrm{Id} - P^*)(\mathrm{Id} - P)\}^{-1}]$. This implies

$$\begin{aligned} P_{\mathrm{ort}}^* &= \{(PP')^* + ((\mathrm{Id} - P^*)(\mathrm{Id} - P))^*\}^{-1}(PP^*)^* \\ &= \{PP^* + (\mathrm{Id} - P^*)(\mathrm{Id} - P)\}^{-1} PP^* = PP^*\{\ldots\}^{-1} = P_{\mathrm{ort}}\,. \end{aligned}$$

Then we find (keep track of the brackets $\{\ldots\}$):

$$\begin{aligned} (P_{\mathrm{ort}})^2 &= \{PP^*\}\,\{PP^* + (\mathrm{Id} - P^*)(\mathrm{Id} - P)\}^{-1}\, P_{\mathrm{ort}} \\ &= \{[PP^* + (\mathrm{Id} - P^*)(\mathrm{Id} - P)] - (\mathrm{Id} - P^*)(\mathrm{Id} - P)\}\,\{\ldots\}^{-1}\, P_{\mathrm{ort}} \\ &= \{[\ldots] - (\mathrm{Id} - P^*)(\mathrm{Id} - P)PP^* \\ &\qquad - (\mathrm{Id} - P^*)(\mathrm{Id} - P)(\mathrm{Id} - P^*)(\mathrm{Id} - P)\}\,\{\ldots\}^{-1}\, P_{\mathrm{ort}} \\ &= \{[\mathrm{Id} - (\mathrm{Id} - P^*)(\mathrm{Id} - P)]\,[PP^* + (\mathrm{Id} - P^*)(\mathrm{Id} - P)]\} \\ &\qquad\qquad \cdot \{PP^* + (\mathrm{Id} - P^*)(\mathrm{Id} - P)\}^{-1}\, P_{\mathrm{ort}} \\ &= \{\mathrm{Id} - (\mathrm{Id} - P^*)(\mathrm{Id} - P)\}\, P_{\mathrm{ort}} \\ &= P_{\mathrm{ort}} - (\mathrm{Id} - P^*)\,\{(\mathrm{Id} - P)P\}\, P^*\,(PP^* + (\mathrm{Id} - P^*)(\mathrm{Id} - P))^{-1} \\ &= P_{\mathrm{ort}} - 0 = P_{\mathrm{ort}}\,. \end{aligned}$$

Now apparently $PP_{\mathrm{ort}} = P_{\mathrm{ort}}$, so we only have to show that $P_{\mathrm{ort}}P = P$. Let $v = Pu$. Then

$$\begin{aligned} P_{\mathrm{ort}}(v) &= PP^*\,\{PP^* + (\mathrm{Id} - P^*)(\mathrm{Id} - P)\}^{-1}\,(Pu) \\ &= Pu - (\mathrm{Id} - P^*)(\mathrm{Id} - P)\,\{PP^* + (\mathrm{Id} - P^*)(\mathrm{Id} - P)\}^{-1}\,(Pu) \\ &= Pu - (\mathrm{Id} - P^*)(\mathrm{Id} - P)P\,\{PP^* + (\mathrm{Id} - P^*)(\mathrm{Id} - P)\}^{-1}\,(u) \\ &= Pu\,. \end{aligned}$$

(b): If P is a pseudo-differential operator, then P_{ort}, too; and if P has an orthogonal symbol, then our construction does not change the principal symbol. □

13. Existence of Traces of Null Space Elements

The establishing of the Calderón projector for Dirac operators over compact manifolds with boundary is rounded off by showing that all elements of $\ker_+(A, s)$, even the distributional solutions, have well-defined traces (boundary values).

We still have not completed the proof of Theorem 12.4a. The argument presented in Chapter 12 works only in case $s > \frac{1}{2}$. To apply the closing argument of the proof of Theorem 12.4a for $s \leq \frac{1}{2}$, we have to show the existence of the trace $\gamma_0 v$ in $H^{s-\frac{1}{2}}(Y; S^+|_Y)$ for $v \in \ker_+(A, s)$. This is the subject of the next theorem which is crucial to the whole theory of elliptic boundary problems. We adopt the notation of the preceding chapters, especially

- the construction of the invertible double $\widetilde{A}$ of chapter 9,
- the parametrization $\{Y_\varepsilon\}_{\varepsilon \in I}$ of a collar neighbourhood of the boundary Y, the restriction operator r^+, and for $s > \frac{1}{2}$ the trace operator $\gamma : H^s(X; S^+) \times I \to H^{s-\frac{1}{2}}(Y; S^+|_Y)$ of Chapter 11,
- and for arbitrary real s the null space $\ker(A, s)$ and the Poisson type operator $K_+ : H^{s-\frac{1}{2}}(Y; S^+|_Y) \to \ker_+(A, s)$ of Chapter 12.

Theorem 13.1. *Let $v \in \ker_+(A, s)$ (then in particular v is smooth in $X \setminus Y$, $Av = 0$ in $X \setminus Y$, and for $\varepsilon > 0$, the trace $\gamma_\varepsilon v$ is a well-defined element of $C^\infty(Y_\varepsilon; S^+|_Y)$). Then, as $\varepsilon \to 0_+$, the sections $\gamma_\varepsilon v$ converge to an element $g \in H^{s-\frac{1}{2}}(Y; S^+|_Y)$.*

Proof. First we extend $v \in H^s(X; S^+)$ to a section $\tilde{v} \in H^s(\widetilde{X}; \widetilde{S^+})$ such that $r^+\tilde{v} = v$ (see Definition 11.7d). Then $w := \tilde{A}\tilde{v}$ is an element of $H^{s-1}(\widetilde{X}; \widetilde{S^-})$ which vanishes on $X \setminus Y$. Therefore it is enough to prove that, for any distribution $w \in H^t(\widetilde{X}; \widetilde{S^-})$ with $w = 0$ in $X \setminus Y$,

$$\text{(13.1)} \qquad \lim_{\varepsilon \to 0_+} \gamma_\varepsilon(\tilde{A}^{-1} w) \text{ exists in } H^{t+\frac{1}{2}}(Y; S^+|_Y).$$

Use a partition of unity to localize the problem. One has to investigate the limit

$$\lim_{\varepsilon\to 0_+} \gamma_\varepsilon(M_\varphi \tilde{A}^{-1} M_\psi w),$$

where φ and ψ are smooth functions with compact support contained in a neighbourhood of a given point $y \in Y$. Hence consider the following problem: Split $\mathbf{R}^n = \mathbf{R} \times \mathbf{R}^{n-1}$, where

$$\mathbf{R}^{n-1} = \{(u, x_1, \ldots, x_{n-1}) \mid u = 0\}.$$

Let Q denote the operator $M_\varphi(\tilde{A})^{-1}M_\psi$. It is a pseudo-differential operator of order -1 in $\mathbf{R}^n$. Then, to prove Theorem 13.1, one must only show:

(13.2) *Assume that* $w \in H^s(\mathbf{R}^n)$ *(with values in* $\mathbf{C}^N$*) vanishes for* $u > 0$. *Then* $\lim_{\varepsilon\to 0_+}\gamma_\varepsilon Qw$ *is a well-defined element of* $H^{s+\frac{1}{2}}(\mathbf{R}^{n-1})$.

To prove (13.2) we use the operator

$$\Lambda_+ := -\partial_u + \sqrt{1+\Delta}\,.$$

Here $\Delta = \Delta_{(n-1)}$ denotes the Laplacian $-\sum_{j=1}^{n-1}(\partial/\partial x_j)^2$ in $\mathbf{R}^{n-1}$, and $\sqrt{1+\Delta}$ is a positive pseudo-differential operator of order 1 with total symbol $a(y,\zeta) = \sqrt{1+\sum_{j=1}^{n-1}\zeta_j^2}$. We have the following classical lemma.

Lemma 13.2. *For any* $s \in \mathbf{R}$, *the operator* Λ_+ *gives an isomorphism of Banach spaces*

$$\Lambda_+ : H^s(\mathbf{R}^n) \longrightarrow H^{s-1}(\mathbf{R}^n).$$

Proof. To prove that Λ_+ extends to a bounded operator from $H^s(\mathbf{R}^n) \to H^{s-1}(\mathbf{R}^n)$, it is enough to show that $\sqrt{1+\Delta}$ extends to such a mapping. Let $f \in H^s(\mathbf{R}^n)$. We may assume $s > 0$ since the general result then follows by duality. We write $(x,\xi) = (u,y;\tau,\zeta) \in \mathbf{R}^n \times T\mathbf{R}^n$ with $y = (x_1,\ldots,x_{n-1})$ and $\zeta = (\zeta_1,\ldots,\zeta_{n-1})$. We have

$$\|f\|_s := \int_{\mathbf{R}^n} |\hat{f}(\xi)|^2(1+|\xi|^2)^s\,d\xi\,;$$

hence

$$\|\sqrt{1+\Delta}\, f\|_{s-1}^2 = \int \left|\sqrt{1+|\zeta|^2}\hat{f}(\tau,\zeta)\right|^2 (1+|\xi|^2)^{s-1}\, d\xi$$
$$= \int |\hat{f}(\tau,\zeta)|^2(1+|\zeta|^2)(1+|\xi|^2)^{s-1}\, d\xi$$
$$= \int |\hat{f}(\xi)|^2(1+|\xi|^2)^s \underbrace{\frac{1+|\zeta|^2}{1+|\xi|^2}}_{\leq 1}\, d\xi \leq \|f\|_s\,,$$

thus proving the continuity. Also the proof that Λ_+ is an isomorphism is trivial. Notice that $(\Lambda_+)^* = \partial_u + \sqrt{1+\Delta} =: \Lambda_-$. Then the operator

$$(\Lambda_+)^*\Lambda_+ = -(\partial_u)^2 + \Delta + 1 = 1 + \Delta_{(n)}$$

is an isomorphism between H^s and H^{s-2}; here $\Delta_{(n)}$ denotes the Laplacian in $\mathbf{R}^n$. □

The reason for using the Λ_+-operator to prove (13.2) is the following statement.

Proposition 13.3. *Let w be a distribution in $\mathcal{S}'(\mathbf{R}^n)$ with support in the half-space $\{u \leq 0\}$. Then for any $t \in \mathbf{R}$, the distribution $\Lambda_+^t(w)$ also has support in $\{u \leq 0\}$.*

Proof. Let $\langle w, f\rangle$ denote the duality pairing between elements of $\mathcal{S}'$ and $\mathcal{S}$. Then for Λ_-, defined as above, we have $\langle \Lambda_+^t w, f\rangle = \langle w, \Lambda_-^t f\rangle$. It suffices to show that $\Lambda_-^t f$ vanishes for $u \leq 0$, if $f \in \mathcal{S}$ vanishes for $u \leq \varepsilon$ for some $\varepsilon > 0$. The point is that for such an f

$$\widehat{f}(\tau,\zeta) = \int_\varepsilon^\infty \int_{\mathbf{R}^{n-1}} e^{-iu\tau}\, e^{-i\langle y,\zeta\rangle}\, f(u,y)\, dy du$$

has a holomorphic extension to the half plane $\{\Im(\tau) < 0\}$. Let $\tau = a - ib$ and $g(u,y) := e^{-ub} f(u,y)$. Then

$$\widehat{f}(a-ib,\zeta) = \widehat{g}(a,\zeta)\,, \tag{13.3}$$

which tends to 0 in $\mathcal{S}$ as $b \to +\infty$. Moreover,

(13.4)

$(\Lambda_-^t f)(u,y) =$

$$= \int_{\mathbf{R}^{n-1}} e^{-i\langle y,\zeta\rangle}\, d\zeta \int_{-\infty}^{+\infty} e^{-iu\tau} \left[(1+\sum_{j=1}^{n-1}\zeta_j^2)^{\frac{1}{2}} + i\tau\right]^t \widehat{f}(\tau,\zeta) d\tau$$

$$= \int_{\mathbf{R}^{n-1}} e^{-i\langle y,\zeta\rangle}\, d\zeta \lim_{R\to\infty} \int_{-R}^{R} e^{-iu\tau} \left[(1+\sum_{j=1}^{n-1}\zeta_j^2)^{\frac{1}{2}} + i\tau\right]^t \widehat{f}(\tau,\zeta) d\tau\,.$$

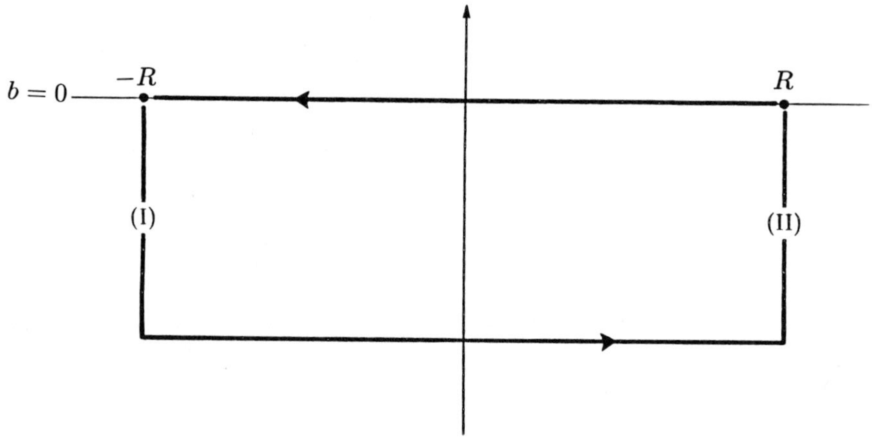

Fig. 13.1 Integration path for Cauchy's theorem

Now apply Cauchy's theorem. Since $\widehat{f}$ is an element of $\mathcal{S}$, the integrals along the sides (I) and (II) disappear as $R \to \infty$. Using (13.3) and (13.4) yields

$$(\Lambda_-^t f)(u,y) = \int_{\mathbf{R}^{n-1}} e^{-i\langle y,\zeta\rangle}\, d\zeta \, \cdot \tag{13.5}$$

$$\cdot \lim_{R\to\infty} \int_{a=R}^{a=-R} da\, e^{iua}\, e^{ub} \left(\sqrt{1+\sum_{j=1}^{n-1} \zeta_j^2} + ia + b \right)^t \widehat{f}(a-ib,\zeta)$$

$$= \int_{\mathbf{R}^{n-1}} \int_{-\infty}^{\infty} e^{-i\langle y,\zeta\rangle}\, e^{iua}\, e^{ub} \left(\sqrt{1+\sum_{j=1}^{n-1} \zeta_j^2} + ia + b \right)^t \widehat{f}(a-ib,\zeta) d\zeta da\,.$$

Taking $u < 0$ and letting $b \to +\infty$ yields $(\Lambda_-^t f)(u,y) = 0$. □

Unfortunately Λ_+ is not quite a pseudo-differential operator, since its symbol $\lambda_+(x;\xi) = \lambda_+(u,y;\tau,\zeta) = -i\tau + a(y,\zeta)$ violates the classical estimate

$$\left| \frac{\partial^\alpha}{\partial \xi^{|\alpha|}} \frac{\partial^\beta}{\partial x^{|\beta|}} \lambda(x,\xi) \right| \overset{!}{\leq} C_{\alpha,\beta}\, (1+|\xi|)^{1-|\alpha|}$$

for any $\alpha = (0,\dots)$ with $|\alpha| > 1$ and $|\tau| \to \infty$. However, as noticed in Lemma 13.2, still Λ_+ is an operator of order 1. We furthermore have the following obvious result which shows that Λ_+ also in other aspects behaves like a pseudo-differential operator:

Lemma 13.4. *Let $m \in \mathbf{Z}$. For any pseudo-differential operator Q of order $\leq m$, the commutator*

$$[Q, \Lambda_+] := Q\Lambda_+ - \Lambda_+ Q$$

(and any iterated commutator $[[.[[Q, \Lambda_+], \Lambda_+], \ldots \Lambda_+], \Lambda_+]$) is an operator of order $\leq m$.

Proof. We are content to prove the lemma in our case of $m = -1$ and

$$Q := M_\varphi(\tilde{A})^{-1} M_\psi\,,$$

where all is easy modulo small technical details. Notice, for example,

$$\begin{aligned}[M_\varphi(\tilde{A})^{-1}M_\psi, \Lambda_+] &= M_\varphi(\tilde{A})^{-1}[\Lambda_+, \tilde{A}](\tilde{A})^{-1}M_\psi \\ &\quad + M_\varphi(\tilde{A})^{-1}[M_\psi, \Lambda_+] + [M_\varphi, \Lambda_+](\tilde{A})^{-1}M_\psi\,.\end{aligned}$$

Clearly for a pseudo-differential operator S of order $k \geq 0$, the commutator $[S, \Lambda_+]$ is an operator of order $\leq k$, hence $[\tilde{A}, \Lambda_+]$ an operator of order ≤ 1 and $[M_\varphi, \Lambda_+]$ an operator of order ≤ 0. □

Resume the proof of Theorem 13.1. The proof goes via induction. Begin with the special, but most interesting case $s = -1$. Let $w \in H^{-1}(\mathbf{R}^n)$ vanish for $u > 0$. We have

$$\begin{aligned}Qw &= (Q\Lambda_+ - \Lambda_+ Q)\Lambda_+^{-1}(w) + \Lambda_+ Q\Lambda_+^{-1}(w) \\ &= [Q, \Lambda_+]\Lambda_+^{-1}(w) + \Lambda_+ Q \Lambda_+^{-1}(w)\,.\end{aligned} \tag{13.6}$$

The operator $[Q, \Lambda_+]\Lambda_+^{-1}$ is of order -2, hence $[Q, \Lambda_+]\Lambda_+^{-1}(w) \in H^1(\mathbf{R}^n)$, and by the trace theorem 11.4

$$\lim_{\varepsilon \to 0_+} \gamma_\varepsilon [Q, \Lambda_+]\Lambda_+^{-1}(w)$$

exists in $H^{\frac{1}{2}}(\mathbf{R}^{n-1})$. The same argument gives the existence of the trace

$$\lim_{\varepsilon \to 0_+} \gamma_\varepsilon Q\Lambda_+^{-1}(w) \in H^{\frac{1}{2}}(\mathbf{R}^{n-1})\,.$$

Now

$$\begin{aligned}\Lambda_+ Q \Lambda_+^{-1}(w) &= (-\partial_u + \sqrt{1+\Delta})Q\Lambda_+^{-1}(w) \\ &= (-\partial_u Q \Lambda_+^{-1})(w) + \sqrt{1+\Delta}Q\Lambda_+^{-1}(w)\,.\end{aligned}$$

Since the operator $\sqrt{1+\Delta}$ acts in tangential direction we have

$$\begin{aligned}\gamma_0\left(\sqrt{1+\Delta}Q\Lambda_+^{-1}(w)\right) &= \lim_{\varepsilon\to 0_+}\gamma_\varepsilon\left(\sqrt{1+\Delta}Q\Lambda_+^{-1}(w)\right)\\ &= \sqrt{1+\Delta}\lim_{\varepsilon\to 0_+}\left(\gamma_\varepsilon Q\Lambda_+^{-1}(w)\right) = \sqrt{1+\Delta}\left(\gamma_0 Q\Lambda_+^{-1}(w)\right).\end{aligned}$$

We are left with the term $-\partial_u Q\Lambda_+^{-1}(w)$. We have

$$\begin{aligned}Q\Lambda_+^{-1}(w) &= M_\varphi \tilde{A}^{-1}M_\psi\Lambda_+^{-1}(w)\\ &= \tilde{A}^{-1}M_\psi\Lambda_+^{-1}(w) + M_{(1-\varphi)}\tilde{A}^{-1}M_\psi\Lambda_+^{-1}(w).\end{aligned}$$

The second summand is smooth (for a suitable choice of φ and ψ). The first summand actually belongs to the null space of A on $X_+\setminus Y$, since

$$\tilde{A}(\tilde{A}^{-1}M_\psi\Lambda_+^{-1}(w)) = M_\psi\Lambda_+^{-1}(w).$$

Here, let $X_\pm$ denote the two halves of the closed double $\widetilde{X}$ of $X = X_+$. It follows from Proposition 13.3 that $\Lambda_+^{-1}(w)$ has support in $u \le 0$, hence $h := M_\psi\Lambda_+^{-1}(w)$ has support in X_-. Then $\tilde{A}^{-1}h$ belongs to $H^1(\widetilde{X};\widetilde{S^+})$ and has a well-defined trace in Y. We also have

$$\begin{aligned}0 &= \lim_{\varepsilon\to 0_+}\gamma_\varepsilon(\tilde{A}(\tilde{A}^{-1}h)) = \lim_{\varepsilon\to 0_+}\gamma_\varepsilon(G(\partial_u+B)(\tilde{A}^{-1}h))\\ &= G\left(\lim_{\varepsilon\to 0_+}\gamma_\varepsilon(\partial_u+B)(\tilde{A}^{-1}h)\right)\\ &= G\left(\lim_{\varepsilon\to 0_+}\gamma_\varepsilon\partial_u(\tilde{A}^{-1}h) + B\gamma_0(\tilde{A}^{-1}h)\right)\end{aligned}$$

which implies

$$\lim_{\varepsilon\to 0_+}\gamma_\varepsilon\partial_u(\tilde{A}^{-1}h) = -B\gamma_0(\tilde{A}^{-1}h), \tag{13.7}$$

and the theorem is proved for $s=-1$. The argument works for any $s > -\frac{3}{2}$. To prove statement (13.2) for arbitrary s we argue by induction, using step by step the lifting argument $Qw \mapsto \Lambda_+ Q\Lambda_+^{-1}(w)$ of (13.6). To explain the induction, let us repeat the arguments:

Let v be a distribution which is a solution of $Av = 0$ in $X\setminus Y$. As in the last part of the proof for $s=-1$, we obtain that the traces $\gamma_0((\partial_u)^k v)$ exist for any $k \ge 1$, if the trace $\gamma_0 v$ exists. The point is that we have

$$\gamma_0(A^k v) = 0, \tag{13.8}$$

and that we can solve that equation to obtain a formula for $\gamma_0((\partial_u)^k v)$ in terms of

$$\{\gamma_0((\partial_u)^j B^j v) \mid 0 \le j < k\}$$

as in (13.7) for $k = 1$. In particular, let $w \in H^s(\mathbf{R}^n)$ with $\operatorname{supp} w \subset \{u \le 0\}$ and let k be sufficiently large, namely $k > -s - \frac{1}{2}$. Then it follows from the preceding argument that the distribution

$$\Lambda_+^k Q \Lambda_+^{-k} w \in H^{s+1}(\mathbf{R}^n)$$

has a well-defined trace $\gamma_0(\Lambda_+^k Q \Lambda_+^{-k} w) \in H^{s+\frac{1}{2}}(\mathbf{R}^{n-1})$.

Now use that result for the induction. To explain the induction, let us begin once again with a special case. Let $w \in H^s(\mathbf{R}^n)$ with $\operatorname{supp} w \subset \{u \le 0\}$, but now $s > -\frac{5}{2}$. Then

$$\begin{aligned} Qw &= [Q, \Lambda_+]\Lambda_+^{-1} w + \Lambda_+ Q \Lambda_+^{-1} w \\ &= [[Q, \Lambda_+]\Lambda_+^{-1}, \Lambda_+]\Lambda_+^{-1} w + \Lambda_+ [Q, \Lambda_+]\Lambda_+^{-2} w + \Lambda_+ Q \Lambda_+^{-1} w \\ &= [[Q, \Lambda_+]\Lambda_+^{-1}, \Lambda_+]\Lambda_+^{-1} w + 2\Lambda_+ Q \Lambda_+^{-1} w - \Lambda_+^2 Q \Lambda_+^{-2} w\,. \end{aligned}$$

By Lemma 13.4, the first summand belongs to $H^{s+3}(\mathbf{R}^n)$, hence it has a trace by the general trace theorem 11.4. By the same argument $Q\Lambda_+^{-2} w$ has a trace. The third summand is equal to

$$-\left((\partial_u)^2 + \Delta_{(n-1)} - 2\partial_u \sqrt{1 + \Delta_{(n-1)}} + 1\right) Q\Lambda_+^{-2} w\,.$$

The traces $\gamma_0(\partial_u(Q\Lambda_+^{-2} w))$ and $\gamma_0((\partial_u)^2(Q\Lambda_+^{-2} w))$, as explained above, are well defined, since $\gamma_0(Q\Lambda_+^{-2} w)$ exists. And, since $\Delta_{(n-1)}$ acts tangentially, it follows that $\gamma_0(\Lambda_+^2 Q \Lambda_+^{-2} w)$ is well defined, and we are left with the trace of $\Lambda_+ Q \Lambda_+^{-1}$. All we need to show is that $Q\Lambda_+^{-1} w$ has a well-defined trace in $H^{s+\frac{3}{2}}(\mathbf{R}^{n-1})$, since then our method of solving (13.8) for the normal derivatives yields that $\partial_u Q \Lambda_+^{-1} w$ has a well-defined trace in $H^{s+\frac{1}{2}}(\mathbf{R}^{n-1})$. The same is true for $\sqrt{1 + \Delta_{(n-1)}} Q \Lambda_+^{-1} w$ and hence for $\Lambda_+ Q \Lambda_+^{-1}$. Now consider

$$\begin{aligned} Q\Lambda_+^{-1} w &= Q\Lambda_+^{-1} w - \Lambda_+ Q \Lambda_+^{-2} w + \Lambda_+ Q \Lambda_+^{-2} w \\ &= [Q, \Lambda_+]\Lambda_+^{-2} w + \Lambda_+ Q \Lambda_+^{-2} w\,. \end{aligned}$$

The first summand belongs to $H^{s+3}(\mathbf{R}^n)$ and hence has a well-defined trace in $H^{s+\frac{5}{2}}(\mathbf{R}^{n-1})$. With regard to the second summand, we notice that $Q\Lambda_+^{-2} w$ has a well-defined trace in $H^{s+\frac{3}{2}}(\mathbf{R}^{n-1})$, hence also

$\Lambda_+(Q\Lambda_+^{-2}w)$ by our (13.8) argument. This shows the existence of the trace of $\Lambda_+ Q\Lambda_+^{-1}w$ in $H^{s+\frac{1}{2}}(\mathbf{R}^{n-1})$.

For general s, use the formula

$$Qw = [[\dots[Q,\Lambda_+]\Lambda_+^{-1},\dots,\Lambda_+]\Lambda_+^{-1},\Lambda_+]\Lambda_+^{-1}w + \sum_{j=1}^{k} c_{jk}\Lambda_+^{j} Q\Lambda_+^{-j}w\,.$$

The first summand is in $H^{s+k+1}(\mathbf{R}^n)$, hence it has for $k > -s-\frac{1}{2}$ a trace, and we argue as before to show that then each $\Lambda_+^{j}Q\Lambda_+^{-j}w$ has a well-defined trace in $H^{s+\frac{1}{2}}(\mathbf{R}^{n-1})$. □

To show that the traces established in the preceding theorem are reasonable with respect to the Poisson type operator K_+ and the Calderón projector $\mathcal{P}_+$, we prove

Proposition 13.5. *Let $v \in \ker_+(A,s)$ and $r := \gamma_0(v)$. Then*

$$v = K_+(r) \quad \text{and} \quad \mathcal{P}_+(r) = r\,. \tag{13.9}$$

Proof. The proposition is proved for $s > \frac{1}{2}$. Now let s be arbitrary and $v \in \ker_+(A,s)$. Let $x \in X\setminus Y$ and choose ε so that

$$x \in X_{2\varepsilon} := X\setminus\{[0,2\varepsilon]\times Y\}\,.$$

The distribution v is C^∞ in X_ε, hence $v = \tilde{A}^{-1}\gamma_\varepsilon^* G(\gamma_\varepsilon v)$ in X_ε. Choose φ and ψ in $C^\infty(X)$ so that

$$\varphi\cdot\psi = 0,\ \varphi(x) = 1,\ \text{and}\ \operatorname{supp}\varphi \subset [0,2\varepsilon]\times Y\,. \tag{13.10}$$

Then

$$v(x) = (\varphi\tilde{A}^{-1}\psi)\gamma_\varepsilon^*(G\gamma_\varepsilon v)(x)\,. \tag{13.11}$$

The operator $\varphi\tilde{A}^{-1}\psi$ is a smoothing operator, and we have a convergence

$$\gamma_\varepsilon^*(G\gamma_\varepsilon v) \to \gamma_0^*(Gr) \qquad \text{in } H^t(\tilde{X};\widetilde{S^-})\,,$$

where t is a negative number $\leq s$. This shows that for any ε and any $x \in X_{2\varepsilon}$

$$v(x) = (\varphi\tilde{A}^{-1}\psi)(\gamma_\varepsilon^* G\gamma_\varepsilon v)(x) \to (\varphi\tilde{A}^{-1}\psi)(\gamma_\varepsilon^* G\gamma_0 r)(x) = K_+ r(x)\,.$$

Therefore we have

$$v = K_+ r \quad \text{and} \quad \mathcal{P}_+ r = \lim_{\varepsilon\to 0_+}\gamma_\varepsilon K_+ r = \lim_{\varepsilon\to 0_+}\gamma_\varepsilon v = r\,.$$ □

To complete the proof of Theorem 12.4a, we need two more results.

Lemma 13.6. *Let* $r \in H^{s-\frac{1}{2}}(Y; S^+|_Y)$ *and* $\mathcal{P}_+ r = r$. *Then there exists* $v \in H^s(X; S^+)$ *such that* $Av = 0$ *in* $X \setminus Y$ *and*

$$\gamma_0 v = \lim_{\varepsilon \to 0_+} \gamma_\varepsilon v = r \qquad \text{in } H^{s-\frac{1}{2}}(Y; S^+|_Y).$$

In other words, $\operatorname{range}(\mathcal{P}_+) \subset \gamma_0(\ker_+(A, s))$.

Proof. Let $\{r_\nu\}$ denote a sequence of smooth sections such that $\mathcal{P}_+ r_\nu = r_\nu$ and $r_\nu \to r$ in $H^{s-\frac{1}{2}}(Y; S^+|_Y)$. Then the previous estimate (12.12) implies that $\{K_+(r_\nu)\}$ is a Cauchy sequence in $H^s(X; S^+)$, hence it has a limit and of course

$$\lim_{\nu \to \infty} K_+(r_\nu) = K_+ r = v,$$

and once again we have

$$r = \mathcal{P}_+ r = \lim_{\varepsilon \to 0_+} \gamma_\varepsilon v. \qquad \square$$

Recall that $H_+(A, s)$ denotes the closure of $H_+(A)$ in $H^{s-\frac{1}{2}}$. Then

Corollary 13.7. *The range of* $\mathcal{P}_+$ *in* $H^{s-\frac{1}{2}}(Y; S^+|_Y)$ *is the space* $H_+(A, s)$, *and the operator* $K_+ : H^{s-\frac{1}{2}}(Y; S^+|_Y) \to H^{s-1}(X; S^+)$ *is injective, when restricted to* $H_+(A, s)$.

Proof. Assume that we are given $r_1 = \mathcal{P}_+ r_1$ and $r_2 = \mathcal{P}_+ r_2$ in $H^{s-\frac{1}{2}}(Y; S^+|_Y)$, such that $K_+ r_1 = v = K_+ r_2$. Then

$$r_1 = \lim_{\varepsilon \to 0_+} \gamma_\varepsilon v = r_2. \qquad \square$$

This corollary ends the proof of Theorem 12.4.

Next we use Theorem 12.4 and Lemma 12.7 to show the existence of traces of other distributional sections not necessarily belonging to the null space.

Theorem 13.8. *Let* $v \in H^t(X; S^+)$ *and* $Av \in H^s(X; S^-)$ *with* t, s *real, and* $s > -\frac{1}{2}$. *Then the trace of* v *on* Y *is well-defined in* $H^{t-\frac{1}{2}}(Y; S^+|_Y)$.

Proof. If $t > s + 1$, we have $t > \frac{1}{2}$, and the assertion simply follows from Theorem 11.4. Assume $t \leq s + 1$. Then $w := r^+ \tilde{A}^{-1} e^+(Av)$ is well-defined in $H^{s+1}(X; S^+)$, and we have $Aw = Av$ in $H^s(X; S^-)$; hence $v - w \in \ker_+(A, t)$. Theorem 12.4 implies that there exists a unique $r = \mathcal{P}_+ r \in H^{t-\frac{1}{2}}(Y; S^+|_Y)$ such that $K_+ r = v - w$ and w has a well-defined trace in $H^{s-\frac{1}{2}}(Y; S^+|_Y) \subset H^{t-\frac{1}{2}}(Y; S^+|_Y)$. Therefore we have

$$\gamma_0 v = \lim_{\varepsilon \to 0_+} \gamma_\varepsilon((v - w) + w) = r + \gamma_0(w) . \qquad \square$$

We have one important case which has already been established in the second part of the proof of Theorem 13.1.

Corollary 13.9. *Let $v \in L^2(X; S^+)$ and $Av \in L^2(X; S^-)$. Then v has a well-defined trace in $H^{-\frac{1}{2}}(Y; S^+|_Y)$.*

Remark 13.10. With Theorem 13.8 and its corollaries we sharpen the trace theorem of Chapter 11, which gave a trace in $H^{t-\frac{1}{2}}(Y)$ for $u \in H^t$ only if $t > \frac{1}{2}$. Theorem 13.8 shows that the assumption $t > \frac{1}{2}$ is dispensible, if $Au \in H^s$ for some $s > -\frac{1}{2}$. In Chapters 19 and 20 below, it turns out that the theorem and its corollaries are decisive for regularity and closedness of (global) elliptic boundary value problems.

14. Spectral Projections of Dirac Operators

We account for the construction and the basic properties of the spectral projections associated with the tangential part of a Dirac operator.

In this chapter we discuss one more pseudo-differential projection which plays an important role in the theory of elliptic boundary problems for Dirac operators. Unlike the Calderón projection, which is defined by global data of a Dirac operator A over a manifold X with boundary Y, we shall employ only the tangential part B for the construction of the spectral projection. In this chapter we may therefore forget about X and A and just consider arbitrary pseudo-differential elliptic self-adjoint operators over a closed manifold Y. Let

$$B : C^\infty(Y;V) \to C^\infty(Y;V)$$

be such an operator acting on sections of a Hermitian vector bundle V over Y. Assume (though it is not necessary) that B is of first order.

It is well-known that B has a discrete spectrum contained in $\mathbf{R}$ numbered like

$$(14.1) \qquad \cdots \leq \lambda_{-j} \leq \cdots \leq \lambda_{-1} < 0 \leq \lambda_0 \leq \lambda_1 \leq \cdots \leq \lambda_j \leq \ldots,$$

and one can find an orthonormal basis $\{e_j\}_{j\in\mathbf{Z}}$ of $L^2(Y;V)$ consisting of (smooth) eigenfunctions of B (i.e. $Be_j = \lambda_j e_j$, $j \in \mathbf{Z}$), see e.g. Gilkey [1984; Lemma 1.6.3]. Such a system $\{\lambda_j; e_j\}_{j\in\mathbf{Z}}$ is called a *spectral decomposition* of $L^2(Y;V)$ generated by B, or, in short, a *spectral resolution* of B.

Definition 14.1. For a given self-adjoint elliptic operator B and for any real a we shall denote by

$$\mathrm{P}_{\geq a} : L^2(Y;V) \longrightarrow L^2(Y;V)$$

the *spectral projection*, that is, the orthogonal projection of $L^2(Y;V)$ onto the subspace spanned by $\{e_j \mid \lambda_j \geq a\}$, where $\{\lambda_j; e_j\}_{j\in\mathbf{Z}}$ is a spectral resolution of B.

Note. We shall write $\mathrm{P}_{\geq a}(B)$, when we want to stress the dependence on B; we shall omit the a when $a = 0$; and we shall adopt the notation $\mathrm{P}_{<} := \mathrm{Id} - \mathrm{P}_{\geq}$. Hence

$$\mathrm{P}_{\geq} := \mathrm{P}_{\geq 0}(B) \quad \text{and} \quad \mathrm{P}_{<} := \mathrm{P}_{<0} = \mathrm{Id} - \mathrm{P}_{\geq} .$$

It is well-known that the spectral projection $\mathrm{P}_{\geq a}$ is a pseudo-differential operator of order zero (see e.g. Atiyah, Patodi & Singer [1976, p. 48], Baum & Douglas [1981, Proposition 2.4], or Booß & Wojciechowski [1985, Lemma 2.2]). We shall repeat the arguments as this proposition is basic for the following:

Proposition 14.2. *The spectral projection $\mathrm{P}_{\geq a}$ is a pseudo-differential operator of order zero for any real number a. The principal symbol of $\mathrm{P}_{\geq a}$ does not depend on a, and for any $(y,\eta) \in T^*Y \setminus \{0\}$ it is equal to the orthogonal projection*

$$p_+(y,\eta) : V_y \longrightarrow V_y$$

of V_y onto the direct sum of the eigenspaces of the automorphism

$$b(y,\eta) : V_y \longrightarrow V_y$$

corresponding to the positive eigenvalues. Here b denotes the principal symbol of the operator B.

Proof. For any pair of real numbers $a < b$, the operator $\mathrm{P}_{\geq a} - \mathrm{P}_{\geq b}$ is the projection onto the direct sum of eigenspaces corresponding to the eigenvalues from the interval $[a, b)$. This is a finite-dimensional space with a basis $\{e_j\}_{a \leq \lambda_j < b}$ consisting of smooth sections; hence $\mathrm{P}_{\geq a} - \mathrm{P}_{\geq b}$ is an operator with smooth kernel. Thus it suffices to prove the proposition for $a = 0$.

Let B' denote the bounded operator (in $L^2(Y;V)$, i.e. operator of order zero)

$$B' := (\mathrm{Id} + B^2)^{-1/2} B . \tag{14.2}$$

It has a spectral decomposition

$$\left\{ \frac{\lambda_j}{\sqrt{1+\lambda_j^2}} ; e_j \right\}_{j \in \mathbf{Z}} . \tag{14.3}$$

The advantage of investigating B' instead of B is due to two facts, namely that

– the new spectrum is contained in the interval $(-1,1)$,

– and there still exists $\varepsilon > 0$ such that

(14.4) $$\operatorname{spec} B' \cap (-2\varepsilon, 2\varepsilon) \subset \{0\}.$$

It is well-known from spectral theory, cf. e.g. Kato [1976, Section III.6.4], that the operators $\mathrm{P}_{\geq} = \mathrm{P}_{\geq 0}(B) = \mathrm{P}_{\geq 0}(B')$ and $\mathrm{P}_{<} = \mathrm{P}_{<0} = \mathrm{Id} - \mathrm{P}_{\geq}$ have the integral representation

(14.5)
$$\mathrm{P}_{\geq} = \frac{1}{2\pi i}\int_{\Gamma_{\geq}} (B'-\lambda)^{-1} d\lambda\,, \text{ and } \mathrm{P}_{<} = \frac{1}{2\pi i}\int_{\Gamma_{<}} (B'-\lambda)^{-1} d\lambda\,,$$

where

(14.6) $$\Gamma_{\geq} := \{(1+\varepsilon)e^{it}+1\}\,, \quad \text{and} \quad \Gamma_{<} := \{(1-\varepsilon)e^{it}-1\}$$

and ε is chosen so small that $[-\varepsilon, 0) \cap \operatorname{spec}(B') = \emptyset$, see Figure 14.1.

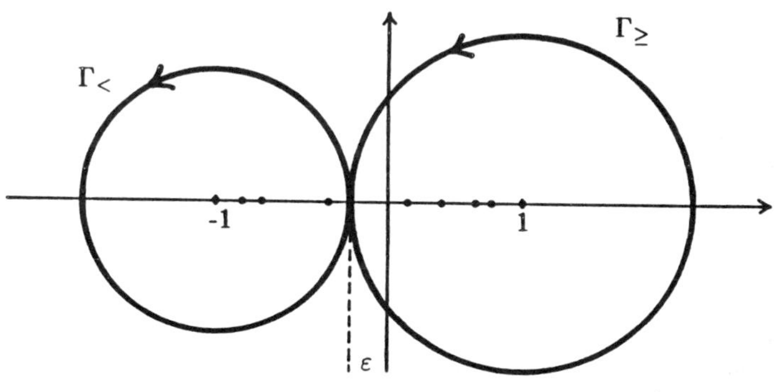

Fig. 14.1 Separation of the spectrum of B'

It is easy to determine the full symbol of the operator (the *resolvent*) $(B'-\lambda)^{-1}$ for $\lambda \in \mathbf{C}\backslash\operatorname{spec}(B')$ in any local chart. Its asymptotic expansion has the form $\sum_{j\geq 0} p_{-j}(y,\eta;\lambda)$, where the p_j are rational functions of λ of order $\leq -j$ with poles at $\lambda = \pm 1$ (see e.g. Gilkey [1984, p. 51f]). In particular we find for the principal symbol of B'

$$b'(y,\eta) = p_+(y,\eta) - (\mathrm{id} - p_+)(y,\eta) = p_+ - p_-\,.$$

This yields

$$
\begin{aligned}
p_0(y,\eta;\lambda) &= (b'-\lambda)^{-1}(y,\eta) = [(b'-\lambda)(y,\eta)]^{-1} \\
&= \frac{1}{1-\lambda^2}(b'(y,\eta)+\lambda) = \frac{1}{1-\lambda^2}(p_+ - p_- + \lambda) \\
&= \frac{1}{1-\lambda^2}((1+\lambda)p_+ - (1-\lambda)p_-) \\
&= \frac{1}{1-\lambda}p_+(y,\eta) - \frac{1}{1+\lambda}p_-(y,\eta)\,.
\end{aligned} \tag{14.7}
$$

This shows that the principal symbol of $\mathrm{P}_{\geq}$ is equal to

$$
\begin{aligned}
\sigma_0(\mathrm{P}_{\geq})(y,\eta) &= \frac{1}{2\pi i}\int_{\Gamma_{\geq}} (b'-\lambda)^{-1}(y,\eta)\,d\lambda \\
&= \frac{1}{2\pi i}\int_{\Gamma_{\geq}} \left\{\frac{1}{1-\lambda}p_+(y,\eta) - \frac{1}{1+\lambda}p_-(y,\eta)\right\} d\lambda \\
&= p_+(y,\eta)\,.
\end{aligned} \tag{14.8}
$$

In the same way we compute

$$
c_k(y,\eta) = \frac{1}{2\pi i}\int_{\Gamma_{\geq}} p_k(y,\eta;\lambda)\,d\lambda\,. \tag{14.9}
$$

Let C_k denote an operator with symbol $\sum_{\iota=0}^{k} c_\iota(y,\eta)$. Then, by the standard argument, one obtains that $\mathrm{P}_{\geq} - C_k$ is an operator of order $-k-1$. □

Corollary 14.3. *Let us assume that the operator B is the tangential part of a Dirac operator A. Then, for any real a, $\mathrm{P}_{\geq a}(B) - \mathcal{P}_+(A)$ is an operator of order -1.*

Proof. The operators $\mathrm{P}_{\geq a}(B)$ and $\mathcal{P}_+(A)$ have the same principal symbol. □

Remarks 14.4. (a) To prove that $\mathrm{P}_{\geq}(B)$ is a pseudo-differential operator, an alternative argument may be summarized as follows:

Let $B = U_B|B|$ be the (unique) polar decomposition of B, see e.g. Kato [1976, Section VI.2.7]. Here $|B|$ denotes the non-negative square root of $B^2 = B^*B$. It is an elliptic pseudo-differential operator of the same order as B by Seeley's theorem on (complex) powers of pseudo-differential operators. The operator U_B is a partial isometry from $L^2(Y;V)$ in itself with a decomposition of $L^2(Y;V)$ into the three eigenspaces m_+, m_-, and m_0 with eigenvalues equal $1, -1$, and 0. These eigenspaces are generated by the eigenfunctions of B corresponding to positive, negative, and zero eigenvalues (the last being the kernel of B).

The space $m_0 = \ker B$ is finite-dimensional and consists of C^∞-sections. Let k denote the $L^2(Y;V)$ orthogonal projection onto $\ker B$,

$$kg := \sum_{j=1}^{l} \langle g, e_j \rangle e_j \,, \tag{14.10}$$

where $e_1, \dots, e_l$ is an orthonormal basis of $\ker B$. This shows that k can be defined by a C^∞-kernel, so k is a pseudo-differential operator of order $-\infty$, i.e. a smoothing operator.

We observe that

$$\mathrm{P}_{\geq}(B) = \frac{1}{2}(\mathrm{Id} + U_B) + k \,. \tag{14.11}$$

Therefore, in order to show that $\mathrm{P}_{\geq}(B)$ is a pseudo-differential operator of order zero, it suffices to show that U_B is a pseudo-differential operator of order zero.

Notice that the operators B and $|B|$ have the same kernel. Hence $|B|+k$ is a pseudo-differential operator of the same order as B, namely of order 1, elliptic and invertible, and $(|B|+k)^{-1}$ is an (elliptic) pseudo-differential operator of order -1. Then we can write

$$U_B = U_B(|B|+k)(|B|+k)^{-1} = U_B|B|(|B|+k)^{-1} = B(|B|+k)^{-1}$$

since $U_B k = 0$. This shows that U_B is a pseudo-differential operator of order 0.

(b) Notice that the Calderón projector $\mathcal{P}_+(A)$ of a Dirac operator over a manifold with boundary and the spectral projection $\mathrm{P}_{\geq}(B)$ of its tangential part coincide for the Cauchy-Riemann operator over the disc (see Example 21.1 below). One might ask whether that coincidence is a peculiarity of dimension two. For higher dimensions and certain symmetric spaces, harmonic analysis e.g. establishes related 1-1 correspondences, namely between the space of *harmonic forms* over a symmetric n-dimensional domain with boundary and the *Hardy space* of harmonic eigenforms to non-negative eigenvalues over the $\frac{n}{2}$-dimensional Šilov boundary.

More specifically, we always have, by definition, the inclusion

$$\ker B \subset \operatorname{range} \mathrm{P}_{\geq}(B)\,,$$

whereas the question

$$\ker B \overset{?}{\subset} \operatorname{range} \mathcal{P}_{+}(A)$$

is non-trivial: Roughly speaking, a harmonic form over the 1-codimensional boundary might have too many conditions to satisfy to be extendable over the whole manifold. Nevertheless, Kori [1993; Theorem 5.10] obtains $\mathcal{P}_{+}(A^{+}) = \mathrm{P}_{\geq}(B)$ for the Dirac operator over the 4-dimensional ball with boundary S^3 in a suitable metric.
(c) A more general variant of spectral projections of elliptic operators is discussed in Wojciechowski [1985a; Sections 2–4], where the index theory for families of operators with two rays of minimal growth is discussed.

15. Pseudo-Differential Grassmannians

The homotopy groups of the space of pseudo-differential projections with given principal symbol are computed. Criteria are given for two projections belonging to the same connected component.

The next tool needed to explain what the index of global elliptic boundary problems determines is the *Grassmannian of pseudo-differential projections.* We shall consider the set Gr_{p_+} of all pseudo-differential projections P with principal symbol equal to p_+, and with the topology given by the standard norm topology in the space of all bounded operators acting on the space $L^2(Y; S|_Y)$. (In this chapter we only assume that $S|_Y$ is a Hermitian bundle and do not care whether $S|_Y$ originates from a full or split bundle of Clifford modules.) We assume that p_+ is an orthogonal projection and that $p_+ \neq 0$ and $p_+ \neq \mathrm{Id}$ or, equivalently, that $\dim\ker P = \infty = \dim\operatorname{range} P$. We calculate the homotopy groups of Gr_{p_+} and it turns out that the closure of this space is a classifying space for the K^0 functor; but we shall only exploit the fact that this space has countably many components.

Remarks 15.1. (a) The homotopy groups of the corresponding *total* Grassmannian in a separable Hilbert space (the closure of Gr_{p_+}) are well-known (see e.g. Wojciechowski [1981], Booß & Wojciechowski [1982/1985], Wojciechowski [1985a], and Pressley & Segal [1986]); they only differ by the choice of equivalence classes, namely, whether one takes classes modulo operators of finite rank, modulo compact operators, or modulo operators of Hilbert-Schmidt type. (This makes no difference to the respective homotopy groups, as shown by Palais [1965b]). Fuglede [1976] observed earlier that the total Grassmannian consists of infinitely many connected components. The point of our presentation is that we have to restrict ourselves to the pseudo-differential Grassmannian.

(b) We assume here that the term *projection* denotes an *idempotent* ($P^2 = P$). However, as base point for our homotopy groups, we choose an *orthogonal* projection ($P_+ = P_+^* = P_+^2$). We can do that without loss of generality, as explained in Lemma 15.11 below. More specifically, we let P_+ be an orthogonal projection onto a subspace of

$L^2(Y;S|_Y)$. In the context of this book, a natural choice of P_+ is the orthogonal projection onto the space $H_+(A)$ of Cauchy data of a given Dirac operator A. As explained in Lemma 12.8, the projection P_+ is a pseudo-differential operator and has the same principal symbol as the Calderón projector constructed in that chapter. The reason is that in our case the principal symbol $p_+(y,\zeta)$ of the Calderón projector is orthogonal.

We introduce a group of invertible elliptic operators on $S|_Y$:

$$(15.1) \quad \mathcal{E}\ell\ell^\times := \{g \mid g \text{ is an elliptic invertible operator of order 0 on } S|_Y \text{ with principal symbol equal to } \mathrm{id}_S\}.$$

The key to the whole problem is the following elementary lemma:

Lemma 15.2. *Let P_0 and P_1 denote two elements of Gr_{p_+}, such that $\|P_1 - P_0\| < 1$. Then the operator*

$$(15.2) \qquad T = \mathrm{Id} + (P_1 - P_0)(2P_0 - \mathrm{Id})$$

belongs to $\mathcal{E}\ell\ell^\times$ and satisfies:

$$(15.3) \qquad TP_0 = P_1 T.$$

Proof. We have

$$\|T - \mathrm{Id}\| = \|(P_1 - P_0)(2P_0 - \mathrm{Id})\| \leq \|P_1 - P_0\| \, \|2P_0 - \mathrm{Id}\| \leq \|P_1 - P_0\| < 1,$$

since $(2P_0 - \mathrm{Id})^2 = \mathrm{Id}$ for any projection P_0 and hence $\|2P_0 - \mathrm{Id}\| = 1$. So T is invertible.

For the principal symbol of T we obtain $\sigma(T) = \mathrm{Id} + 0 = \mathrm{Id}$. Finally, an elementary check yields $TP_0 = P_1 P_0 = P_1 T$ which proves (15.3). □

Corollary 15.3. *The space Gr_{p_+} is locally contractible and, in particular, the connected components are path-connected.*

Proof. A deformation retraction of $\{P_1 \mid \|P_1 - P_0\| < 1\|\}$ onto P_0 is given by the homotopy

$$H_t(P_1) := (T(P_1,t))^{-1} P_0 T(P_1,t), \quad \text{for } 0 \leq t \leq 1,$$

where

$$T(P_1, t) := \mathrm{Id} + t(P_1 - P_0)(2P_0 - \mathrm{Id}).$$

Clearly $H_0(P_1) = P_0$, $H_1(P_1) = P_1$, and $\{H_t(P_1)\}$ remains in Gr_{p_+}.

□

Later on we shall use this lemma to show that any single connected component of our Grassmannian is a base space of a certain principal fibre bundle with the total space $\mathcal{E}\ell\ell^\times$ (Theorem 15.5). Hence, in order to compute the homotopy groups $\pi_k(\mathrm{Gr}_{p_+})$ for $k > 0$, we have to compute the homotopy groups of $\mathcal{E}\ell\ell^\times$. They are well-known, but there are no computations published to establish them.

Proposition 15.4. *The homotopy groups of $\mathcal{E}\ell\ell^\times$ (with base point* Id*) are given by*

$$\pi_k(\mathcal{E}\ell\ell^\times) \cong \begin{cases} 0 & \text{for } k \text{ even} \\ \mathbf{Z} & \text{for } k \text{ odd}. \end{cases}$$

Proof. First recall that these are exactly the homotopy groups of the groups GL_c and GL^∞ given by Bott periodicity (see Karoubi [1978; I.3.14 and II.3.19]) and Palais' results (see Palais [1965b]). Here GL_c denotes the group of invertible operators acting on $L^2(Y; S|_Y)$, which differ from the identity by a compact operator, and GL^∞ denotes the group of invertible operators acting on $L^2(Y; S|_Y)$, which differ from the identity by an operator of finite rank.

Now it is enough to show that any family of operators of the form $f : S^k \to \mathcal{E}\ell\ell^\times$ can be deformed within $\mathcal{E}\ell\ell^\times$ to a family of operators from GL^∞. This will be done in four steps:

1st step. Let $f : S^k \to \mathcal{E}\ell\ell^\times$ be a continuous family. Deform $f_0(x) := f(x)$ into the unitary operator $f_1(x) := f(x)\,|f(x)|^{-1}$, where $|f(x)|$ denotes the unique positive square root of the operator $f(x)f^*(x)$, $x \in S^k$. The deformation is provided by combination with the standard retraction

$$g_t(x) := t\,\mathrm{Id} + (1-t)|f(x)|, \quad 0 \le t \le 1$$

of the space of positive operators onto the identity which rests within $\mathcal{E}\ell\ell^\times$.

2nd step. The unitary operator $f_1(x)$ belongs to $\mathcal{E}\ell\ell^\times$. This means that its spectrum is contained in the circle S^1 and that 1 is the only point of the essential spectrum of $f_1(x)$. Thus we have a (possibly

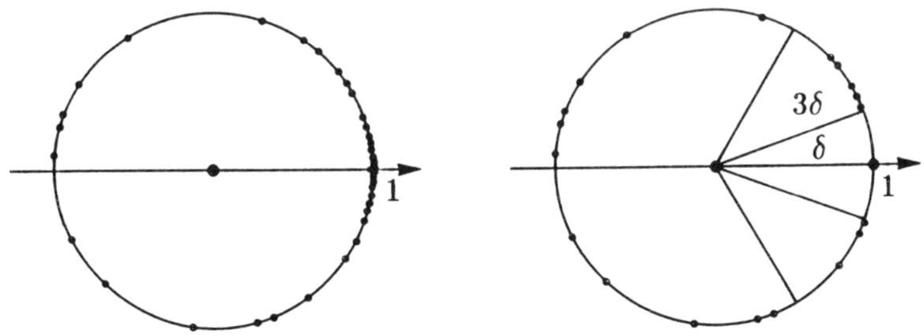

Fig. 15.1 Spectral deformation of a unitary $f_1(x)$ into H_1 $(f_1(x)) \in GL^\infty$

infinite) set $\{\lambda_m\}$ of eigenvalues of finite multiplicity contained in S^1, and 1 is the only limiting point of this set (it may be an eigenvalue of infinite multiplicity). Then for any small $\varepsilon > 0$ we can find a deformation $f_t : S^k \to GL_c$, $1 \le t \le 2$ with $f_2(x) \in GL^\infty$ and $\|f_t(x) - f_1(x)\| < \frac{\varepsilon}{2}$ for all $x \in S^k$ and $1 \le t \le 2$. In fact, we can choose a small $\delta > 0$ and a related "deformation retraction" of S^1 onto S^1 in such a way that

$$\begin{aligned} H_0 &= \mathrm{Id} \\ H_1(e^{i\theta}) &= 1 \quad \text{for } -\delta < \theta < \delta \\ H_t(e^{i\theta}) &= e^{i\theta} \quad \text{for } 3\delta < \theta < 2\pi - 3\delta \end{aligned}$$

for $0 \le t \le 1$ and $e^{i\theta} \in S^1$; hence

$$x \mapsto H_t(f_1(x)), \quad x \in S^k$$

is a continuous deformation of the family $\{f_1(x)\}$ into a family

$$\{f_2(x) := H_1(f_1(x))\}_{x \in S^k}$$

of operators belonging to GL^∞; for δ sufficiently small, we get

$$\|H_t(f_1(x)) - f_1(x)\| < \frac{\varepsilon}{2} \quad \text{for } 0 \le t \le 1$$

as required. (See also Figure 15.1).

3rd step. The technical difficulty here is that GL^{∞} is not contained in $\mathcal{E}\ell\ell^{\times}$. An element g of GL^{∞} is a pseudo-differential operator, if and only if $g - \mathrm{Id}$ is an operator with a smooth kernel. However, it is obvious that we can approximate any family $h : S^k \to GL^{\infty}$ by a family of operators h_1, where $h_1 - \mathrm{Id}$ is a family of operators of finite rank with smooth kernels. We shall sketch the proof.

For any $x \in S^k$, the image of the operator

$$\mathrm{Id} - h(x) : L^2(Y; S|_Y) \to L^2(Y; S|_Y)$$

is a finite-dimensional vector space $V(x)$. Since S^k is compact, one can find a finite-dimensional subspace V of $L^2(Y; S|_Y)$ such that for any x, the image of $h(x) - \mathrm{Id}$ is contained in V. By a small deformation, one can change h to $\tilde{h}$ and V to $\tilde{V}$ such that $\mathrm{Id} - \tilde{h}(x) : L^2(Y; S|_Y) \to \tilde{V}$ for all $x \in S^k$ and $\tilde{V}$ consists of smooth sections. Let $\{s_k\}_{k=1}^{N}$, $N = \dim V$ denote an orthonormal basis of V consisting of smooth spinors which may be completed to an orthonormal (smooth) basis of L^2. Then for $s \in L^2$:

$$(\mathrm{Id} - h(x))\, s = \sum_{i=1}^{N} \sum_{j=1}^{\infty} k_{i,j}(x) \langle s; s_j \rangle\, s_i\ .$$

It is obvious that the continuous functions $k_{i,j}$ can be approximated by smooth functions up to an arbitrary order.

This implies that to determine the set $[S^k, GL^{\infty}] = \pi_k(GL^{\infty})$, we may use a family $h : S^k \to GL^{\infty}$ of maps, where the range of h consists of operators belonging to $\mathcal{E}\ell\ell^{\times}$.

Thus we finally have a family $f_3 : S^k \to GL^{\infty} \cap \mathcal{E}\ell\ell^{\times}$ with

$$\|f_3(x) - f_2(x)\| < \frac{\varepsilon}{2}\,.$$

4th step. The family f_3 was obtained by continuous deformations within the space GL_c. But only the first homotopy and the results of the second and third homotopy are confined to $\mathcal{E}\ell\ell^{\times}$. However, since $f_3(x)$ belongs to $\mathcal{E}\ell\ell^{\times}$ and $\|f_3(x) - f_1(x)\| < \varepsilon$ for sufficiently small ε, we can choose the linear path

$$t f_3(x) + (1-t) f_1(x)$$

which is totally contained in $\mathcal{E}\ell\ell^{\times}$ as a new homotopy. Thus we have obtained a deformation of $f = f_0$ within $\mathcal{E}\ell\ell^{\times}$ into a family f_3 which consists of operators belonging to GL^{∞} (see Figure 15.2). Therefore

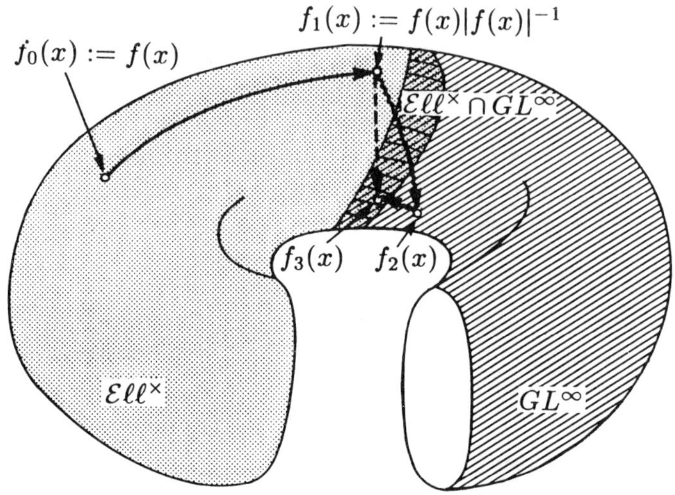

Fig. 15.2 Deformation of $f_0(x)$ within $\mathcal{E}\ell\ell^\times$ into $f_3(x) \in GL^\infty$

the homotopy groups of the group $\mathcal{E}\ell\ell^\times$ and of the group GL^∞ are the same. This ends the proof of the proposition. □

Now we calculate the higher homotopy groups of the connected components of Gr_{p_+}.

Theorem 15.5. *Fix a projection* $P_+ \in \mathrm{Gr}_{p_+}$ *and let* Gr_{P_+} *denote the connected component of* Gr_{p_+} *which contains* P_+. *Let* $\Phi : \mathcal{E}\ell\ell^\times \to \mathrm{Gr}_{P_+}$ *denote the continuous map given by the formula* $\Phi(g) = gP_+g^{-1}$. *Then* Φ *is a principal fibre bundle with fibre and structure group equal to*

$$\mathcal{E}\ell\ell^\times{}_{P_+} := \{g \in \mathcal{E}\ell\ell^\times \mid gP_+ = P_+g\},$$

where $\mathcal{E}\ell\ell^\times{}_{P_+}$ *acts from the right on* $\mathcal{E}\ell\ell^\times$.

Proof. The result follows easily from standard fibre bundle theory: For any $g \in \mathcal{E}\ell\ell^\times$, the operator $\Phi g = gP_+g^{-1}$ is a projection with principal symbol equal to p_+ and belongs to the same connected component as P_+ since $\mathcal{E}\ell\ell^\times$ is connected. So Φg is well-defined and belongs to Gr_{P_+}. Moreover, Φ is well-defined on the left cosets $\{g \cdot \mathcal{E}\ell\ell^\times{}_{P_+}\}$ of $\mathcal{E}\ell\ell^\times{}_{P_+}$ in $\mathcal{E}\ell\ell^\times$, since

$$\Phi(g \cdot h) = ghP_+h^{-1}g^{-1} = gP_+g^{-1} = \Phi g$$

for $g \in \mathcal{E}\ell\ell^{\times}$ and $h \in \mathcal{E}\ell\ell^{\times}{}_{P_+}$. In fact, Φ provides a homeomorphism between the homogeneous space $\mathcal{E}\ell\ell^{\times} / \mathcal{E}\ell\ell^{\times}{}_{P_+}$ and the base space Gr_{P_+}: To show the injectivity of the mapping, check that for any g, $g_0 \in \mathcal{E}\ell\ell^{\times}$ with $\Phi g = \Phi g_0$, the operator $h = g_0^{-1} g$ belongs to $\mathcal{E}\ell\ell^{\times}{}_{P_+}$, i.e.

$$hP_+h^{-1} = g_0^{-1} g P_+ g^{-1} g_0 = g_0^{-1}\Phi(g) g_0 = g_0^{-1}\Phi(g_0) g_0 = P_+ \,.$$

To show that the map Φ is surjective, take an arbitrary $P \in \mathrm{Gr}_{P_+}$. Since Gr_{P_+} is path-connected (Corollary 15.3), we have a finite sequence $\{P_i\}_{i=1,\dots,N} \in \mathrm{Gr}_{P_+}$ such that $P_1 = P_+$, $P_N = P$, and $\|P_{i+1} - P_i\| < 1$ for all $i = 1, \dots, N-1$. From Lemma 15.2 we then obtain operators $T_i \in \mathcal{E}\ell\ell^{\times}$ with $P_{i+1} = T_i P_i T_i^{-1}$, hence $P = TP_+T^{-1} = \Phi(T)$ with $T := T_{N-1} \cdot \ldots \cdot T_1 \in \mathcal{E}\ell\ell^{\times}$.

To prove that

$$\Phi : \mathcal{E}\ell\ell^{\times} \to \mathrm{Gr}_{P_+} \cong \mathcal{E}\ell\ell^{\times} / \mathcal{E}\ell\ell^{\times}{}_{P_+}$$

is a (principal) fibre bundle with fibre and structure group $\mathcal{E}\ell\ell^{\times}{}_{P_+}$, it only remains to assure us that $\mathcal{E}\ell\ell^{\times}{}_{P_+}$ has a local cross-section in $\mathcal{E}\ell\ell^{\times}$, i.e. the existence of a function $T : P \mapsto T_P$ mapping a neighbourhood U of $\Phi(\mathcal{E}\ell\ell^{\times}{}_{P_+}) = \{P_+\}$ continuously into $\mathcal{E}\ell\ell^{\times}$ and such that $\Phi T_P = P$ for each $P \in U$. Choose

$$U := U_{P_+} = \{P \mid P \in \mathrm{Gr}_{P_+} \text{ and } \|P - P_+\| < 1\} \,.$$

Set $T_P := \mathrm{Id} + (P - P_+)(2P - \mathrm{Id})$, the operator from Lemma 15.2. The function T is continuous in the operator norm and we have $\Phi(T_P) = T_P P_+ (T_P)^{-1} = P$. Then our theorem follows from the bundle structure theorem; see Steenrod [1965, p. 30f], see also Figure 15.3. □

Remark 15.6. Notice that a local trivialization of the total space $\mathcal{E}\ell\ell^{\times}$ over the neighbourhood U_{P_+} is given by the formula

(15.4)
$$\Phi^{-1}(U_{P_+}) \ni g \mapsto \big(\Phi(g) = gP_+g^{-1}; (T_{\Phi(g)})^{-1} g\big) \in U_{P_+} \times \mathcal{E}\ell\ell^{\times}{}_{P_+},$$

where $T : U_{P_+} \to \mathcal{E}\ell\ell^{\times}$ is the local section from the preceding proof. In fact

$$\begin{aligned}\Phi((T_{\Phi(g)})^{-1} g) &= (T_{\Phi(g)})^{-1} g \cdot P_+ \cdot ((T_{\Phi(g)})^{-1} g)^{-1} \\ &= (T_{\Phi(g)})^{-1} \cdot gP_+g^{-1} \cdot T_{\Phi(g)} \\ &= (T_{\Phi(g)})^{-1} \Phi(g) T_{\Phi(g)} = P_+ \,,\end{aligned}$$

since $T_P P_+ (T_P)^{-1} = P$ for all $P \in U_{P_+}$ and so also for $P = \Phi(g)$. This proves $(T_{\Phi(g)})^{-1} g \in \mathcal{E}\ell\ell^\times{}_{P_+}$ as claimed in (15.4). Actually (15.4) defines a connection on the bundle $\mathcal{E}\ell\ell^\times$. A parallel transport of an element $g \in \Phi^{-1}(P_1)$ to a fibre over P_2 (where $P_1, P_2 \in U_{P_+}$) is given by $T_{P_2} \cdot (T_{P_1})^{-1} \cdot g$, and there is an obvious construction of a global connection on Gr_{P_+}. We shall not discuss this topic further here.

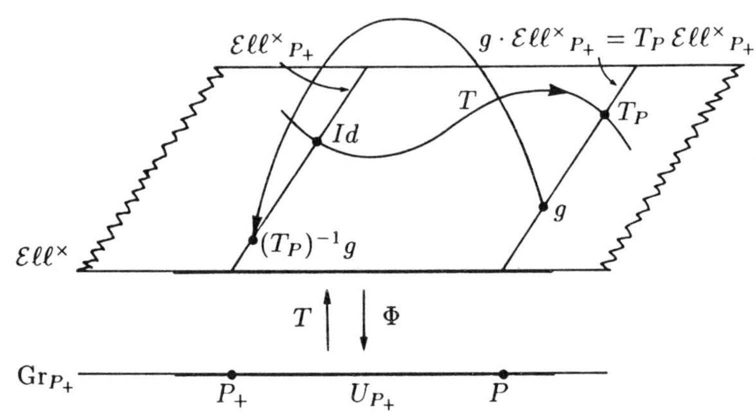

Fig. 15.3 Local section and parallel transport

Recall the assumption that neither the symbol p_+ nor $\mathrm{Id} - p_+$ vanish identically. Then we have $\dim \operatorname{range} P_+ = \dim \operatorname{range}(\mathrm{Id} - P_+) = \infty$, and the operator P_+ decomposes the space $L^2(Y; S|_Y)$ into the direct sum of $\operatorname{range}(P_+)$ and $\operatorname{range}(\mathrm{Id} - P_+)$. With respect to this splitting, the group $\mathcal{E}\ell\ell^\times{}_{P_+}$ decomposes into two factors such that each of them has the homotopy groups of $GL_c(L^2)$ as in the proof of Proposition 15.4. Therefore, the homotopy groups of $\mathcal{E}\ell\ell^\times{}_{P_+}$ are given by

$$\pi_k(\mathcal{E}\ell\ell^\times{}_{P_+}) = \begin{cases} 0 & \text{for } k \text{ even} \\ \mathbf{Z} \oplus \mathbf{Z} & \text{for } k \text{ odd}. \end{cases} \tag{15.5}$$

Now compute the homotopy groups of Gr_{P_+}. Apply the exact homotopy sequence of the fibre bundle $\Phi : \mathcal{E}\ell\ell^\times \to \mathrm{Gr}_{P_+}$ with fibre and structure group $\mathcal{E}\ell\ell^\times{}_{P_+}$ based at P_+, see Steenrod [1965, p. 90f]:

$$\ldots \to \pi_{k+1}(\mathrm{Gr}_{P_+}) \to \pi_k(\mathcal{E}\ell\ell^\times{}_{P_+}) \xrightarrow{j_*} \pi_k(\mathcal{E}\ell\ell^\times) \to \pi_k(\mathrm{Gr}_{P_+})$$
$$\to \pi_{k-1}(\mathcal{E}\ell\ell^\times{}_{P_+}) \to \ldots$$

Here $j : \mathcal{E}\ell\ell^\times{}_{P_+} \hookrightarrow \mathcal{E}\ell\ell^\times$ denotes the natural inclusion. The homotopy groups are relative to a fixed base point, say P_+. The induced map j_* on the homotopy groups is 0 for even k and $j_*(m,n) = m+n$ for odd k. This gives the short exact sequence

$$0 \to \pi_{2i}(\mathrm{Gr}_{P_+}) \to \mathbf{Z} \oplus \mathbf{Z} \xrightarrow{+} \mathbf{Z} \to \pi_{2i-1}(\mathrm{Gr}_{P_+}) \to 0$$

which proves the following theorem:

Theorem 15.7. *For a fixed projection P_+, the homotopy groups of the connected component* Gr_{P_+} *of* Gr_{p_+} *are given by*

$$\pi_k(\mathrm{Gr}_{P_+}, P_+) \cong \begin{cases} \mathbf{Z} & \text{for } k \text{ even and } k > 0 \\ 0 & \text{for } k \text{ odd.} \end{cases}$$

The remaining part of this chapter is devoted to showing that Gr_{p_+} has countably many components classified by the index of certain natural Fredholm operators (Theorem 15.12 below). Following Brown, Douglas & Fillmore [1973] we define:

Definition 15.8. Let P_1, P_2 be pseudo-differential projections with the same principal symbol. The integer

$$\mathbf{i}(P_2, P_1) := \mathrm{index}\,\{P_2 P_1 : \mathrm{range}\, P_1 \to \mathrm{range}\, P_2\} \tag{15.6}$$

is called the *virtual codimension* of P_2 in P_1.

Note. Clearly $P_2P_1 : \mathrm{range}\, P_1 \to \mathrm{range}\, P_2$ is a Fredholm operator, since

$$\ker\{P_2P_1 : \mathrm{range}\, P_1 \to \mathrm{range}\, P_2\} = \ker T \cap \mathrm{range}\, P_1$$

and

$$\mathrm{range}\{P_2P_1 : \mathrm{range}\, P_1 \to \mathrm{range}\, P_2\} = \mathrm{range}\, T \cap \mathrm{range}\, P_2\,,$$

where the 0-order pseudo-differential operator

$$T := P_2P_1 + (\mathrm{Id} - P_2)(\mathrm{Id} - P_1) \tag{15.7}$$

has principal symbol equal id, hence is elliptic.

The following lemma is an obvious corollary from Lemma 15.2:

Lemma 15.9. *Let P_1 and P_2 belong to the same connected component of* Gr_{p_+}. *Then the virtual codimension* $\mathbf{i}(P_2, P_1)$ *vanishes.*

Proof. Lemma 15.2 implies the existence of $g \in \mathcal{E}\ell\ell^\times$ such that $gP_1g^{-1} = P_2$, and it is obvious that:
(15.8)

$$\begin{aligned}\operatorname{index} P_2P_1 &= \operatorname{index} gP_1g^{-1}P_1 \\ &= \operatorname{index}\{g : \operatorname{range} P_1 \xrightarrow{\cong} \operatorname{range} P_2\} + \operatorname{index} P_1g^{-1}P_1 \\ &= 0 \ + \ \operatorname{index} P_1(\mathrm{Id} + \text{compact})P_1 \ = \ 0\,,\end{aligned}$$

since, for $g \in \mathcal{E}\ell\ell^\times$, g^{-1} belongs to $\mathcal{E}\ell\ell^\times$ and can be written as Id + compact. □

Note. The operators g and g^{-1} can be continuously deformed into the identity within the path-connected space $\mathcal{E}\ell\ell^\times$. This follows from formula (15.2).

Now we want to show the opposite direction, namely, that P_1 and P_2 belong to the same connected component, if $\mathbf{i}(P_2, P_1)$ is equal 0. First we prove this for orthogonal projections and afterward we will explain why it works in the general case. We need a stronger variant of Lemma 15.9.

Lemma 15.10. *Two orthogonal projections P_1, $P_2 \in \mathrm{Gr}_{p_+}$ belong to the same connected component of the Grassmannian, if and only if*

$$\mathbf{i}(P_2, P_1) = 0\ .$$

Proof. The *only if* was shown in Lemma 15.9. We show the *if.* Consider the operator

$$\begin{aligned}T &= \mathrm{Id} + (P_2 - P_1)(2P_1 - \mathrm{Id}) \\ &= P_2P_1 + (\mathrm{Id} - P_2)(\mathrm{Id} - P_1) \\ &=: T_+ + T_-\end{aligned}$$

investigated before in Lemma 15.2 for the special situation of $\|P_2 - P_1\| < 1$. Now investigate the general situation. Then T is not necessarily invertible, but still elliptic as observed earlier after equation (15.7). Clearly

$$\ker T \cong \ker T_+ \oplus \ker T_- \qquad \text{and} \qquad \operatorname{coker} T \cong \operatorname{coker} T_+ \oplus \operatorname{coker} T_-\ .$$

We find

(15.9)
$$\begin{aligned}\ker T_+ &= \ker\{P_2P_1 : \operatorname{range} P_1 \to \operatorname{range} P_2\}\\ &= \{s \mid P_1 s = s \text{ and } P_2 s = 0\}\\ &= \{s \mid (\mathrm{Id} - P_2)s = s \wedge (\mathrm{Id} - P_1)s = 0\}\\ &= \ker\{(\mathrm{Id} - P_1)(\mathrm{Id} - P_2) : (\operatorname{range} P_2)^{\perp} \to (\operatorname{range} P_1)^{\perp}\}\\ &= \operatorname{coker}\{(\mathrm{Id} - P_2)(\mathrm{Id} - P_1) : (\operatorname{range} P_1)^{\perp} \to (\operatorname{range} P_2)^{\perp}\}\\ &= \operatorname{coker} T_- \ .\end{aligned}$$

Of course here we exploit that P_1 and P_2 are *orthogonal* projections, so that e.g.

$$\begin{aligned}(15.10)\quad &\ker\{P_1P_2 : \operatorname{range} P_2 \to \operatorname{range} P_1\}\\ &= \operatorname{coker}\{(P_1P_2)^* = P_2^*P_1^* = P_2P_1 : \operatorname{range} P_1 \to \operatorname{range} P_2\} \ .\end{aligned}$$

One can check (15.10) directly:

$$\begin{aligned}&\operatorname{coker}\{P_2P_1 : \operatorname{range} P_1 \to \operatorname{range} P_2\}\\ &= \{s \mid P_2 s = s \text{ and } s \perp \operatorname{range} P_2P_1\}\\ &= \{s \mid P_2 s = s \text{ and } \langle s; P_2P_1 f\rangle = 0 \text{ for all } f\}\\ &= \{s \mid P_2 s = s \text{ and } \langle P_1P_2 s; f\rangle = 0 \text{ for all } f\}\\ &= \{s \in \ker P_1P_2\} \ .\end{aligned}$$

Similar to the derivation of (15.9), by interchanging P_1 and P_2, one gets

$$\ker T_- = \operatorname{coker} T_+ \ . \tag{15.11}$$

Now assume $\mathbf{i}(P_2, P_1) = 0$. It suffices to construct an operator T' which belongs to $\mathcal{E}\ell\ell^{\times}$ and satisfies $T'P_1 = P_2T'$. Then the lemma is proved, since $\mathcal{E}\ell\ell^{\times}$ is connected. First construct T':

Since $\mathbf{i}(P_2, P_1) = 0$, we have

$$\dim\ker T_+ = \dim\left(\ker T_- \ (\overset{(15.11)}{=} \operatorname{coker} T_+)\right) \ .$$

Hence one can fix a unitary morphism $V : \ker T_+ \longrightarrow \ker T_-$ which defines an involution $\Psi : \ker T \longrightarrow \ker T$ by the formula

$$\Psi := \begin{pmatrix} 0 & V^* \\ V & 0 \end{pmatrix} \tag{15.12}$$

with respect to the decomposition

$$\ker T = \ker T_+ \oplus \ker T_- \ = \ker P_2 P_1 \oplus \ker P_1 P_2 \ . \tag{15.13}$$

Extend Ψ (by 0) to the whole Hilbert space $L^2(Y; S|_Y)$ and define

$$T' := T + \Psi$$

to be the perturbation of T by the finite rank operator Ψ. Then also the operator T' is a pseudo-differential elliptic operator of order 0 with principal symbol id_S. Moreover $\ker T' = \{0\} = \operatorname{coker} T'$; hence $T' \in \mathcal{E}\ell\ell^\times$. An elementary computation shows

$$T' P_1 = P_2 T'$$

which implies that P_1 and P_2 are elements of the same connected component. □

To extend this result to the general case of not necessarily orthogonal projections, we need the following lemma.

Lemma 15.11. *For any $P \in \mathrm{Gr}_{p_+}$, let $\widetilde{P}$ denote the orthogonal projection onto the range of P. Then P and $\widetilde{P}$ belong to the same connected component of* Gr_{p_+}.

Proof. It follows from Lemma 12.8 that $\widetilde{P} \in \mathrm{Gr}_{p_+}$. We have to show that P and $\widetilde{P}$ belong to the same connected component of Gr_{p_+}. Once more we argue with the elliptic operator construction

$$T := \mathrm{Id} + (\widetilde{P} - P)(2P - \mathrm{Id}) = \widetilde{P}P + (\mathrm{Id} - \widetilde{P})(\mathrm{Id} - P) \ .$$

Again, T is a pseudo-differential operator of order 0 which differs from the identity only by a compact operator, so it has principal symbol equal id_S and is elliptic. Therefore the index of T vanishes. We show that T is invertible and hence belongs to the connected space $\mathcal{E}\ell\ell^\times$: Let $s \in \ker T$. Then

$$\widetilde{P}Ps + (\mathrm{Id} - \widetilde{P})(\mathrm{Id} - P)s = 0 \ ,$$

hence

$$\widetilde{P}Ps = 0 \text{ and } (\mathrm{Id} - \widetilde{P})(\mathrm{Id} - P)s = 0 \ . \tag{15.14}$$

Since $\widetilde{P}P = P$ and $(\mathrm{Id} - \widetilde{P})(\mathrm{Id} - P) = \mathrm{Id} - P$, we get from (15.14):

$$s = Ps + (\mathrm{Id} - P)s = 0 + 0 = 0 \ .$$

Therefore $\ker T = \{0\}$, hence also $\operatorname{coker} T = \{0\}$, and so $T \in \mathcal{E}\ell\ell^\times$. Our assertion follows from $\widetilde{T} = \widetilde{P}P = TP$. □

Theorem 15.12. *Two projections,* P_1, $P_2 \in \mathrm{Gr}_{p_+}$, *belong to the same connected component of* Gr_{p_+}, *if and only if*

$$\mathbf{i}(P_2, P_1) = 0 .$$

Proof. All we must know is

$$\begin{aligned}
\mathbf{i}(P_2, P_1) &= \mathrm{index}\{P_2P_1 : \mathrm{range}\, P_1 \to \mathrm{range}\, P_2\} \\
&\overset{\text{Lemma 15.10}}{=} \mathrm{index}\{T_2^{-1}\widetilde{P}_2T_2\, T_1^{-1}\widetilde{P}_1T_1 : \mathrm{range}\, P_1 \to \mathrm{range}\, P_2\} \\
&= \mathrm{index}\{\widetilde{P}_2T_2\, T_1^{-1}\widetilde{P}_1 : \mathrm{range}\, P_1 \to \mathrm{range}\, P_2\} \\
&= \mathrm{index}\{\widetilde{P}_2(\mathrm{Id} + \mathrm{compact})\widetilde{P}_1 : \mathrm{range}\, P_1 \to \mathrm{range}\, P_2\} \\
&= \mathbf{i}(\widetilde{P}_2, \widetilde{P}_1) .
\end{aligned}$$

Then the theorem follows from Lemma 15.10. □

Corollary 15.13. *The Grassmannian* Gr_{p_+} *has enumerable many connected components:*

$$\pi_0(\mathrm{Gr}_{p_+}) \cong \mathbf{Z} .$$

Proof. Fix an orthogonal projection $P_+ = P_0$ which belongs to Gr_{p_+}. Then the proof has two parts. First *(1)* construct a projection $P \in \mathrm{Gr}_{p_+}$ with non-trivial virtual codimension $\mathbf{i}(P, P_0)$ of P in P_0 and get, more precisely,

$$\pi_0(\mathrm{Gr}_{p_+}) \overset{\mathbf{i}(.,P_0)}{\longrightarrow} \mathbf{Z} \qquad \text{surjective.}$$

Then *(2)* show that this codimension is sufficient for distinguishing the connected components, so

$$\pi_0(\mathrm{Gr}_{p_+}) \overset{\mathbf{i}(.,P_0)}{\longrightarrow} \mathbf{Z} \qquad \text{injective.}$$

(1) Let $\{e_j\}_{j \in \mathbf{Z}}$ be an orthonormal basis of $L^2(Y; S|_Y)$ which consists of smooth sections of $S|_Y$, such that

$$H_0 := \mathrm{range}\, P_0 = \mathcal{L}_{\mathbf{C}}\{e_j\}_{j \geq 0} . \tag{15.15}$$

More generally, let $H_k := \mathcal{L}_{\mathbf{C}}\{e_j\}_{j \geq k}$ for $k \in \mathbf{Z}$, and let P_k denote the orthogonal projection onto H_k. Then $P_k \in \mathrm{Gr}_{p_+}$ for any k and $\mathbf{i}(P_k, P_0) = k$.

(2) Now let $P_1,\ P_2 \in \mathrm{Gr}_{p+}$ be orthogonal projections such that

$$\mathbf{i}(P_1, P_0) = k = \mathbf{i}(P_2, P_0)\ . \tag{15.16}$$

We shall show that then P_1 and P_2 belong to the same connected component of Gr_{p+}. We shall reduce the various cases $k \neq 0$ to the case $k = 0$ which was already explicitly solved by the preceding theorem. Actually, we only have to distinguish two situations, $k < 0$ and $k > 0$.

Case $k < 0$: This implies that

$$\dim \operatorname{coker}\{P_1 P_0 : \operatorname{range} P_0 \longrightarrow \operatorname{range} P_1\} > -k\ ,$$

and the same holds for the operator $P_2 P_0$. We have

$$\operatorname{coker} P_1 P_0 = \{s \in L^2(Y; S|_Y) \mid P_1 s = s \text{ and } P_0 s = 0\}\,;$$

notice that it consists of smooth sections since it is a subspace of the cokernel of the elliptic operator $P_1 P_0 + (\mathrm{Id} - P_1)(\mathrm{Id} - P_0)$. Let $\{s_1, \dots, s_l\}$ $(l \geq -k)$ be an orthonormal basis of $\operatorname{coker} P_1 P_0$. Introduce

$$\widetilde{P}_1 s := P_1 s - \sum_{j=1}^{-k} \langle s; s_j \rangle s_j\ .$$

The operator $\widetilde{P}_1$ is a projection onto the orthogonal complement of the space $W_1 := \mathcal{L}_{\mathbf{C}}\{s_1, \dots, s_{-k}\}$ in $\operatorname{range} P_1$. In the same way remove the space W_2 from $\operatorname{range} P_2$ and obtain $\widetilde{P}_2$. We write $P_i = \widetilde{P}_i + Q_i$ where Q_i denotes the orthogonal projection onto W_i, $i = 1, 2$.

Then $\widetilde{P}_1,\ P_2 \in \mathrm{Gr}_{p+}$, and it is obvious that

$$\mathbf{i}(\widetilde{P}_1, P_0) = 0 = \mathbf{i}(\widetilde{P}_2, P_0)\ .$$

Hence $\widetilde{P}_1$ and $\widetilde{P}_2$ belong to the same connected component of Gr_{p+} as P_0, therefore by the preceding theorem $\mathbf{i}(\widetilde{P}_2, \widetilde{P}_1) = 0$.

So we also have

$$\begin{aligned}
\mathbf{i}(P_2, P_1) &= \operatorname{index}\{P_2 P_1 : \operatorname{range} P_1 \to \operatorname{range} P_2\} \\
&= \operatorname{index}\{\widetilde{P}_2 \widetilde{P}_1 + \widetilde{P}_2 Q_1 + Q_2 \widetilde{P}_1 + Q_2 Q_1 : \operatorname{range} P_1 \to \operatorname{range} P_2\} \\
&= \operatorname{index}\{\widetilde{P}_2 \widetilde{P}_1 : \operatorname{range} P_1 \to \operatorname{range} P_2\} \\
&= \operatorname{index}\{\widetilde{P}_2 \widetilde{P}_1 : \operatorname{range} \widetilde{P}_1 \to \operatorname{range} \widetilde{P}_2\} = \mathbf{i}(\widetilde{P}_2, \widetilde{P}_1) = 0\,,
\end{aligned}$$

which shows by Theorem 15.12 that P_1 and P_2 belong to the same connected component of Gr_{p_+}.

Case $k > 0$: Now remove some subspaces V_i, $i = 1, 2$ from

$$\ker\{P_i P_0 : \operatorname{range} P_0 \longrightarrow \operatorname{range} P_1\}$$

which are found in quite the same way as the spaces W_i. One obtains two perturbated projections

$$\widetilde{P}_{0,i} := P_0 - R_i \,,$$

where R_i denotes the orthogonal projection of range P_0 onto V_i. Hence

$$\mathbf{i}(P_1, \widetilde{P}_{0,1}) = 0 = \mathbf{i}(P_2, \widetilde{P}_{0,2}) \,. \tag{15.17}$$

Moreover

(15.18)

$$\begin{aligned}
\mathbf{i}(\widetilde{P}_{0,2}, \widetilde{P}_{0,1}) &= \operatorname{index}\{\widetilde{P}_{0,2}\widetilde{P}_{0,1} : \operatorname{range} \widetilde{P}_{0,1} \to \operatorname{range} \widetilde{P}_{0,2}\} \\
&= \operatorname{index}\{\widetilde{P}_{0,2}\widetilde{P}_{0,1} : \operatorname{range} \widetilde{P}_{0,1} \oplus V_2 \to \operatorname{range} \widetilde{P}_{0,2} \oplus V_1\} \\
&= \operatorname{index}\{\widetilde{P}_{0,2}\widetilde{P}_{0,1} : \operatorname{range} P_0 \to \operatorname{range} P_0\} \\
&= \operatorname{index}\{(P_0 - R_2)(P_0 - R_1) : \operatorname{range} P_0 \to \operatorname{range} P_0\} \\
&= \mathbf{i}\,((P_0 - R_2) + R_2, (P_0 - R_1) + R_1) \\
&= \mathbf{i}(P_0, P_0) = 0 \,.
\end{aligned}$$

Because of (15.17) and (15.18), P_1 and P_2 belong to the same connected component. □

Note. For later use we repeat the construction of an operator with nontrivial index for the special case $k = 1$ as an exercise: Let $f \in L^2(Y; S|_Y)$ be a unit section belonging to range $P_+ =: H_+$. We define P_1 by the formula:

$$P_1(s) := P_+(s) - \langle s; f\rangle f. \tag{15.19}$$

It is immediate that $\ker P_1 P_+$ is spanned by f and that range $P_1 P_+ = \{cf\}^{\perp H_+}$, the orthogonal complement of $\{cf\}_{c \in \mathbf{C}}$ in H_+. Since the orthogonal complement is the whole range of P_1, we get for the operator $P_1 P_+ : \operatorname{range} P_+ \to \operatorname{range} P_1$ a vanishing cokernel and hence $\mathbf{i}(P_1, P_+) = \operatorname{index} P_1 P_+ = 1$.

Remark 15.14. As shown in Theorem 15.12, the obstruction for $P_0, P_1 \in \mathrm{Gr}_{p_+}$ belonging to the same connected component is the integer $\mathbf{i}(P_1, P_0)$, and as shown by Corollary 15.13, this integer is in fact crucial for distinguishing the connected components of the Grassmannian Gr_{p_+}. Based on the notion of the spectral flow, which is explained in Section 17, we get an alternative expression which is sometimes easier to calculate. Assume that P_0, P_1 are orthogonal. Then

$$\{B_t := t(2P_1 - \mathrm{Id}) + (1-t)(2P_0 - \mathrm{Id})\}_{t\in I}$$

is a family of elliptic self-adjoint operators of order 0 parametrized over the unit interval $I = [0,1]$. Compare also the different families constructed for the proof of Corollary 15.3. The family $\{B_t\}$ joins the operators $B_i = (\mathrm{Id} - P_i) - P_i$ $(i = 0,1)$ with spectrum equal to $\{-1, +1\}$, and an easy computation shows that

$$\mathrm{sf}\{B_t\} = \mathbf{i}(P_1, P_0)\ . \tag{15.20}$$

From Corollary 15.13 follows the basic property of Gr_{p_+}, that the diagram of Figure 15.4 is *index-commutative*:

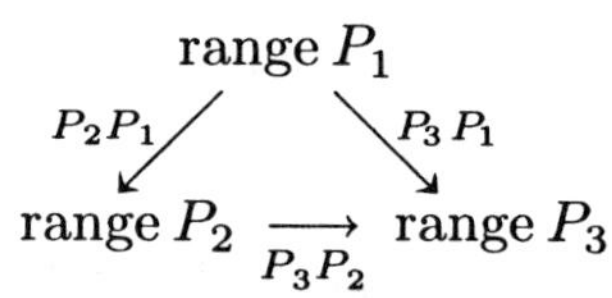

Fig. 15.4 $\mathbf{i}(P_3, P_1) = \mathbf{i}(P_3, P_2) + \mathbf{i}(P_2, P_1)$

Proposition 15.15. *Let $P_1, P_2, P_3 \in \mathrm{Gr}_{p_+}$. Then*

$$\begin{aligned}\mathbf{i}(P_3, P_1) &= \mathbf{i}(P_3, P_2) + \mathbf{i}(P_2, P_1)\\ &= \mathrm{index}\{P_3P_2P_1 : \mathrm{range}\, P_1 \to \mathrm{range}\, P_3\}.\end{aligned}$$

Proof. We may assume

$$P_j = P_{k_j} \qquad \text{for suitable } k_j \in \mathbf{Z};\ j = 1,2,3\ , \tag{15.21}$$

where P_{k_j} denotes the orthogonal projection of $L^2(Y; S|_Y)$ onto the subspace H_{k_j} introduced in the proof of Corollary 15.13. We have to consider different possibilities

$$k_1 \le k_2 \le k_3\,, \quad k_1 \le k_3 \le k_2\,, \quad k_2 \le k_1 \le k_3\,, \qquad \text{and so on}$$

to analyse $\mathrm{index}\{P_{k_3}P_{k_2}P_{k_1} : H_{k_1} \to \mathrm{range}\, H_{k_3}\}$. Elementary computation shows that the proposition is true in each case. $\square$

16. The Homotopy Groups of the Space of Self-Adjoint Fredholm Operators

We offer an elementary and relatively simple calculation of the homotopy groups of the space of self-adjoint Fredholm operators.

We determine the topology of the space of self-adjoint Fredholm operators in Hilbert space. This was done in Atiyah & Singer [1969], but though transparent, their computations are quite long and complicated. We make three simplifications: We do not bother to classify properties in K-theory, but calculate only the homotopy groups; we do not consider skew-adjoint Fredholm operators, but (complex) self-adjoint ones; and we do not repeat all arguments of Atiyah & Singer [1969] *ab ovo*, but use only a simple argument from the beginning of their work which permits us to apply the concepts and methods of infinite-dimensional Grassmannians provided in Chapter 15 of this book.

Actually the story is even more simple: In Chapter 15 we restricted ourselves to Grassmannians of *pseudo-differential* projections with given principal symbol. This was differential analysis. Now we need not be concerned with all the intricacies we had to surmount at that time, since we make our deformations in the *full* Grassmannian of all (orthogonal) projections which differ from a given non-trivial projection by a compact operator (and the corresponding Grassmannian of self-adjoint involutions in the Calkin algebra). This is functional analysis. As such, this chapter solves a problem posed by Singer [1971; p. 951], namely to *find an "elementary" or "direct" proof* of the periodicity of the homotopy groups of $\widehat{\mathcal{F}}_*$ (see Proposition 16.2 and Theorem 16.10 below).

16A. Elementary Decompositions and Deformations

First a few rather generally known observations about the spectrum of Fredholm operators and self-adjoint Fredholm operators.

Proposition 16.1. *Let H be a separable infinite-dimensional complex Hilbert space and let $\mathcal{F}$ denote the space of all Fredholm operators*

on H. *Then*
(a) $T \in \mathcal{F} \Longrightarrow T + \lambda \in \mathcal{F}$ *for small* λ.
(b) *If* $T \in \mathcal{F}$ *is self-adjoint, then 0 belongs to the discrete spectrum of* T*; more precisely there exists an* $\varepsilon > 0$ *such that* $(-\varepsilon, \varepsilon) \cap \operatorname{spec}(T)$ *is empty or equal to* $\{0\}$ *which is an eigenvalue of finite multiplicity.*

Proof. *(a)* is obvious since $\mathcal{F}$ is open in the normed space $\mathcal{B}$ of all bounded operators. To prove *(b)* we recall that the range of a Fredholm operator is closed, hence by Proposition 9.4 the inverse operator $(T|_{\ker T)^\perp})^{-1}$ is bounded. Therefore there exists $\varepsilon > 0$ so that

$$\|T\varphi\| \geq \varepsilon\|\varphi\| \qquad \text{for } \varphi \in (\ker T)^\perp .$$

□

The first remarkable fact about the topology of the space of self-adjoint Fredholm operators is that it decomposes into three connected components which are distinguished by the essential spectrum. This will be proved in the following proposition. For a better understanding of this result, recall from ordinary Fredholm theory:

The space $\mathcal{F}$, equipped with the operator norm, decomposes into $\mathbf{Z}$ connected components which are distinguished by the index. More generally, for any continuous family $\{A_m\}_{m\in M}$ of Fredholm operators parametrized by a compact topological space M, we can define a homotopy invariant, the index bundle

$$\operatorname{index}\{A_m\}_{m\in M} \in K(M),$$

where $K(M)$ is the Grothendieck group of complex vector bundles over M. The Atiyah-Jänich theorem states that the index bundle defines a bijection $[M, \mathcal{F}] \to K(M)$, where $[..,..]$ denotes the homotopy classes of mappings. Hence the homotopy type of the space $\mathcal{F}$ is completely identified: it is a classifying space for the functor K.

It turns out that one can develop an analogous theory for the space $\widehat{\mathcal{F}}$ of self-adjoint Fredholm operators, not based on the index which vanishes on $\widehat{\mathcal{F}}$, but on spectral invariants, namely the essential spectrum and the spectral flow. The most profound result is the following:

Proposition 16.2. *Let* $\widehat{\mathcal{F}}$ *denote the space of self-adjoint Fredholm operators on a fixed complex separable Hilbert space* H. *It has three* *connected components $\widehat{\mathcal{F}}_+$, $\widehat{\mathcal{F}}_-$, and $\widehat{\mathcal{F}}_*$ characterized by*

$$\begin{aligned} T \in \widehat{\mathcal{F}}_\pm :&\Longleftrightarrow T - k \textit{ is a positive (negative) operator} \\ &\qquad \textit{for some compact operator } k \\ &\Longleftrightarrow \operatorname{spec}_{\text{ess}}(T) \subset \mathbf{R}_\pm \\ T \in \widehat{\mathcal{F}}_* :&\Longleftrightarrow T \textit{ not in } \widehat{\mathcal{F}}_\pm \\ &\Longleftrightarrow T \textit{ has both positive and negative essential spectrum.} \end{aligned}$$

Moreover, the components $\widehat{\mathcal{F}}_+$ and $\widehat{\mathcal{F}}_-$ are contractible.

Proof. First check the equivalence of the two alternative descriptions of the sets $\widehat{\mathcal{F}}_+$, $\widehat{\mathcal{F}}_-$, and $\widehat{\mathcal{F}}_*$ given in the statement of the proposition:

If $T - k$ is a positive operator for some compact operator k (one also says T is *essentially positive*), we have

$$\operatorname{spec}_{\text{ess}}(T) \quad \overset{k \text{ compact}}{=} \quad \operatorname{spec}_{\text{ess}}(T-k) \quad \overset{T-k \text{ positive}}{\subset} \quad (0,+\infty)\ .$$

Vice versa, if $\operatorname{spec}_{\text{ess}}(T) \subset (0,+\infty)$, it follows from Proposition 16.1 that there exists $\varepsilon > 0$, such that there are only finitely many discrete eigenvalues in the interval $(-\infty, \varepsilon)$; hence T is a positive operator on the complementary space of the finite direct sum of these finite-dimensional eigenspaces.

The same arguments apply for $\widehat{\mathcal{F}}_-$ and $\widehat{\mathcal{F}}_*$ so that also the spectral characterization of these spaces are established.

To proceed further it is suitable to turn to the images of $\mathcal{F}$, $\widehat{\mathcal{F}}$, $\widehat{\mathcal{F}}_\pm$, and $\widehat{\mathcal{F}}_*$ in the Calkin algebra $\mathcal{C} := \mathcal{B}/\mathcal{K}$ of the Hilbert space H. Let $\mathcal{G}$ denote the group of units of $\mathcal{C}$ and $\widehat{\mathcal{G}}$ the subspace of self-adjoint invertible elements in $\mathcal{C}$ (with respect to the natural C^*-algebra structure for $\mathcal{C}$). Let $\mathbf{p} : \mathcal{B} \to \mathcal{C}$ denote the canonical projection. A widely exploited fact is that $\mathcal{F} = \mathbf{p}^{-1}(\mathcal{G})$ which is obvious, since the Fredholm operators on H can be characterized as those which are invertible modulo $\mathcal{K}$. We also see that $\mathbf{p}$ maps $\widehat{\mathcal{F}}$ into $\widehat{\mathcal{G}}$, and the mapping $\widehat{\mathcal{F}} \to \widehat{\mathcal{G}}$ is also surjective: for $g \in \widehat{\mathcal{G}}$ and $T \in \mathcal{F}$ with $\mathbf{p}(T) = g = g^*$ we have $\mathbf{p}(T^*) = g$ and hence $\mathbf{p}(\frac{T+T^*}{2}) = g$ and $\frac{T+T^*}{2} \in \widehat{\mathcal{F}}$.

By definition

$$\operatorname{spec}_{\text{ess}} T = \operatorname{spec} \mathbf{p}(T) := \{\lambda \mid \mathbf{p}(T) - \lambda \notin \mathcal{G}\}. \tag{16.1}$$

Therefore the subsets $\widehat{\mathcal{F}}_\pm$ and $\widehat{\mathcal{F}}_*$ are the inverse images of corresponding subsets $\widehat{\mathcal{G}}_\pm$, $\widehat{\mathcal{G}}_*$ of $\widehat{\mathcal{G}}$ under the projection $\mathbf{p}$. Of course, here the

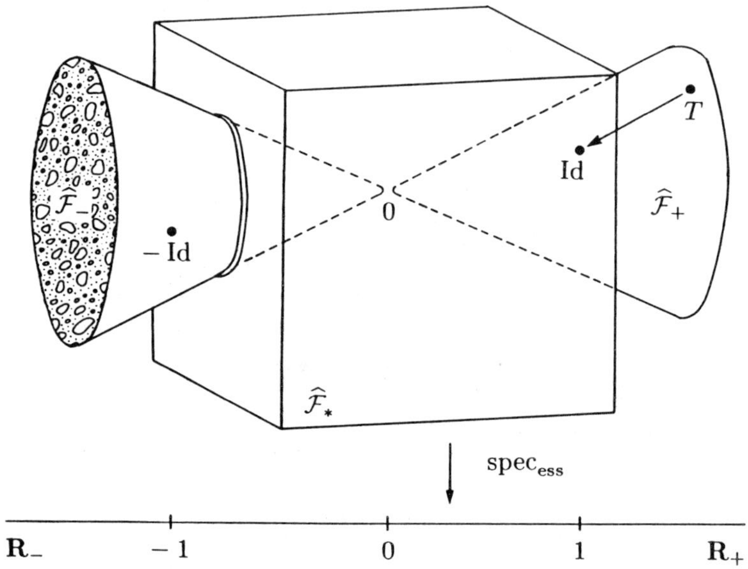

Fig. 16.1 The decomposition of $\widehat{\mathcal{F}}$ into its three connected components

point is that the $\widehat{\mathcal{F}}$-subspaces are defined via the essential spectrum and the $\widehat{\mathcal{G}}$-subspaces only via the spectrum.

These subspaces of $\widehat{\mathcal{G}}$ are both open and closed and hence also its inverse images $\widehat{\mathcal{F}}_\pm$ and $\widehat{\mathcal{F}}_*$. Therefore the space $\widehat{\mathcal{F}}$ is not connected and decomposes into the union of these three mutually disjoint subsets. It turns out that each of these subsets is connected. This is obvious for $\widehat{\mathcal{F}}_\pm$. Let $T \in \widehat{\mathcal{F}}_+$: Then

$$(16.2) \qquad T_t := (1-t)T + t\,\mathrm{Id}, \qquad 0 \le t \le 1$$

does not lead out of $\widehat{\mathcal{F}}_+$, since by (16.1)

$$\begin{aligned}\mathrm{spec}_{\mathrm{ess}}\, T_t &= \mathrm{spec}\,\mathbf{p}(T_t) = (1-t)(\mathrm{spec}\,\mathbf{p}(T)) + t \\ &= (1-t)(\mathrm{spec}_{\mathrm{ess}}\, T) + t \subset \mathbf{R}_+\end{aligned}$$

for $\mathrm{spec}_{\mathrm{ess}}\, T \subset \mathbf{R}_+$. Hence (16.2) provides a contraction of $\widehat{\mathcal{F}}_+$ to the operator Id, cf. also Figure 16.1. Similarly $\widehat{\mathcal{F}}_-$ contracts to the operator $-\,\mathrm{Id}$. We postpone the proof that $\widehat{\mathcal{F}}_*$ is also connected. □

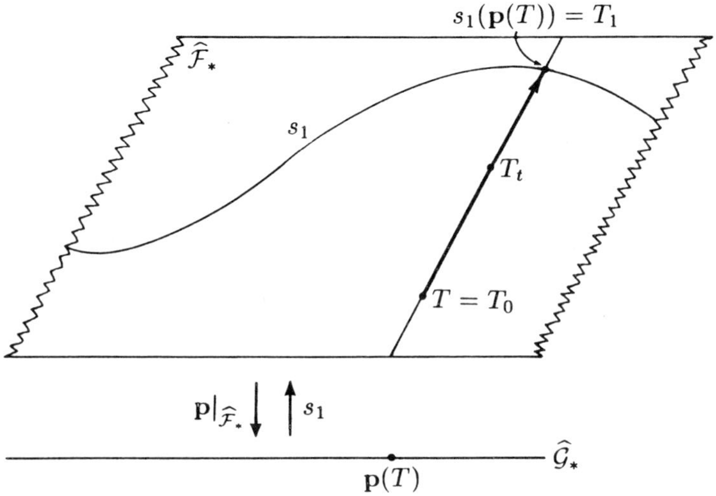

Fig. 16.2 The retraction of $\widehat{\mathcal{F}}_*$ onto $s_1(\widehat{\mathcal{G}}_*)$

Lemma 16.3. *The natural map* $\mathbf{p}|_{\widehat{\mathcal{F}}_*} : \widehat{\mathcal{F}}_* \to \widehat{\mathcal{G}}_*$ *is a homotopy equivalence.*

Proof. Recall the theorem of Bartle and Graves (see e.g. Bessaga & Pelczynski [1975; p. 86]) which states that each surjective, continuous, linear operator from one Frechet space into another Frechet space possesses a right (not necessarily linear) continuous inverse. Since the projection $\mathbf{p} : \mathcal{B} \to \mathcal{C}$ is surjective, continuous, and linear, we get from the theorem of Bartle and Graves a *section* s in the *bundle* $\mathbf{p} : \mathcal{B} \to \mathcal{C}$ over $\mathcal{C} = \mathcal{B}/\mathcal{K}$ with *total space* $\mathcal{B}$ and *fibre* $\mathcal{K}$, namely a continuous (in our case definitely *not* linear) right inverse s of $\mathbf{p}$. Then

$$s_1 : g \mapsto \frac{s(g) + (s(g))^*}{2}$$

is a section (continuous right inverse) for the mapping

$$\mathbf{p}|_{\widehat{\mathcal{F}}_*} : \widehat{\mathcal{F}}_* \to \widehat{\mathcal{G}}_* \, .$$

Modulo homotopy, the section s_1 is also a right inverse of $\mathbf{p}|_{\widehat{\mathcal{F}}_*}$, since we can retract $\widehat{\mathcal{F}}_*$ back onto $s_1(\mathcal{G}_*)$ linearly by

$$T_t := (1-t)T + t\, s_1(\mathbf{p}(T)), \qquad 0 \le t \le 1$$

for arbitrary $T \in \widehat{\mathcal{F}}_*$. Note that all T_t belong to $\widehat{\mathcal{F}}_*$ and in fact to the same *fibre* $\mathbf{p}^{-1}(\mathbf{p}(T))$ of $\widehat{\mathcal{F}}_*$. This is clear since $\mathbf{p}s_1 = \mathrm{id}$, so $s_1\mathbf{p}(T) - T \in \mathcal{K}$ and therefore also $T_t - T \in \mathcal{K}$. See also Figure 16.2.

□

Thus the topology of the $\widehat{\mathcal{F}}$-spaces has been replaced by the topology of the $\widehat{\mathcal{G}}$-spaces. However, the $\widehat{\mathcal{G}}$-spaces are still rather large and their spectra are not easily grasped. For further topological investigations, it is convenient to replace these spaces by suitable deformation retracts, which only consist of single points, or for which at least the spectrum only consists of single points. We apply the standard C^*-algebra retraction

$$g_t := g\left((1-t)\,\mathrm{id} + t\sqrt{g^*g}^{\,-1}\right), \qquad 0 \le t \le 1. \tag{16.3}$$

As a special case, we have that the group G of unitary elements of the Calkin algebra $\mathcal{C}$ is a deformation retract of (and hence homotopy equivalent with) $\mathcal{C}$. Fortunately the $\widehat{\mathcal{G}}$-spaces are stable under the retraction (16.3) because

$$g^* = g \quad \Longrightarrow \quad g_t^* = g_t\,. \tag{16.4}$$

Therefore

$$\mathcal{S} \cap G \text{ is a deformation retract of } \mathcal{S} \tag{16.5}$$

for $\mathcal{S} \in \{\widehat{\mathcal{G}}, \widehat{\mathcal{G}}_+, \widehat{\mathcal{G}}_-, \widehat{\mathcal{G}}_*\}$.

Denote the corresponding unitary retracts by Roman letters. Then

$$\begin{aligned} \widehat{G} &= \{g \in \mathcal{C} \mid g^* = g \text{ and } g^*g = \mathrm{id} \text{ hence } g^2 = \mathrm{id}\}, \\ \widehat{G}_+ &= \{\mathrm{id}\}, \qquad \widehat{G}_- = \{-\mathrm{id}\}, \qquad \text{and} \\ \widehat{G}_* &= \{g \in \mathcal{C} \mid g^* = g \text{ and } \mathrm{spec}(g) = \{+1,-1\}\}\,. \end{aligned} \tag{16.6}$$

From Lemma 16.3 we get with (16.5) the following theorem:

Theorem 16.4. *The space $\widehat{\mathcal{F}}_*$ is homotopy equivalent with the space $\widehat{G}_*$.*

Now we shall determine the homotopy groups of $\widehat{G}_*$.

16B. The Homotopy Groups of $\widehat{G}_*$

Let H be a complex separable infinite-dimensional Hilbert space. We have seen in (16.6) that the space $\widehat{G}_*$ consists of the self-adjoint involutions in the Calkin algebra $\mathcal{C}(H) = \mathcal{B}(H)/\mathcal{K}(H)$ of H which are not equal to the identity:

$$\widehat{G}_* = \{j \in \mathcal{C}(H) \mid j^* = j\,,\ j^2 = \mathrm{id}\,,\ \text{and } j \neq \pm\,\mathrm{id}\}\,.$$

Thus it can be identified with a suitable Grassmannian

$$\text{(16.7)} \qquad \mathrm{Gr}(\mathcal{C}(H)) := \{q \in \mathcal{C}(H) \mid q^2 = q = q^* \neq 0,\ \mathrm{id}\}.$$

The isomorphism is given by

$$\text{(16.8)} \qquad \widehat{G}_* \ni j \mapsto \frac{1}{2}(\mathrm{id} + j) \in \mathrm{Gr}(\mathcal{C}(H))\,.$$

Consider the case of the corresponding Grassmannian in $\mathcal{B}(H)$:

$$\text{(16.9)} \qquad \mathrm{Gr}(H) := \{P \in \mathcal{B}(H) \mid P = P^2 = P^* \\ \text{and } \dim\mathrm{range}\, P = \infty = \dim\mathrm{range}(\mathrm{Id} - P)\}.$$

If we fix a $P_0 \in \mathrm{Gr}(H)$, then $\mathrm{Gr}(H)$ is isomorphic with the homogeneous space

$$\text{(16.10)} \quad \mathcal{U}(H)/\mathcal{U}(P_0(H)) \oplus \mathcal{U}((\mathrm{Id} - P_0)H) \cong \mathcal{U}(H)/\mathcal{U}(H) \oplus \mathcal{U}(H)\,,$$

where $\mathcal{B}(X)$ denotes the C^*-algebra of bounded operators on X for

$$X \in \{H,\ P_0(H),\ (\mathrm{Id} - P_0)(H)\}$$

and $\mathcal{U}(X)$ its unitary retract. The isomorphism is given by the map

$$\text{(16.11)} \qquad \mathbf{R}_{P_0} : \mathcal{U}(H) \ni U \mapsto U P_0 U^{-1} \in \mathrm{Gr}(H)\,.$$

Repeating an argument from Chapter 15 (cf. especially the proof of Theorem 15.5 and Remark 15.6, see also Remark 24.5b below) we obtain the following theorem:

Theorem 16.5. *The mapping* $\mathbf{R}_{P_0} : \mathcal{U}(H) \longrightarrow \mathrm{Gr}(H)$ *of (16.11) defines a principal fibre bundle with fibre and structure group* $\mathcal{U}(H) \oplus \mathcal{U}(H)$.

Note. By Kuiper's theorem, it follows that $\mathrm{Gr}(H)$ is the image of a contractible space and therefore itself contractible.

Now repeat this argument in the case of $\widehat{G}_*$. The only essential difference is that now neither the total space, nor the structure group of the bundle is contractible. As before, let G denote the unitary group of $\mathcal{C}(H)$. Similar to how we proved Lemma 16.3, we see that G has the homotopy type of the space of Fredholm operators and in particular

$$\pi_k(G, j_0) = \begin{cases} \mathbf{Z} & \text{for } k \text{ even} \\ 0 & \text{for } k \text{ odd} \end{cases} \tag{16.12}$$

for any base point $j_0 \in G$. We fix $j_0 \in \widehat{G}_*$. As already noticed (in the proof of Lemma 15.2), $P_0 - (\mathrm{Id} - P_0)$ is an involution for any projection P_0. Now the next elementary observation is proved:

Lemma 16.6. *There exists* $P_0 \in \mathrm{Gr}(H)$ *such that* $\mathbf{p}(P_0 - (\mathrm{Id} - P_0)) = j_0$, *where* $\mathbf{p} : \mathcal{B}(H) \to \mathcal{C}(H) = \mathcal{B}(H)/\mathcal{K}(H)$ *denotes the natural projection.*

Proof. Consider the following diagram which is not necessarily commutative and contains a semi-continuous mapping $\mathrm{P}_{\geq}$ (because of jumps of dimensions):

$$\begin{array}{ccc} \widehat{\mathcal{F}}_* & \xrightarrow{\ \mathbf{p}|_{\widehat{\mathcal{F}}_*}\ } & \widehat{\mathcal{G}}_* \\ {\scriptstyle \mathrm{P}_{\geq}}\downarrow & & \uparrow{\scriptstyle \mathbf{i}} \\ \mathrm{Gr}(H) & \xrightarrow[\ \widehat{\mathbf{p}}\ ]{} & \widehat{G}_* \end{array} \tag{16.13}$$

Here $\mathrm{P}_{\geq}(T)$ denotes the spectral projection of T onto $[0, \infty)$ and

$$\widehat{\mathbf{p}} : \mathrm{Gr}(H) \to \widehat{G}_*$$

is defined by

$$\widehat{\mathbf{p}}(P) := \mathbf{p}(2P - \mathrm{Id}). \tag{16.14}$$

Begin with the fixed class $j_0 \in \widehat{G}_* \subset \widehat{\mathcal{G}}_*$. Since $\mathbf{p}|_{\widehat{\mathcal{F}}_*} : \widehat{\mathcal{F}}_* \to \widehat{\mathcal{G}}_*$ is surjective, there exists $T \in \widehat{\mathcal{F}}_*$ such that $\mathbf{p}(T) = j_0$. The operator T is a self-adjoint Fredholm operator with $\operatorname{spec}_{\mathrm{ess}}(T) = \{\pm 1\}$, and T is *essentially* an involution:

$$T^2 \in \mathrm{GL}_c := \mathbf{p}^{-1}\{\mathrm{id}\} = \{\mathrm{Id} + \mathcal{K}\}. \tag{16.15}$$

On the other hand $2\,\mathrm{P}_{\geq}(T) - \mathrm{Id}$ is a true involution and $\widehat{\mathbf{p}}(\mathrm{P}_{\geq}(T)) = j_0$.

□

The following corollary completes the proof of Proposition 16.2.

Corollary 16.7. *The spaces $\widehat{G}_*$ and $\widehat{\mathcal{F}}_*$ are connected.*

Proof. First we show that $\widehat{G}_*$ is connected. Fix a $P_0 \in \mathrm{Gr}(H)$ and consider the following composition of maps

$$\mathcal{U}(H) \xrightarrow{\mathbf{R}_{P_0}} \mathrm{Gr}(H) \xrightarrow{\widehat{\mathbf{p}}} \widehat{G}_* .$$

The first map is clearly continuous and surjective by Theorem 16.5. The second map is continuous by construction and surjective by the preceding lemma. Hence $\widehat{G}_*$ is the image of the connected (even contractible) space $\mathcal{U}(H)$ under a continuous map and therefore connected. Since, by Lemma 16.3, $\widehat{\mathcal{F}}_*$ is homotopy equivalent with $\widehat{G}_*$, it is connected as well. □

As seen below, the continuous image $\widehat{G}_*$ does not inherit the contractibility from $\mathcal{U}(H)$.

Similar to (16.10) above and Theorem 16.5 we now define a map $\mathbf{r}_{j_0} : G \to \widehat{G}_*$ on the unitary group G of the Calkin algebra by the formula (for fixed j_0)

$$\mathbf{r}_{j_0}(u) := u j_0 u^{-1}. \tag{16.16}$$

We want

(1) to show that $\mathbf{r}_{j_0} : G \to \widehat{G}_*$ defines a principal fibre bundle,
(2) to derive its long exact homotopy sequence,
(3) and to determine the homotopy groups of $\widehat{G}_*$.

We begin with the following result:

Lemma 16.8. *The map* $\mathbf{r}_{j_0} : G \to \widehat{G}_*$ *is surjective.*

Proof. Let $j \in \widehat{G}_*$. It follows from Lemma 16.6 that $j = \mathbf{p}(P - (\mathrm{Id} - P))$ for an appropriate $P \in \mathrm{Gr}(H)$, and it follows from Theorem 16.5 that there exists a unitary operator $U \in \mathcal{U}(H)$ such that $UP_0U^{-1} = P$, where $P_0 \in \mathrm{Gr}(H)$ was fixed such that $\mathbf{p}(P_0 - (\mathrm{Id} - P_0)) = j_0$. Set $u := \mathbf{p}(U)$. Then

$$j = \mathbf{p}\left(P - (\mathrm{Id} - P)\right) = \mathbf{p}\left(U(P_0 - (\mathrm{Id} - P_0))U^{-1}\right) = uj_0u^{-1}\,.$$

□

Now investigate the fibre

$$G_0 := \mathbf{r}_{j_0}^{-1}(j_0) = \{u \in G \mid uj_0u^{-1} = j_0\}.$$

It is a subgroup of G. Moreover:

Proposition 16.9. *We have* $G_0 \cong G \oplus G$.

Note. This result can also be formulated as follows:

$$G_0 = \mathbf{p}(\mathcal{U}_{P_0}(H))\,, \tag{16.17}$$

where

$$\mathcal{U}_{P_0}(H) := \{U \in \mathcal{U}(H) \mid UP_0 - P_0U \in \mathcal{K}(H)\} \tag{16.18}$$

and $P_0 \in \mathrm{Gr}(H)$ with $\mathbf{p}(P_0 - (\mathrm{Id} - P_0)) = j_0$. Then Proposition 16.9 states that any class $u \in G_0$ can be expressed (in the Calkin algebra) as a class of direct sums of unitary operators on $P_0(H)$ and on $(\mathrm{Id} - P_0)(H)$.

Proof. Define $q^0_\pm := \frac{1}{2}(\mathrm{id} \pm j_0)$. Assume that a $P_0 \in \mathrm{Gr}(H)$ is given with $j_0 = \mathbf{p}(P_0)$. Then for any class $u \in G_0$, we have a decomposition

$$u = u_+ \cdot u_- = u_- \cdot u_+\,, \tag{16.19}$$

where

$$u_\pm := q^0_\pm u q^0_\pm \; + \; q^0_\mp\,. \tag{16.20}$$

The map $u \mapsto (u_+, u_-)$ is an isomorphism of the group G_0 of unitary elements of the Calkin algebra onto

$$G_+ \oplus G_- \cong G \oplus G\,,$$

where $G_\pm$ denotes the unitary part of the group $\mathbf{p}(\mathcal{F}(P_0H))$ (or $\mathbf{p}(\mathcal{F}((\mathrm{Id}-P_0)(H)))$ resp.) which is a deformation retract of it. □

To follow the line of arguments in proving Theorem 15.7, we now construct a local section of the bundle $\mathbf{r}_{j_0} : G \to \widehat{G}_*$:

Let $j \in \widehat{G}_*$ with $\|j - j_0\| < 1$. Then, following Lemma 15.2, the class

$$t(j) := \mathrm{id} + (j - j_0)(2j_0 - \mathrm{id}) \tag{16.21}$$

is invertible in $\mathcal{C}(H)$ and we have

$$t(j)j_0 = jt(j)\,, \tag{16.22}$$

therefore

$$j = t(j)\, j_0\, t(j)^{-1}\,, \qquad \text{hence} \qquad \mathbf{r}_{j_0}(t(j)) = j\,. \tag{16.23}$$

So t defines a local section. To be more precise, we have to take the unitary part of $t(j)$ in order to get an element of G, since a priori only $t \in \mathbf{p}(\mathcal{F}(H))$. This can always be done.

Now following the proof of Theorem 15.7 we get the short exact sequence

$$0 \to \pi_{2k+1}(\widehat{G}_*) \to \mathbf{Z} \oplus \mathbf{Z} \xrightarrow{+} \mathbf{Z} \to \pi_{2k}(\widehat{G}_*) \to 0 \tag{16.24}$$

exploiting that

$$\pi_{2k}(G \oplus G) \cong \mathbf{Z} \oplus \mathbf{Z} \qquad \text{and} \qquad \pi_{2k}(G) \cong \mathbf{Z}\,.$$

Here as usual the homotopy groups are defined relatively to fixed base points. From (16.24) the main result of this chapter is obtained:

Theorem 16.10. *Let $T_0 = P_0 - (\mathrm{Id} - P_0) \in \widehat{\mathcal{F}}_*$ and $j_0 \in \widehat{G}_*$ be fixed elements. Then the homotopy groups of the non-trivial component $\widehat{\mathcal{F}}_*$ of the space of self-adjoint Fredholm operators and of the corresponding space $\widehat{G}_*$ of self-adjoint non-trivial involutions of the Calkin algebra are given by*

$$\pi_i(\widehat{\mathcal{F}}_*, T_0) = \pi_i(\widehat{G}_*, j_0) = \begin{cases} 0 & \textit{for } i \textit{ even} \\ \mathbf{Z} & \textit{for } i \textit{ odd}. \end{cases}$$

17. The Spectral Flow of Families of Self-Adjoint Operators

The spectral flow is investigated for families of self-adjoint elliptic operators with non-vanishing symbol class over an oriented closed smooth manifold.

Our main tool for the treatment of global elliptic boundary problems for operators of Dirac type is the concept of the spectral flow, this peculiar blend of classical analysis and differential topology, which will be explained in detail in the present chapter.

17A. Continuity of Eigenvalues

From a topological point of view, the most fundamental property of the spectrum of self-adjoint operators is its continuity under continuous changes of the operator:

Lemma 17.1. *Let A be a self-adjoint bounded linear operator in a Hilbert space H and let a be a positive real number such that the intersection* $\operatorname{spec} A \cap (-a, a)$ *consists only of a finite system of eigenvalues*

$$-a < \lambda_k \le \lambda_{k+1} \le \cdots < 0 \le \cdots \le \lambda_m < a$$

(all the eigenvalues repeated according to their multiplicity). Then, for all self-adjoint operators A' sufficiently close to A, the intersection $\operatorname{spec} A' \cap (-a, a)$ *consists of the same number of eigenvalues*

$$-a < \lambda'_k \le \lambda'_{k+1} \le \cdots \le \lambda'_m < a$$

and one has

$$|\lambda_j - \lambda'_j| \le \|A - A'\| \textit{ for each } j \in \{k, \dots, m\}.$$

Proof. First decompose A into the difference $A_+ - A_-$ of two non-negative operators. The operator A_+ has the eigenvalues $\lambda_0, \dots, \lambda_m$ and A_- has the eigenvalues $-\lambda_k, \dots, -\lambda_{-1}$. Since $\|A_+ - A'_+\| \le \|A - A'\|$, it suffices to consider the case when A is non-negative.

Recall that then the j-th eigenvalue λ_j can be characterized by the minimum-maximum principle (see e.g. Reed & Simon [1972] or Kato [1976; Section I.6.10]) through

$$\lambda_j = \sup\{U_A(v_1, .., v_j) \mid v_1, \dots, v_j \in H\},$$

where

$$U_A(v_1, \dots, v_j) = \inf\{\langle Aw, w\rangle \mid \|w\| = 1 \text{ and } w \in \mathcal{L}_{\mathbf{C}}(v_1, .., v_j)^{\perp}\}.$$

Now for each choice of $v_1, \dots, v_j$ and for each $\varepsilon > 0$ choose a vector w with $\|w\| = 1$, orthogonal to $v_1, \dots, v_j$ such that

$$\langle Aw, w\rangle - U_A(v_1, \dots, v_j) < \varepsilon.$$

It follows that

$$\begin{aligned} U_{A'}(v_1, \dots, v_j) &\leq \langle A'w, w\rangle \\ &= \langle Aw, w\rangle + \langle (A' - A)w, w\rangle \\ &\leq U_A(v_1, \dots, v_j) + \|A - A'\| + \varepsilon; \end{aligned}$$

hence

$$U_{A'}(v_1, \dots, v_j) \leq U_A(v_1, \dots, v_j) + \|A - A'\|.$$

In the same way we get

$$U_A(v_1, \dots, v_j) \leq U_{A'}(v_1, \dots, v_j) + \|A - A'\|,$$

which proves the lemma. □

Remarks 17.2. (a) Recall that self-adjoint elliptic operators of positive order over a closed manifold have a discrete spectrum $\{\lambda_j\}_{j\in\mathbf{Z}}$ of finite multiplicity. Moreover, there exists no essential spectrum, and the eigenfunctions span the whole L^2E. We get the continuity of the eigenvalues immediately from Lemma 17.1. As usual, operators of order r can be reduced to operators of order 0 by the substitution

$$B \mapsto (1 + B^2)^{-1/2}B =: \widehat{B}.$$

More precisely, let $\{\lambda_j, e_j\}_{j\in\mathbf{Z}}$ be a spectral decomposition of the Hilbert space of square Lebesgue-integrable sections in a Hermitian vector bundle E generated by B, i.e., the $\{\lambda_j\}_{j\in\mathbf{Z}}$ are the eigenvalues of $B : H^r E \to L^2 E$, where H^r is the r-th Sobolev space, with the corresponding eigenfunctions $\{e_j\}_{j\in\mathbf{Z}}$ spanning the whole L^2E. Then

$$\{\lambda_j/\sqrt{1+\lambda_j^2},\, e_j\}_{j\in\mathbf{Z}}$$

is a spectral decomposition of L^2E generated by $\widehat{B}$ with all eigenvalues bounded by $+1$. Hence the continuity of the eigenvalues of $\widehat{B}$ carries over to the eigenvalues of B.
(b) The continuity of a finite system of eigenvalues can be obtained for all closed, not necessarily bounded or self-adjoint operators as shown in Kato [1976, IV.3.5].

Now we could provide the definition of the spectral flow directly by exploiting the perturbation theory of the spectrum of closed, not necessarily bounded operators, as mentioned in the preceding remark. However, the topological meaning of the spectral flow will be more transparent, if instead, we recall the homotopy investigations from Chapter 16 and first work with self-adjoint (bounded) Fredholm operators in Hilbert space. Later we shall carry the arguments over to elliptic operators of positive order.

17B. The Spectral Flow on Loops in $\widehat{\mathcal{F}}_*$

Now we define the spectral flow of a family $\{B_t\}_{t\in[0,1]} \subset \widehat{\mathcal{F}}_*$. We begin with operator families belonging to the subspace

$$\hat{\mathrm{F}}^{\infty} := \{B \in \widehat{\mathcal{F}}_* \mid \operatorname{spec}_{\mathrm{ess}} B = \{1,-1\} \quad \text{and } B \text{ has only finitely many eigenvalues}\}.$$

We can argue in a similar way as in the proof of Proposition 15.4 (second step) in order to show that $\hat{\mathrm{F}}^{\infty}$ is a deformation retract of $\widehat{\mathcal{F}}_*$. However, we do not need the full argument, since we only want to explain the notion of the spectral flow. Let $\{B_t\}_{t\in S^1}$ be a periodic family of operators of $\hat{\mathrm{F}}^{\infty}$. Then the spectrum of $\{B_t\}$ can be described in the following way.

Lemma 17.3. *Let $B : S^1 \to \hat{\mathrm{F}}^{\infty}$ be a continuous map. Then the graph of the spectrum of B is given by a finite monotone sequence of continuous functions $j_i : I \to [-1,1]$,*

$$-1 = j_0 \le j_1 \le \cdots \le j_m = 1\,,$$

i.e., $\operatorname{spec} B_t = \{j_0(t), j_1(t), \ldots, j_m(t)\}, \quad t \in I.$

Definition 17.4. Let ℓ be the integer such that $j_k(1) = j_{k+\ell}(0)$ for all k, where we eventually define

$$j_{-\ell} = j_{-\ell+1} = \cdots = j_0 = -1 \quad \text{and} \quad j_{m+\ell} = j_{m+\ell-1} = \cdots = j_m = 1\,.$$

The integer ℓ is called the *spectral flow* of the family B.

Remarks 17.5. (a) The spectral flow is therefore just the difference between the number of eigenvalues, which change the sign from $-$ to $+$ as t goes from 0 to 1, and the number of eigenvalues, which change the sign from $+$ to $-$. In Definition 17.4, we rearranged the eigenvalues to get a monotone sequence.

(b) Consider a family $B : S^1 \to \widehat{\mathcal{F}}_*$ in $\widehat{\mathcal{F}}_*$. Now we might deform B to a family with values in $\hat{\mathrm{F}}^{\infty}$. However, this is not needed since the discussion in Section 17A showed that for any family B with values in $\widehat{\mathcal{F}}_*$, the graph of the spectrum of B near 0 can be parametrized by a finite set of continuous functions $j_1 \leq \cdots \leq j_m : I \to \mathbf{R}$. This yields once again a well-defined integer ℓ such that $j_k(1) = j_{k+\ell}(0)$.

Proposition 17.6. *The spectral flow defines an isomorphism*

$$\mathrm{sf} : \pi_1(\widehat{\mathcal{F}}_*) \to \mathbf{Z}\,.$$

Note. Since $\widehat{\mathcal{F}}_*$ is connected, the choice of a base point in $\widehat{\mathcal{F}}_*$ is irrelevant to the definition of the homotopy group $\pi_1(\widehat{\mathcal{F}}_*)$.

Proof. First the spectral flow is a homotopy invariant of the families: Let $A = \{A_t\}$ and $B = \{B_t\}$ be two periodic families parametrized by $t \in I$ which can be deformed into each other, i.e., there exists a two-parameter family $\{F_{t,u}\}_{t,u\in I}$ such that $F_{t,0} = A_t$, $F_{t,1} = B_t$ for each t, and $F_{0,u} = F_{1,u}$ for each u. Let $C = \{C_s\}_{s\in I}$ be the boundary of $\{F_{t,u}\}$, i.e.

$$C_s = \begin{cases} A_{4s} & \text{for } 0 \leq s \leq 1/4 \\ F_{1,4s-1} & \text{for } 1/4 < s \leq 1/2 \\ B_{3-4s} & \text{for } 1/2 < s \leq 3/4 \\ F_{0,4-4s} & \text{for } 3/4 < s \leq 1\,. \end{cases}$$

Then $\mathrm{sf}\, C = \mathrm{sf}\, A - \mathrm{sf}\, B$, and C can be deformed into a constant family C'. Let $\{G_{t,u}\}_{t,u\in I}$ be such a two-parameter family with $G_{t,0} = C_t, G_{t,1} = C'_t$ and $G_{0,u} = G_{1,u}$. Let m be the spectral flow of the family $\{G_{0,u}\}_{u\in I}$, and let $\{j_k(t,u)\}_k$ be the sequence of functions parametrizing the spectrum. Then

$$j_k(0,0) = j_{k-m}(0,1) = j_{k-m}(1,1) = j_k(1,0)\,,$$

hence $\operatorname{sf} C = 0$. It is clear that the spectral flow defines a homomorphism

$$\operatorname{sf} : \pi_1(\widehat{\mathcal{F}}_*) \to \mathbf{Z}\,.$$

In fact it is an isomorphism: We construct a periodic family with spectral flow equal to 1. Let $\{e_k\}_{k\in\mathbf{Z}}$ be an orthonormal basis of H. Take

$$B_t := \sum_{k=1}^{\infty}(P_k - P_{-k}) + (2t-1)P_0\,,$$

where P_k is the orthogonal projection onto $\mathcal{L}_{\mathbf{C}}\{e_k\}$; hence

$$\operatorname{spec} B_t = \{-1, 2t-1, 1\}, \quad t \in I\,.$$

□

The preceding proposition gives the main result of this chapter.

Theorem 17.7. *Let $\widehat{\mathrm{Ell}}(Y;E)$ be the space of elliptic self-adjoint operators of positive order over a Riemannian manifold Y on sections of a Hermitian bundle E, and let $B : I \to \widehat{\mathrm{Ell}}(Y;E)$ be a family of operators, where the coefficients depend smoothly on the parameter $t \in I$.*
(a) Then the graph of the spectrum of B can be parametrized near the 0-line through a finite set of continuous functions

$$j_1 \le j_2 \le \cdots \le j_m : I \to \mathbf{R}\,.$$

(b) If B_0 and B_1 have the same spectrum, we get $j_k(1) = j_{k+\ell}(0)$ for some integer ℓ which is the same for each k. That ℓ will be called the spectral flow $\operatorname{sf}\{B_t\}$ *of the family $\{B_t\}$.*
(c) The spectral flow is a C^∞ homotopy invariant of periodic (i.e. $B_1 = B_0$) families of elliptic self-adjoint operators of positive order, which means that the spectral flow does not change under C^∞-deformations of the coefficients of the operators involved.

The situation may be illustrated by Figure 17.1, where e.g. $j_1(1) = j_2(0)$, $j_2(1) = j_3(0)$, $j_3(1) = j_4(0)$ etc., hence $\operatorname{sf}\{B_t\} = 1$.

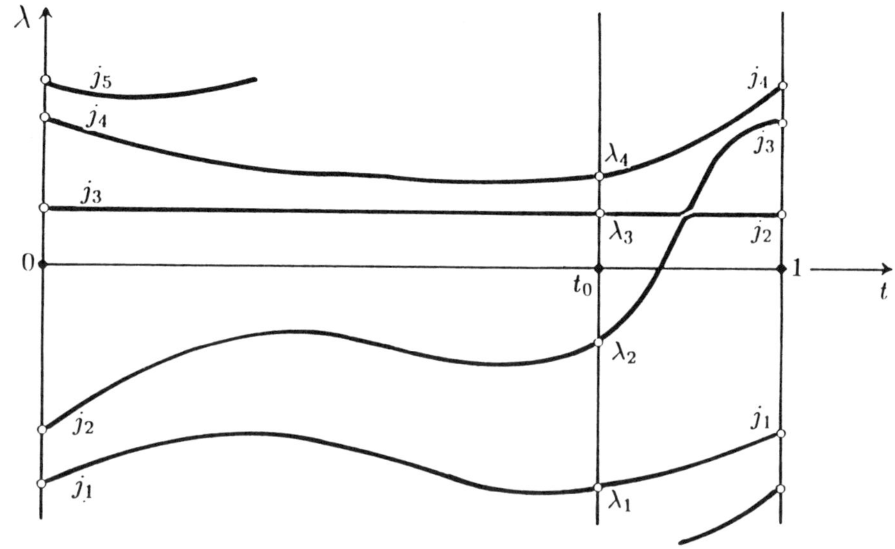

Fig. 17.1 Spectrum of a family with spectral flow = 1

Examples 17.8. (a) Let H be a complex separable Hilbert space and let P_+, P_- be complementary projections, i.e. $P_+ + P_- = \mathrm{Id}$, both having infinite-dimensional range. Then, for any $v \in \operatorname{range} P_-$ and $t \in \mathbf{R}$, the operator $P_+ - P_- + tP_v$ belongs to $\widehat{\mathcal{F}}_*$. Here P_v denotes the projection of H onto $\mathcal{L}_{\mathbf{C}}\{v\}$. The spectrum of $P_+ - P_- + tP_v$ consists of the essential spectrum $\{1, -1\}$ and the eigenvalue $\lambda = -1 + t$ of multiplicity 1. The spectrum of the family

$$\{B_{v,t} := P_+ - P_- + 2tP_v\}_{t \in I}$$

is given by the left graph of Figure 17.2 and hence $\operatorname{sf}\{B_{v,t}\}_t = 1$. This family was already considered in the proof of Proposition 17.6. Taking P_V instead of P_v, where V is a subspace of range P_- of finite dimension N, we get the same graph and a spectral flow equal to N by multiplicity argument.

(b) Now consider the family

$$\left\{ B_t := -i\frac{d}{dx} + t \right\}_{t \in S^1}$$

of ordinary differential operators over the circle S^1, parametrized by $t \in S^1 = I/\{0, 1\}$. We have a spectral decomposition of $H := L^2(S^1)$

by the system $\{e^{ikx}\}_{k\in\mathbf{Z}}$ of eigenfunctions with corresponding eigenvalues $\{k+t\}_{k\in\mathbf{Z}}$, i.e. the spectrum of B is given by the right graph of Figure 17.2 and hence $\operatorname{sf} B = 1$.

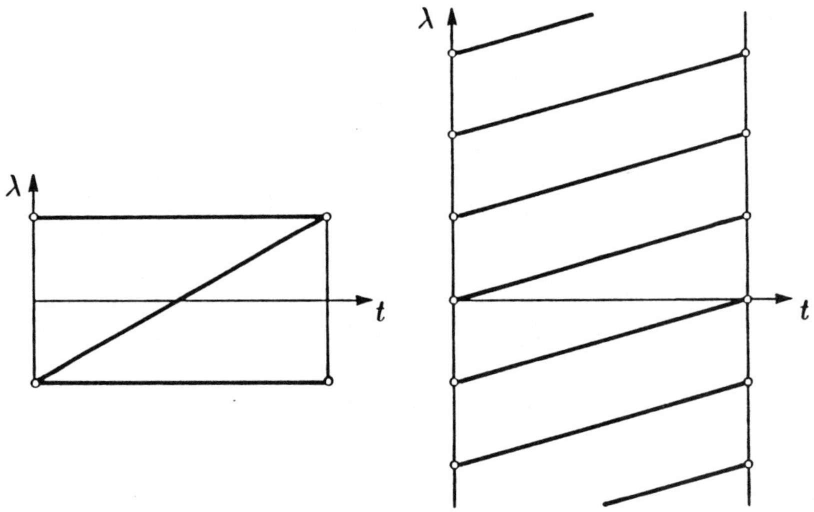

Fig. 17.2 The spectrum of $\{P_+ - P_- + 2tP_v\}$(*left*) and of $\{-id/dx + t\}$ (*right*)

Note that the reduction $B \mapsto B' := (Id + B^2)^{-1/2}B$ leads to a family $B'_t \in \widehat{\mathcal{F}}_*$. This family can be deformed further into the family B_v of (a), where P_+, P_- denote the projections onto $\mathcal{L}_{\mathbf{C}}\{e^{i2\pi kx}\}$ with $k > 0$ and $k \leq 0$ respectively, and v denotes the constant function equal to 1 (corresponding to $k = 0$).

The preceding Example 17.8b raises two natural questions:

(1) Instead of the family $\{B_t = -id/dx + t\}$ of self-adjoint elliptic operators over S^1 parametrized by $t \in S^1$, we may consider its *suspension*, namely the (non-self-adjoint) elliptic operator

$$A := \frac{\partial}{\partial t} - i\frac{\partial}{\partial x} + t$$

over the torus $\mathbf{T} = S^1 \times S^1$ acting on the space

$$C^\infty(\mathbf{T}; E) := \{f \in C^\infty(I \times S^1) \mid f(1,x) = e^{-i2\pi x} f(0,x) \quad \text{for all } x \in S^1\}.$$

How are the index of the operator A and the spectral flow of the family $\{B_t\}$ interrelated? In Theorem 17.13 below we shall explain why these two integers must coincide.

(2) We noticed that the preceding examples (a) and (b) are interrelated. More generally, instead of the operator $A \sim -\partial/\partial t + B_t$, or (equivalently) the family $\{B_t\}$, we may consider the spectral projections $\mathrm{P}_{\geq}$ and $\mathrm{P}_{<}$ belonging to B_0, and the *gauge transformation* $B_1 = g^{-1} B_0 g$ for a suitable morphism g. But how are the spectral flow of the family $\{B_t\}$ (or the index of A) and the index of the *desuspension* $\mathrm{P}_{\geq} - g\,\mathrm{P}_{<}$ related to each other? In Theorem 17.17 it will be explained why these two integers also must coincide.

As we shall show in Chapter 24–26, both questions are fundamental for the understanding of the cutting and pasting of elliptic operators. Especially the second question seems to be central to understanding the geometry of the Grassmannian of all elliptic boundary problems and to solving single boundary problems. See also Remark 15.14 above, which shows that two pseudo-differential projections with the same principal symbol naturally define a family of self-adjoint elliptic operators, and that the spectral flow of that family is the obstruction for the projections belonging to the same connected component in the Grassmannian.

17C. Spectral Flow and Index

Let $\{B_u\} : S^1 \to \widehat{\mathcal{F}}_*$ be a family of self-adjoint Fredholm operators, where the circle S^1 is parametrized by $u \in [0,1]$. By Proposition 17.6, such a family defines an element of the homotopy group $\pi_1(\widehat{\mathcal{F}}_*) \cong \mathbf{Z}$, and two families define the same element if and only if they have the same spectral flow. We shall relate the *spectral* invariant $\mathrm{sf}\{B_u\}$ with *topological* invariants: the indices of related single operators (of the *suspension*, Theorem 17.13, and of the *desuspension*, Theorem 17.17). To that purpose it is suitable to introduce an intermediate, typically *differential* invariant, the *Chern character* of the *analytical index*, which is a certain vector bundle canonically associated with the family $\{B_u\}$. We shall make all constructions as explicit as possible.

The basic notations of K-theory may be found e.g. in Karoubi [1978], Gilkey [1984], or Booß & Bleecker [1985].

Recall that the elements of

$$K^{-1}(S^1) := K\left(S^1 \times S^1 \,/\, (\{0\} \times S^1 \cup S^1 \times \{1\})\right)$$

are determined by the homotopy classes of the clutching mappings $g : S^1 \to U(N)$ of the bundles, and that the generator is the class of the map $g(u) = e^{iu}$. In this case the Chern character is the isomorphism

$$\mathrm{ch} : K^{-1}(S^1) \to H^1(S^1; \mathbf{Z}) \; (\cong \mathbf{Z})$$

given by the formula

$$\mathrm{ch}(g) := \frac{1}{2\pi i} \int_{S^1} \mathrm{tr}\, g^{-1} dg = \deg g\,, \tag{17.1}$$

where g is assumed to be differential.

Consider the diagram

$$\begin{array}{ccc} \pi_1(\widehat{\mathcal{F}}_*) & \xrightarrow{\text{a-index}} & K^{-1}(S^1) \\ & {\scriptstyle \mathrm{sf}} \searrow \quad \swarrow {\scriptstyle \mathrm{ch}} & \\ & \mathbf{Z} & \end{array}$$

To describe the horizontal map a-index which gives the identification, we recall the construction of Example 17.8a: For fixed complementary projections $P_\pm$, both having infinite-dimensional range and a non-negative integer k, choose a subspace V_k of range(P_-) of dimension k. (For negative k one may correspondingly choose a subspace V_k of range(P_+) of dimension $-k$). Then any element of $\pi_1(\widehat{\mathcal{F}}_*)$ is defined by a family

$$\{B_{k,u} := P_+ - P_- + 2uP_k\}$$

with $\mathrm{sf}\{B_{k,u}\} = k$.

Definition 17.9. With the preceding notations the *analytical index* denotes the homomorphism

$$\text{a-index} : \pi_1(\widehat{\mathcal{F}}_*) \to K^{-1}(S^1)$$

given by a-index$\{B_{k,u}\} := [h]$, where $\{h_u\}$ denotes the family of unitary matrices defined by

$$h_u = -\exp i\pi B_{k,u} := \begin{cases} \mathrm{Id} & \text{on } (V_k)^\perp \\ e^{2\pi i u} & \text{on } V_k\,. \end{cases} \tag{17.2}$$

The preceding definition of the analytical index gives at once the identification of ch(a-index$\{B_u\}$) and $\mathrm{sf}\{B_u\}$:

Proposition 17.10. *Let* $\{B_u\} : S^1 \to \widehat{\mathcal{F}}_*$ *be a continuous family. Then*

$$\mathrm{sf}\{B_u\} = \mathrm{ch}(\text{a-index}\{B_u\})\,.$$

Proof. Assume $\mathrm{sf}\{B_u\} = k$ and represent the class of $\{B_u\}$ in $\pi_1(\widehat{\mathcal{F}}_*)$ by the standard family $B_{k,u}$. Then by (17.1) and (17.2)

$$\begin{aligned}\mathrm{ch}(\text{a-index}\{B_u\}) &= \mathrm{ch}(-\exp i\pi B_{k,u}) = \frac{1}{2\pi i}\int_0^1 \mathrm{tr}\, h_u^{-1}\, dh_u \\ &= \frac{1}{2\pi i}\int_0^1 \mathrm{tr}\, e^{-2\pi i u}\, d(e^{2\pi i u})\, \mathrm{Id}_{V_k} = \dim V_k \int_0^1 du = k\,.\end{aligned}$$

□

The remaining part of this chapter is less elementary since we shall exploit the full strength of the index theorem including the index theorem for families.

We proceed with the topological study of families of self-adjoint elliptic operators in the realm of symbolic calculus. Recall that the homotopy class of an elliptic operator B over a closed manifold Y only depends on the homotopy class of its principal symbol σ_B. For any such symbol, the usual theory of elliptic symbols constructs a *symbol class* $[\sigma_B]$, which is an element of $K(TY)$, and which leads to the computation of the index of B by the *topological index* homomorphism $K(TY) \to \mathbf{Z}$.

If B is a self-adjoint elliptic operator over Y, then again its homotopy class only depends on the homotopy class of its principal symbol, that is, B can be deformed into another self-adjoint elliptic operator C through a smooth 1-parameter family $\{B_t\}$ of self-adjoint elliptic operators if and only if σ_B can be deformed into σ_C in the class of self-adjoint and (for each non-zero cotangent vector) invertible symbols. This is a standard exercise. However, the usual symbol class $[\sigma_B] \in K(TY)$ is vanishing. In Atiyah, Patodi & Singer [1976, §3], it was found that the characteristic element of a self-adjoint elliptic operator lies in the group $K^{-1}(TY)$:

Definition 17.11. (a) Let B be a self-adjoint elliptic operator acting on the C^∞-sections of a Hermitian vector bundle E over a closed Riemannian C^∞-manifold Y, and let

$$\sigma_B(y,\eta) : E_y \xrightarrow{\cong} E_y, \quad y \in Y,\ \eta \in TY_y \setminus 0, \text{ and } |\eta| = 1\,,$$

be the principal symbol of B. Consider the family

$$\sigma'_t(y,\eta) = \begin{cases} \cos t\,\mathrm{Id} + i \sin t\,\sigma_B(y,\eta) & \text{for } 0 \le t \le \pi \\ e^{it}\,\mathrm{Id} & \text{for } \pi < t < 2\pi \end{cases}$$

of elliptic symbols over Y parametrized by a point t on the circle S^1. This yields an automorphism of the bundle π^*E, where $\pi : S^1 \times SY \to Y$ is the projection ($SY \subset TY$ is the sphere bundle of cotangent vectors) and hence defines an element $[\sigma'_B] \in K(S^1 \times TY)$. Its restriction to $\{0\} \times TY$ is trivial and so $[\sigma'_B]$ can be regarded as an element of $K^{-1}(TY)$. It will be called the *stable symbol class* of the operator B.
(b) Let Y and E be as in (a). Then, in the same way as in (a), we obtain, for a family $B = \{B_m\}_{m \in M}$ of elliptic self-adjoint operators on $C^\infty(Y;E)$ with a compact parameter space M, a two-parameter family of elliptic symbols $\sigma'_{m,t}$ over Y parametrized by $m \in M$ and $t \in S^1$ and hence an automorphism of the bundle E lifted to $S^1 \times M \times SY$. This yields an element in $K(S^1 \times M \times TY)$, trivial on $\{0\} \times M \times TY$ and hence an element in $K^{-1}(M \times TY)$, the *stable symbol class* $[\sigma'_B]$ of the family B.

Lemma 17.12. *Let $B = \{B_t\}_{t \in S^1}$ be a family of elliptic self-adjoint first-order operators acting on sections of a Hermitian bundle E over a closed manifold Y and let*

$$B' := \{-\frac{\partial}{\partial t} + B_t\}_{t \in S^1}$$

be the associated first-order elliptic operator over the torus $S^1 \times Y$ acting on sections $u(t,y)$ of the vector bundle ρ^*E (= E lifted to $S^1 \times Y$ by the canonical projection $\rho : S^1 \times Y \to Y$). Then we have

$$K^{-1}(S^1 \times TY) \ni [\sigma'_B] = [\sigma_{B'}] \in K(T(S^1 \times Y)).$$

Proof. Clearly B' is an elliptic differential operator, but in general, B' is not self-adjoint since its formal adjoint is $B'^* = \{\frac{\partial}{\partial t} + B_t\}_{t \in S^1}$. Since the tangent space of $S^1 \times Y$ is diffeomorphic to the product $S^1 \times \mathbf{R} \times TY$ by the mapping $(t, y; s\frac{d}{d\theta} + \eta) \mapsto (t, s, (y;\eta))$, we get a natural isomorphism

$$K(T(S^1 \times Y)) \cong K(S^1 \times \mathbf{R} \times TY) \cong K^{-1}(S^1 \times TY).$$

Now consider

$$\sigma_{B'}(t,y;\tau,\eta) = -i\tau + \sigma_{B_t}(y,\eta)\,,$$

where $t \in S^1$, $y \in Y$, $(\tau,\eta) \in T(S^1 \times Y)$ with $|\tau|^2 + |\eta|^2 = 1$. Since B_t is an operator of first order, the symbol σ_B is homogeneous in η and we can rewrite

$$\begin{aligned}\sigma_{B'}(t,y;\tau,\eta) &= -i\left(\tau + i|\eta|\,\sigma_{B_t}(y,\frac{\eta}{|\eta|})\right)\\ &= -i\left(\cos r + i\sin r\,\sigma_{B_t}(y,\frac{\eta}{|\eta|})\right)\end{aligned}$$

by substituting $\tau = \cos r$ for $0 \le r \le \pi$. As usual we join $-i$ with the identity and so we get a homotopy between the clutching isomorphism

$$-i\tau + \sigma_{B_t}(y,\eta), \quad \tau \in S^1,\, t \in S^1,\, (y,\eta) \in SY$$

which defines the symbol class $[\sigma_{B'}]$ of the operator B' and the clutching isomorphism

$$\begin{cases} \cos r + i\sin r\,\sigma_{B_t}(y,\frac{\eta}{|\eta|}) & \text{for } 0 \le r < \pi \\ e^{ir} & \text{for } \pi \le r \le 2\pi\,, \end{cases}$$

where $r \in S^1$, $t \in S^1$, $(y,\eta) \in SY$, which defines the stable symbol class $[\sigma'_B]$ of the family B. Since glueing by homotopic isomorphisms gives the same classes in K-theory, the lemma is proved. □

Theorem 17.13. *Let $\{B_t\}_{t\in S^1}$ be a family of self-adjoint elliptic operators of first order over a closed Riemannian manifold Y parametrized by $t \in S^1$. Then we have*

$$\operatorname{sf} B = \operatorname{index} B',$$

where $B' = \{-\frac{\partial}{\partial t} + B_t\}$ is the elliptic operator on $Y \times S^1$ naturally associated with the family.

Proof. By Proposition 17.10 we have $\operatorname{sf} B = (\operatorname{ch}\text{a-index}\, B)[S^1]$, where $[S^1]$ is the fundamental cycle of S^1 in standard orientation, and a-index $B \in K^{-1}(S^1 \times Y)$, the analytical index of B which equals its topological index; hence

$$\operatorname{sf} B = (\operatorname{ch}\text{t-index}[\sigma'_B])[S^1]\,.$$

Recall the *double* character of $[\sigma'_B] = [\sigma_{B'}]$ by Lemma 17.12 and consider the corresponding diagram

$$\begin{array}{ccccc} K^{-1}(S^1 \times TY) & \xrightarrow{\text{t-index}} & K^{-1}(S^1) & & \\ & & & \searrow^{\text{ch}} & \\ \downarrow \wr\!\wr & & & & \mathbf{Z} \\ & & & \nearrow_{\text{t-index}} & \\ K(S^1 \times (R \times TY)) & \xrightarrow{\cong} & K(T(S^1 \times Y)) & & \end{array}$$

The diagram is commutative (check the commutativity for the periodic family $\{-i\frac{d}{dt} + a\}_{a \in S^1}$ of operators on S^1 of Example 17.8b); hence

$$(\operatorname{ch} \text{t-index}[\sigma'_B])[S^1] = \text{t-index}[\sigma_{B'}]$$

which proves the theorem by the index theorem applied to the single operator B'. □

As is usual in index theory, we want to relate the analytical index which is obtained globally, namely from the space of *solutions*, with the symbol class which is obtained locally, namely from the *coefficients* of the differential equations involved. In the case of one single elliptic operator B over a closed manifold Y, this relation is given by the Atiyah-Singer index theorem:

$$\text{a-index}\, B = \text{t-index}[\sigma_B]\,,$$

where

$$\text{a-index}\, B := [\ker B] - [\operatorname{coker} B] \in K(\mathit{point})$$

and

$$\text{t-index} : K(TY) \to K(\mathit{point}) \cong \mathbf{Z}$$

denotes the topological index given by an embedding $j : Y \to \mathbf{R}^k$, by the excision of a tubular neighbourhood N of $j(Y)$ and by the Bott periodicity $K(T\mathbf{R}^k) \cong K(\mathbf{R}^{2k}) \cong K(\mathit{point})$. In the language of characteristic classes, this becomes

$$\text{t-index}[\sigma_B] = (-1)^{n-1}(\varphi(\operatorname{ch}[\sigma_B]) \cup \tau(Y))[Y]\,,$$

where it is supposed that Y is oriented, $[Y] \in H_n(Y)$ the fundamental cycle of the orientation, n the dimension of Y, $\tau(Y) \in H^*(Y)$, the Todd class of Y,

$$\operatorname{ch} : K(TY) \to H^{\text{even}}_{\text{comp}}(TY; \mathbf{Q})$$

the Chern character, and $\varphi : H^{\text{even}}_{\text{comp}}(TY; \mathbf{Q}) \to H^*(Y)$ the Thom isomorphism.

Corollary 17.14. *Under the assumptions of the preceding theorem we get the following topological formula for the spectral flow:*

$$\operatorname{sf} B = (-1)^{n+1} \int_{SY} \operatorname{ch}[\sigma_{B'}]\, \tau(Y), \quad \textit{where } n = \dim Y.$$

Next we consider a *gauge invariant* family of self-adjoint elliptic operators, i.e. a family $\{B_t : C^\infty(Y;E) \to C^\infty(Y;E)\}_{t\in I}$ with $B_1 = g^{-1}B_0 g$ for a suitable unitary automorphism g of the Hermitian bundle E. Then we shall express the spectral flow of the family $\{B_t\}_{t\in I}$ in terms of the *gauge transformation* g and the spectral projections $\mathrm{P}_\geq$ and $\mathrm{P}_<$ of the initial operator B_0. (For the definition and the fundamental properties of the spectral projections we refer to Definition 14.1 and Proposition 14.2 above.)

Lemma 17.15. *Let B be an elliptic self-adjoint operator of non-negative order acting on the smooth sections of a Hermitian bundle E over a smooth closed Riemannian manifold Y. If the principal symbol p_+ (or p_-) of its spectral projection $\mathrm{P}_\geq$ (or $\mathrm{P}_<$) vanishes, then B is half-bounded as an operator in $L^2(Y;E)$.*

Note. The conclusion is in fact that for an operator B of order zero, the essential spectrum of B lies on only one half line of $\mathbf{R}$, i.e. B is essentially positive (or essentially negative). This is the case, if and only if the principal symbol of B is positive (or negative).

Proof. Let the principal symbol p_+ of the pseudo-differential operator $\mathrm{P}_\geq$ vanish. Then $\mathrm{P}_\geq$ is a pseudo-differential operator of order -1 and hence compact as an operator in L^2. As a projection it has only eigenvalues $\{0,1\}$. Since $\mathrm{P}_\geq$ is compact, the multiplicity of the eigenvalue 1 has to be finite; hence it must have finite-dimensional range. This proves that B only has a finite number of positive eigenvalues and hence

$$\langle Bu, u\rangle \leq \max\{\lambda_j \langle u,u\rangle \mid j \in \mathbf{Z}\} \quad \text{for all } u \in L^2(Y;E)\,.$$

If $\mathrm{P}_<$ vanishes, then similarly we only have a finite number of negative eigenvalues and we get

$$\langle Bu, u\rangle \geq \min\{\lambda_j \langle u,u\rangle \mid j \in \mathbf{Z}\}\,. \qquad \square$$

Lemma 17.15 shows that $p_\pm$ provides a *measure* for the spectral asymmetry of B. More generally, let $E_\pm$ denote the bundles over

SY which consist of the ranges of p_+ and p_-. What happens if the bundles E_+ and E_- are pullbacks of vector bundles F_+, F_- over Y by the projection $\pi : SY \to Y$, that is, if $E_\pm = \operatorname{range} p_\pm = \pi^*(F_\pm)$? This is the case when $p_\pm$ are functions of $y \in Y$ alone. The situation is then more complicated than in Lemma 17.15. However, we find

$$B = p_+ B p_+ + p_- B p_- + (p_+ B p_- + p_- B p_+),$$

where the last term $p_+ B p_- + p_- B p_+$ is an operator of order -1. Hence, up to compact operators, the operator B is a direct sum of half-bounded (essentially positive or essentially negative) operators

$$B \sim \begin{pmatrix} B|_{F_+} & 0 \\ 0 & B|_{F_-} \end{pmatrix} : C^\infty(Y; E) \to C^\infty(Y; E)$$

with $E \cong F_+ \oplus F_-$.

We see that B is non-trivial from our topological point of view if $p_\pm$ defines, as its image, a vector bundle over SY which is not a lifting from Y. Then B has infinitely many eigenvalues on both sides of 0, and it cannot be reduced to the sum of half-bounded (essentially positive or essentially negative) operators.

Note. The range bundle E_+ of p_+ is generally a vector bundle only over a connected component of SY, since the dimension of the range of p_+ may change when we pass to another component.

Now we give a topological formulation of our results: The bundle E_+ defines an element of the group $K(SY)$. Neglecting the topologically trivial case discussed above, we actually get an element of the group $K(SY)/\pi^* K(Y) \cong K^{-1}(TY)$. This last isomorphism can be derived, cf. Atiyah, Patodi & Singer [1976, §3] from the Gysin sequence

$$\begin{array}{ccc} K^*(TY) & \longrightarrow & K^*(Y) \cong K^*(BY) \\ & {\scriptstyle \partial}\nwarrow \quad \swarrow & \\ & K^*(SY) & \end{array}$$

and it follows that the element $\partial[E_+]$ is equal to the stable symbol class of B introduced in Definition 17.11. This proves the following proposition.

Proposition 17.16. *The stable symbol class $[\sigma'_B] \in K^{-1}(TY)$ of an elliptic self-adjoint operator B of non-negative order vanishes, if and only if the operator decomposes into the sum of half-bounded (essentially positive or essentially negative) operators. This is the case if the principal symbols $p_\pm$ of the spectral projections $\mathrm{P}_\geq$ and $\mathrm{P}_<$ of B depend (modulo deformations within the class of idempotent symbols) only on $y \in Y$. If on the contrary $[\sigma'_B] \neq 0$, then we have* $\dim \operatorname{range} \mathrm{P}_\geq = \dim \operatorname{range} \mathrm{P}_< = \infty$.

We are now able to prove the following main theorem on spectral flow and *gauge transformation* (unitary equivalence). When it was first proved in Booß & Wojciechowski [1982] it was called the *desuspension theorem.*

Theorem 17.17. *Let $\{B_t\}$ be a smooth family of elliptic self-adjoint operators of non-negative order acting on the sections of a Hermitian bundle E over a closed manifold Y and having the same principal symbol σ. Let g be an automorphism of E such that $B_1 = g^{-1}B_0 g$. Let $\{e_j\}_{j\in\mathbf{Z}}$ be a spectral decomposition of the Hilbert space $L^2(Y;E)$, i.e. an orthonormal basis of $L^2(Y;E)$ consisting of eigenfunctions e_j of B_0: $B_0 e_j = \lambda_j e_j$. Let $\mathrm{P}_\geq$ and $\mathrm{P}_<$ be the projections onto the subspace $\mathcal{L}_\mathbf{C}\{e_j \mid \lambda_j \geq 0\}$ and $\mathcal{L}_\mathbf{C}\{e_j \mid \lambda_j < 0\}$, respectively. Then $\mathrm{P}_\geq - g\,\mathrm{P}_<$ is an elliptic pseudo-differential operator of order zero over Y and we have*

$$\operatorname{sf}\{B_t\}_{t\in I} = \operatorname{index} \mathrm{P}_\geq - g\,\mathrm{P}_< .$$

Remark 17.18. Recall that the space $\widehat{\mathrm{Ell}}_\sigma$ consists of all elliptic self-adjoint operators of the same order acting on sections of a fixed Hermitian bundle and having the same principal symbol. Hence it is convex. This is an important difference from the case of the space $\widehat{\mathcal{F}}$ of all self-adjoint Fredholm operators having the topological non-trivial component $\widehat{\mathcal{F}}_*$, as seen above in Theorem 16.10 and Proposition 17.6. Hence every true periodic family in $\widehat{\mathrm{Ell}}_\sigma$ could be deformed into the constant family. In order to get a topologically meaningful family, it is necessary to admit that B_1 is different from B_0. We assume that all B_t have the same principal symbol. That permits us to reconnect B_1 with $B_2 := B_0$ in the whole space $\widehat{\mathcal{F}}$ via

$$B_{t+1} := g_t^{-1} B_0 g_t, \quad t \in [0,1],$$

where $\{g_t\}_{t\in I}$ is a retraction of $g_1 := g$ to $g_1 := \mathrm{Id}$ in the contractible group $GL(L^2E)$ of all invertible operators on the Hilbert space L^2E.

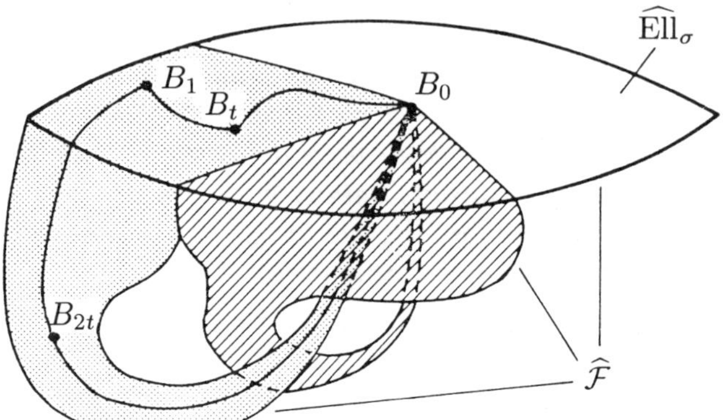

Fig. 17.3 The closing in $\widehat{\mathcal{F}}$ of a family in $\widehat{\mathrm{Ell}}_\sigma$

Of course it is not certain that one can choose a retraction of the operator g to the identity within the subgroup of invertible pseudo-differential operators or even in the group of automorphisms of the vector bundle E. In general, the path $\{B_{t+1}\}_{t\in I}$ will leave the convex space $\widehat{\mathrm{Ell}}_\sigma$. That explains why we can get a topologically non-trivial family by connecting B_0 and B_1. It is obvious that the spectral flow of the family $\{B_{2t}\}_{t\in I}$ only depends on the first half of the path. The choice of the retraction $\{g_t\}$ does not affect the spectrum and hence has no influence at all on the spectral flow. The situation may be illustrated by Figure 17.3. For a more general discussion of the geometry of $\widehat{\mathcal{F}}$ we refer to Fujii [1978], where it is shown that a certain subspace of involutions can in fact serve as classifying space for K^{-1}.

Moreover, it turns out that the spectral flow of a smooth family of elliptic self-adjoint operators only depends on the location of the start and end operators B_0 and B_1. So it is no loss of generality to assume that all operators between B_0 and B_1 have the same principal symbol. Finally, it should be noticed that elliptic boundary problems and the general linear conjugation problem (see Part III below) automatically lead to families with the same principal symbol and admit a unitary equivalence between the start and the end operator.

Proof of Theorem 17.17. To make it more simple, we only consider families of order zero: If $\{B_t\}_{t\in I}$ is a family of order m, then the family

$$\{\widehat{B}_t\}_{t\in I} := \{(\mathrm{Id} + B_t^2)^{-1/2} B_t\}_{t\in I}$$

has not necessarily the same principal symbol over the cotangent sphere bundle SY as $\{B_t\}_{t\in I}$. However, the spectral flow is unchanged (and with it the class of the symbol in $K^{-1}(TY)$).

Assume that

$$\dim \operatorname{range} \mathrm{P}_{\geq} = \infty = \dim \operatorname{range} \mathrm{P}_{<} ,$$

since otherwise $[\sigma'_{B_0}] = 0$ by Proposition 14.2; hence $\operatorname{sf}\{B_t\}_{t\in I} = 0$ and obviously so does $\operatorname{index} \mathrm{P}_{\geq} - g\,\mathrm{P}_{<} = 0$. Now set $S := \mathrm{P}_{\geq} - \mathrm{P}_{<}$. We show that the family $\{B_t\}_{t\in I}$ can be deformed into the family

$$\{B'_t\}_{t\in I} := \{(1-t)S + t(g^{-1}Sg)\}_{t\in I}$$

without changing the spectral flow. Since $\widehat{\mathrm{Ell}}_\sigma$ (the space of all elliptic self-adjoint operators of the same order — here of order zero — acting on sections of a fixed Hermitian bundle and having the same principal symbol σ) is convex, we may assume that $\{B_t\}_{t\in I}$ already has the form

$$B_t = (1-t)B_0 + tB_1, \quad t \in I.$$

Note that the retraction of the original path $\{B_t\}_{t\in I}$ onto the straight line connecting B_0 and B_1 does not change the spectral flow since it is a homotopy invariant by Proposition 17.6. By Proposition 14.2, we get for the principal symbol of S

$$\sigma_S = \sigma_{B_0} = \sigma\,;$$

hence the operator S belongs to the same space $\widehat{\mathrm{Ell}}_\sigma$ as B.

Let $\{C_s\}_{s\in I}$ be the straight line connecting $C_0 := B_0$ with $C_1 := S$. Then the *parallel move*

$$C_{s,t} := (1-t)C_s + t(g^{-1}C_s g), \quad s,t \in I$$

defines a homotopy between the families $\{B_t\}_{t\in I}$ and $\{B'_t\}_{t\in I}$ (and its *invisible* closings $\{B_{1+t}\}_{t\in I} := \{g_t^{-1}B_0 g_t\}$ and $\{B'_{1+t}\}_{t\in I} := \{g_t^{-1}Sg_t\}$). Thus we have $\operatorname{sf}\{B'_t\}_{t\in I} = \operatorname{sf}\{B_t\}_{t\in I}$.

Since B'_0 and B'_1 have the same principal symbol, their difference is an operator of order -1 and hence is compact. This proves that $Sg - gS \in \mathcal{K}(L^2(Y;E))$, since $B'_1 - B'_0 = g^{-1}Sg - S = g^{-1}(Sg - gS)$. Straightforward calculation gives, for $H := L^2(Y;E)$,

$$\begin{aligned}\ker(\mathrm{P}_{\geq} - g\,\mathrm{P}_{<}) &= \{u \in H \mid \mathrm{P}_{\geq} u = g\,\mathrm{P}_{<} u\} \\ &= \{w \in \mathrm{P}_{<} H \mid gw \in \mathrm{P}_{\geq} H\}\end{aligned}$$

and

$$\operatorname{coker}(\mathrm{P}_{\geq} - g\,\mathrm{P}_{<}) = \{w \in \mathrm{P}_{\geq} H \mid gw \in \mathrm{P}_{<} H\},$$

hence

$$\begin{aligned}\operatorname{index}(\mathrm{P}_{\geq} - g\,\mathrm{P}_{<}) &= \dim\{w \in \mathrm{P}_{<} H \mid gw \in \mathrm{P} \geq H\} \\ &\qquad - \dim\{w \in \mathrm{P}_{\geq} H \mid gw \in \mathrm{P}_{<} H\} \\ &= \operatorname{tr}(g^{-1}\,\mathrm{P}_{\geq}\, g - \mathrm{P}_{\geq}).\end{aligned}$$

We determine the discrete spectrum of $B_t' = (1-t)S + tg^{-1}Sg$ by the point set $\{-1, 1, 1-2t, 2t-1\}$ from the elementary calculation

$$B_t'(v) = \begin{cases} v & \text{if } v \in \mathrm{P}_{\geq} H \text{ and } gv \in \mathrm{P}_{\geq} H \\ -v & \text{if } v \in \mathrm{P}_{<} H \text{ and } gv \in \mathrm{P}_{<} H \\ (1-2t)v & \text{if } v \in \mathrm{P}_{\geq} H \text{ and } gv \in \mathrm{P}_{\geq} H \\ (2t-1)v & \text{if } v \in \mathrm{P}_{\geq} H \text{ and } gv \in \mathrm{P}_{\geq} H\,. \end{cases}$$

This gives a graph of the spectrum of the family $\{B_t'\}$, as shown in Figure 17.4, where μ_0, μ_1 parametrize the eigenvalues $2t-1$ and $1-2t$ with multiplicity m_0 and m_1; hence $\operatorname{sf}\{B_t'\} = m_0 - m_1$. As seen above, we have

$$m_0 = \dim\{v \in \mathrm{P}_{<} H \mid gv \in \mathrm{P}_{\geq} H\} = \dim\ker(\mathrm{P}_{\geq} - g\,\mathrm{P}_{<})$$

and

$$m_1 = \dim\{v \in \mathrm{P}_{\geq} H \mid gv \in \mathrm{P}_{<} H\} = \dim\operatorname{coker}(\mathrm{P}_{\geq} - g\,\mathrm{P}_{<})$$

which proves the theorem. □

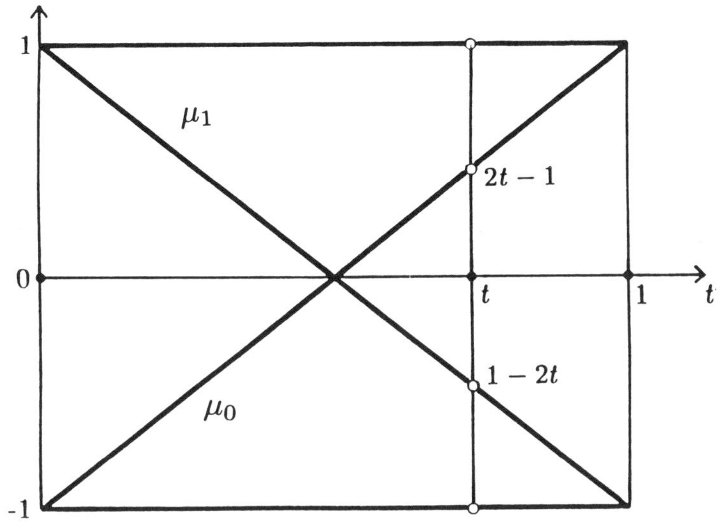

Fig. 17.4 Spectrum parametrization of the family $\{B'_t\}$

17D. Non-Vanishing Spectral Flow

At the end of this chapter about the spectral flow, we present an argument for non-vanishing spectral flow given by physicists. It shows that any solution $s = s(u, y)$ of the equation

$$(\partial_u + B_u)s = 0 \qquad \text{on } S^1 \times Y$$

gives rise to an eigenvalue $\lambda_i(u)$ of B_u, which changes its sign as u goes from 0 to 1. Here $\{B_u\}_{u\in[0,1]}$ is a family of self-adjoint elliptic differential operators of first order acting on sections of a Hermitian bundle E with unitary equivalence $B_1 = gB_0g^{-1}$ for a suitable (non-trivial) choice of g.

It is not difficult to fill the missing details into the chain of arguments below and to obtain the equality $\operatorname{index}(\partial_u + B_u) = \operatorname{sf}\{B_u\}$ in this way. We leave that as an exercise for the reader, who is sufficiently familiar with the needed material involving the Atiyah-Patodi-Singer problem (see Chapter 22 below).

The argument is *adiabatic.* Take a family $\{B_u\}_{u\in[0,1]}$ such that $B_u = B_i$ for u close to $i = 0,\ 1$. Assume that B_i is invertible (we may replace B_u by $B_u + \varepsilon$, where $\varepsilon > 0$ is a small real number which does not belong to the spectrum of B_i). Then extend this family to the whole real line

$$B_u := \begin{cases} B_0 & \text{for } u \in (-\infty, 0) \\ B_u & \text{on } [0,1] \\ B_1 & \text{for } u \geq 1 \end{cases}$$

and consider the operator $A = \partial_u + B_u$ on $\mathbf{R} \times Y$. *A priori* it is not quite obvious that this operator is Fredholm, but actually it is and its index is given by the L^2-index formula of Corollary 23.9 below, which in our case gives:

$$\begin{aligned} L^2\text{-index}(A) &= \int_{\mathbf{R}\times Y} \alpha(x) - \frac{1}{2}\eta_{B_{-\infty}}(0) + \frac{1}{2}\eta_{B_{+\infty}}(0) \\ &= \int_{\mathbf{R}\times Y} \alpha(x) \end{aligned}$$

because of the unitary equivalence of $B_{-\infty}$ and B_∞. Here $\alpha(x)$ denotes the index density which has finite integral.

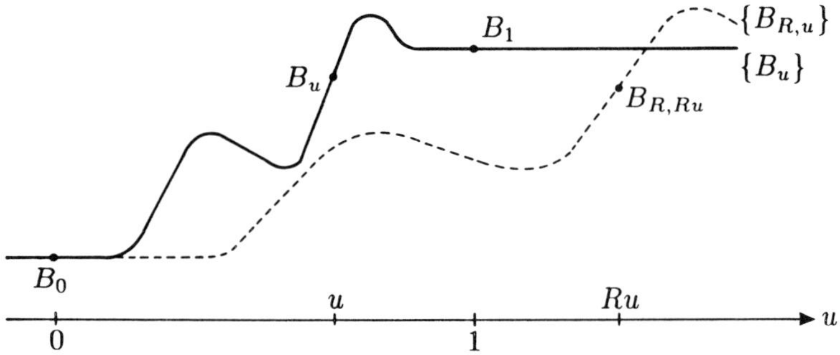

Fig. 17.5 Adiabatic limit of the original family $\{B_u\}$

Now we blow the normal variable $u \to Ru$ and obtain a corresponding family $\{B_{R,u}\} := \{B_{u/R}\}$ and an operator

$$A_R := \partial_u + B_{R,u} \,.$$

We are looking for solutions of $A_R(s) = 0$ of the form

$$s(u,y) = \alpha_i(u)\varphi_{i,u/R}(y)\,,$$

where $\varphi_{i,u/R}$ is a normalized eigensection of $B_{R,u}$ corresponding to an eigenvalue $\lambda_{i,u/R}$ of $B_{u/R}$.

In reality such solutions do not exist; but when $R \to \infty$ we can solve

$$A_R(s) = 0 \tag{17.3}$$

modulo *small* errors. Equation (17.3) is equivalent to

$$\frac{\partial}{\partial u}(\alpha_i \varphi_{i,u/R}) = -B_{u/R}(\alpha_i \varphi_{i,u/R}),$$

hence

$$\frac{\partial \alpha_i}{\partial u}\varphi_{i,u/R} + \alpha_i(u)\frac{1}{R}\frac{\partial \varphi_{i,u/R}}{\partial u}(\frac{u}{R}) = -\alpha_i(u)\lambda_{i,u/R}\,\varphi_{i,u/R}\,.$$

The second summand on the left side is of size $\frac{1}{R}$. We neglect this term and are left with the equation

$$\frac{\partial \alpha_i}{\partial u}\varphi_{i,u/R} = -\alpha_i \lambda_i\, \varphi_{i,u/R}$$

or

$$\frac{\partial \alpha_i}{\partial u} = -\alpha_i \lambda_{i,u/R}$$

with a solution

$$\alpha_i(u) = \exp\left\{-\int_0^u \lambda_{i,u/R}(\tau)\,d\tau\right\},$$

hence

$$s(u,y) = \exp\left\{-\int_0^u \lambda_{i,u/R}(\tau)d\tau\right\}\,\varphi_{i,u/R}(y)\,.$$

Now s must belong to L^2

$$+\infty > \|s\|_{L^2}^2 = \int_{-\infty}^{\infty} \exp\left\{-2\int_0^u \lambda_{i,u/R}(\tau)d\tau\right\}\,du\,.$$

For u negative we have $\int_0^u \lambda_{i,u/R}(\tau)d\tau = -\int_u^0 \lambda_{i,u/R}(\tau)d\tau$ and this term must be positive for large $|u|$. Hence the L^2-norm of s is finite, if and only if

$$\lambda_{i,u/R} \quad \text{is} \quad \begin{cases} \text{positive} & \text{for } u \gg 0 \\ \text{negative} & \text{for } -u \gg 0\,. \end{cases}$$

This shows the following proposition.

Proposition 17.19. *An L^2-solution of the operator $\partial_u + B_{R,u}$ (for large values of R) exists, if and only if there is a family $\{\lambda_R(u)\}_u$ of eigenvalues of $B_{R,u}$ which change the sign from $-$ to $+$ as u goes from $-\infty$ to $+\infty$. Similarly the solution of the adjoint operator $-\partial_u + B_{R,u}$ corresponds to a family of eigenvalues which change sign from $+$ to $-$.*

Part III
Applications

18. Elliptic Boundary Problems and Pseudo-Differential Projections

We explain what it means to pose global elliptic boundary problems and how pseudo-differential projections appear naturally in this context.

Now we can start the discussion of boundary problems for first-order differential operators, say for a Dirac operator

$$A : C^\infty(X; S^+) \to C^\infty(X; S^-)$$

over a manifold X with boundary Y. (Here and in the following we write A instead of A^+ for notational simplicity). In general, A has a closed range of finite codimension, but an infinite-dimensional kernel, which varies depending on the choice of the Sobolev space. Hence, to get regularity and a finite index, we have to restrict the class of admissible sections. The natural way of doing this is to just ignore *almost* all solutions of A.

Definition 18.1. Let $R : C^\infty(Y; S^+|_Y) \to C^\infty(Y; V)$ be a pseudo-differential operator of order 0. We call R a *(global) elliptic boundary condition* for A if the following conditions are satisfied:

(1) For any real s, the extension $R^{(s)} : H^s(Y; S^+|_Y) \to H^s(Y; V)$ of R has a closed range.

(2) Let r denote the principal symbol of R. Then $\operatorname{range}(r) = \operatorname{range}(rp_+) \cong \operatorname{range}(p_+)$. Hence the restriction $r|_{\operatorname{range}(p_+)} : \operatorname{range}(p_+) \to \operatorname{range}(r)$ is an isomorphism of vector bundles.

Recall that $p_+(y,\zeta) : S_y^+ \to S_y^+$ for $(y,\zeta) \in T^*(Y) \setminus 0$ denotes the principal symbol of the Calderón projector $\mathcal{P}_+(A)$ which is the orthogonal projection of S_y^+ onto the direct sum $M_{y,\zeta}^>$ of the eigenspaces of the symbol $b(y,\zeta)$ corresponding to the eigenvalues with positive real part, see Chapter 12. As before, we assume that A is written in the form $G(y)(\partial_t + B)$ close to Y with unitary G and self-adjoint *tangential* operator B over Y with principal symbol b.

Remarks 18.2. (a) Definition 18.1 covers *global* Atiyah-Patodi-Singer conditions as well as standard local elliptic conditions. We get the *Atiyah-Patodi-Singer boundary condition* when $V := S^+|_Y$ and $R := \mathrm{P}_{\geq}(B)$ the spectral projection of B for the interval $[0, +\infty)$. We have seen (in Proposition 14.2) that $\mathrm{P}_{\geq}(B)$ is a pseudo-differential operator of order 0 like the Calderón projector and with the same principal symbol. The range of the extension

$$\mathrm{P}^s_{\geq}(B) : H^s(Y; S^+|_Y) \to H^s(Y; S^+|_Y)$$

is the closure $\overline{\mathcal{L}_{\mathbf{C}_{\geq}}(B)}^s$ of the space spanned by the eigensections of B corresponding to the non-negative eigenvalues, and the principal symbol of $\mathrm{P}_{\geq}(B)$ restricted to range(p_+) is the identity.

(b) For mathematical reasons and for the sake of applications, we define a subclass of the class of all elliptic boundary conditions which contains the original Atiyah-Patodi-Singer condition as well as compact perturbations of this condition. This subclass is distinguished by requiring that R is a projection with $r = p_+$. We shall call such boundary conditions *generalized Atiyah-Patodi-Singer boundary conditions.*

(c) We get a *local elliptic condition* when the range of r can be written as the lifting of the vector bundle V under the natural projection $\pi : T^*Y \setminus 0 \to Y$. Then condition (1) of Definition 18.1 is automatically fulfilled. We shall discuss this in greater detail below (cf. Theorem 20.13 and Theorem 21.5). Note that on odd-dimensional manifolds, the decomposition of the spinors over the (even-dimensional) boundary defines a local elliptic boundary condition. We shall exploit this below in Chapter 21 to prove the cobordism theorem for Dirac operators. On even-dimensional manifolds we have a different situation, since for all Dirac operators there are topological obstructions which prevent the admission of local elliptic boundary conditions (see e.g. Booß & Bleecker [1985, Section II.7.B] or Lawson & Michelsohn [1989, Section II.6]). Nevertheless, *systems* of Dirac operators may admit local elliptic boundary problems. This is always the case for systems associated with transmission problems, see especially Chapter 25 below.

(d) If, as in Remark 12.5b, we identify the space $M^{>}_{y,\zeta} = \operatorname{range} p_+(y,\zeta)$ of *initial values* with the function space

$$\mathcal{M}^{>}_{y,\zeta} = \{w : \mathbf{R} \to S^+_y \mid a^+(0, y; \frac{d}{du}, \zeta)w = 0 \quad \text{and } w(u) \to 0 \text{ as } u \to \infty\},$$

we can reformulate condition (2) of Definition 18.1 in the following way:

(2a) The equation $r(y;\zeta)w|_{u=0} = 0$ has no non-trivial solutions in $\mathcal{M}^{>}_{y,\zeta}$ for each $y \in Y$ and $\zeta \in T^*Y_y$ with $\zeta \neq 0$. This means that $r|_{\mathrm{range}\, p_+}$ is injective, and boundary conditions satisfying this condition are sometimes called *injectively elliptic.*

(2b) For any $s \in r(y,\zeta)(S^+_y)$, the equation $r(y;\zeta)w|_{u=0} = s$ has a solution in $\mathcal{M}^{>}_{y,\zeta}$. This means that $r|_{\mathrm{range}\, p_+} : \mathrm{range}\, p_+ \longrightarrow \mathrm{range}\, r(y,\zeta)$ is surjective.

If, moreover, $\mathrm{range}\, r(y,\zeta) = V_y$, then the pair (A,R) is sometimes called *surjectively elliptic.* Boundary conditions, which are both injectively and surjectively elliptic, we call *local elliptic* conditions, whereas we require less for *(global) elliptic boundary conditions,* namely, that they only be injectively elliptic (2a) and that they satisfy (2b) and the technical condition (1).

What then is the meaning and strength of the concept of global elliptic boundary conditions? Such boundary problems present no serious analytical problems, but its index theory requires a different emphasis, namely somewhere between the broader class of injectively elliptic boundary problems (leading only to semi-Fredholm operators) and the smaller class of local elliptic conditions, where it suffices to concentrate on the symbolic calculus with its approximative statements; in *our* situation of global elliptic boundary problems, as defined above, small changes of the boundary operator R without change of its principal symbol r can lead to jumps of the index (see below Remark 22.25).

Our main interests are the relations between the various expressions of a boundary problem, namely

- the *full mapping pair*

$$(A,R) : C^\infty(X;S^+) \longrightarrow C^\infty(X;S^-) \oplus C^\infty(Y;V)\ , \tag{18.1}$$

given by the formula

$$(A,R)(u) := (Au, R(u|_Y))\ ;$$

- the *realization* of A

$$A_R : \mathrm{dom}\, A_R \longrightarrow L^2(X;S^-) \tag{18.2}$$

as an unbounded operator in $L^2(X;S^+)$ with

$$\mathrm{dom}\, A_R := \{u \in H^1(X;S^+) \mid R(u|_Y) = 0\}\ ;$$

here $R(u|_Y) = 0$ means that u belongs to the kernel of the composite continuous map

$$H^1(X;S^+) \xrightarrow{\gamma_0} L^2(Y;S^+|_Y) \xrightarrow{R_0} L^2(Y;V),$$

where H^1 denotes the first Sobolev space of sections of the spinor bundle S^+ over X, γ_0 denotes the restriction to the boundary, and R_0 is the well-defined continuous extension of R to $H^0(Y;S^+|_Y) = L^2(Y;S^+|_Y)$. Note that γ_0 is even continuous $H^1(X) \to H^{1/2}(Y)$ (Theorem 11.4 and Corollary 11.8);

- and the *boundary integral*

$$(18.3) \qquad R\mathcal{P}_+ : H_+(A) = \operatorname{range}\mathcal{P}_+ \longrightarrow \operatorname{range} R.$$

We have several questions:

(1) Under which restrictions on the operator R does the realization A_R become a Fredholm operator with smooth solutions and smooth solutions of the adjoint operator $(A_R)^*$?
(2) Under which additional conditions for the operator R does the full mapping pair (A, R) become a Fredholm operator with smooth solutions and smooth solutions of the corresponding adjoint problem? The question is whether the boundary conditions R, which provide a *nice* operator A_R, are the same which give a *nice* operator (A, R).
(3) Under which conditions does the *boundary integral* $R\mathcal{P}_+$ become a Fredholm operator, and what about the stability of its index (which was called the *virtual codimension* $\mathbf{i}(R,\mathcal{P}_+)$ of R in $\mathcal{P}_+$ and discussed in Chapter 15 above)?
(4) How can the general (global) and the conventional (local type) elliptic boundary problems be reformulated in the language of projections (pseudo-differential Grassmannians)?

One of the reasons why we can answer these questions is that the projections, induced from the boundary condition R, are pseudo-differential operators, as we shall see now.

Definition 18.3. Let $R : C^\infty(Y;S^+|_Y) \to C^\infty(Y;V)$ be a pseudo-differential operator of order 0 such that the extension

$$R^{(s)} : H^s(Y;S^+|_Y) \to H^s(Y;V)$$

of R has a closed range for any real s. Then $\mathcal{N}_R$, resp. $\mathcal{I}_R$ shall denote the L^2-orthogonal *projections* onto the *kernel* ($\mathcal{N}$ull space), resp. the *range* ($\mathcal{I}$mage) of R defined on $L^2(Y;S^+|_Y)$, resp. $L^2(Y;V)$.

Note. When appropriate, we shall denote the restrictions of the projections onto $C^\infty(Y; S^+|_Y)$, resp. $C^\infty(Y; V)$ by the same symbol.

Clearly we have

Lemma 18.4. *With the preceding notation, the projection onto* $\operatorname{coker} R \cong \ker R^*$ *can be written as*

$$\mathcal{N}_{R^*} = \operatorname{Id} - \mathcal{I}_R .$$

The main purpose of this chapter is to prove the following result of Seeley [1968; Theorem IV.7 and Theorem VI.6]:

Theorem 18.5. *Let R be a (global) elliptic boundary condition in the sense of Definition 18.1. Then the L^2-orthogonal projections $\mathcal{N}_R$, $\mathcal{I}_R$, and $\mathcal{N}_{R^*}$ onto the kernel, the range, and the cokernel of R are pseudo-differential operators.*

The proof of Theorem 18.5 is divided into two parts. First we use condition (1) of Definition 18.1 to show that 0 is an isolated point of the spectrum of the self-adjoint operator R^*R (and of the principal symbol of this operator as well). Then we apply the path integral representation from the proof of Proposition 14.2 to show that the orthogonal projection $\mathcal{N}_R$ onto $\ker R$ is a pseudo-differential operator. We begin with some elementary functional analysis which is part of operator algebra *folklore*, but hard to find in written form in one place.

Proposition 18.6. *Let $T : H_1 \to H_2$ be a bounded operator acting between two Hilbert spaces H_1, H_2. The following conditions are equivalent:*

(1) $\operatorname{range}(T)$ *is closed*

(2) $\operatorname{range}(T) = (\ker T^*)^\perp$

(3) $\operatorname{range}(TT^*)$ *is closed*

(4) $\operatorname{range}(T^*)$ *is closed*

(5) $\operatorname{range}(T^*) = (\ker T)^\perp$

(6) $\operatorname{range}(T^*T)$ *is closed.*

Proof. Since $\operatorname{range} T$ is dense in $(\ker T^*)^\perp$, we clearly have

$$(1) \iff (2) \iff (3) \quad \text{and} \quad (4) \iff (5) \iff (6) .$$

Now it only remains to prove the equivalence (3) $\iff$ (4). It follows from Proposition 9.4 that range TT^* is closed if and only if the self-adjoint operator

$$TT^*|_{(\ker T^* = \ker TT^*)^\perp}$$

has a bounded inverse, i.e. if and only if there exists a positive real number c_1 such that

$$\frac{\langle TT^*v, v\rangle}{\langle v, v\rangle} \geq c_1 \qquad \text{for all } v \in (\ker T^*)^\perp,$$

or equivalently,

$$c_1\|v\|^2 \leq \langle TT^*v, v\rangle = \|T^*v\|^2. \tag{18.4}$$

The estimate (18.4) means that $T^*|_{(\ker T^*)^\perp}$ has a bounded inverse which, once again by Proposition 9.4, is equivalent to closed range T^*.

□

Corollary 18.7. *Let $T : H_1 \to H_2$ be a bounded operator of closed range. Then 0 is an isolated point of the spectrum of the operator T^*T.*

Proof. It follows from Proposition 18.6.(6) that T^*T has a closed range and from Proposition 9.4 that

$$T^*T|_{(\ker T)^\perp} > \varepsilon$$

for some $\varepsilon > 0$. □

Now we can prove Theorem 18.5: The results proved up to now show that there exists $\gamma > 0$ such that the operator $R^*R - \gamma$ is an invertible (elliptic) operator with

$$\ker R^*R = V_{-\gamma} \quad \text{and} \quad \operatorname{spec}(R^*R - \gamma) \cap (-\infty, 2\gamma) \subset \{-\gamma\},$$

where V_a denotes the eigenspace of $R^*R - \gamma$ corresponding to the real eigenvalue a.

Next we essentially repeat the argument given in the proof of Proposition 14.2. The projection $\mathcal{N}_R$ onto $\ker R = \ker R^*R$ is given by the path integral

$$\mathcal{N}_R = \frac{1}{2\pi i}\int_{\lambda=\gamma} (R^*R - \lambda)^{-1}\, d\lambda\,. \tag{18.5}$$

To show that $\mathcal{N}_R$ is a pseudo-differential operator we must compute the full symbol of $\mathcal{N}_R$. The only difficulty is to show that the principal symbol of R^*R has no eigenvalues on the circle $|\lambda| = \gamma$. This follows from the next proposition.

Proposition 18.8. *Let $T : C^\infty(Y;V) \to C^\infty(Y;V)$ be a pseudo-differential operator of order 0, acting on sections of a Hermitian vector bundle over a closed Riemannian manifold Y, and let σ_T denote the principal symbol of T. Then the union*

$$\operatorname{spec}(\sigma_T) := \bigcup_{(y,\zeta)\in T^*Y\setminus 0} \operatorname{spec}\sigma_T(y,\zeta) \tag{18.6}$$

of the symbol spectra, taken for all non-vanishing cotangent vectors of Y, is contained in the spectrum $\operatorname{spec}(T)$ *of T.*

Proof. Let $y_0 \in Y$, $\zeta \in T^*Y_{y_0}$, $\zeta \neq 0$, and let $\lambda \in \operatorname{spec}(\sigma_T(y_0,\zeta))$. Hence there is $v_0 \in V_{y_0}$ with $\langle v_0; v_0\rangle = 1$ and $\lambda v_0 - \sigma_T(y_0,\zeta)v_0 = 0$. Extend v_0 to a section v with $\langle v; v\rangle = 1$ in a neighbourhood of y_0 and extend ζ to a vector field $e \in C^\infty(Y;T^*Y)$ with $e_y \neq 0$ for all y. Let θ_{ε,y_0} denote an approximation of the distribution δ_{y_0} by smooth functions. From the expansion of the total symbol of T we obtain

$$\|T(\theta_{\varepsilon,y_0}v) - \sigma_T(y,e)(\theta_{\varepsilon,y_0}v)\|_{L^2} \to 0 \qquad \text{for } \varepsilon \to 0\,. \tag{18.7}$$

By continuity we get $\lambda \in \operatorname{spec}(T)$. □

The picture of the spectral correspondence between operator and principal symbol can be made even more clear.

Theorem 18.9. *Let T be a self-adjoint pseudo-differential operator of order 0. Then* $\operatorname{spec}(T)\setminus\operatorname{spec}(\sigma_T)$ *consists of isolated points of finite multiplicity.*

Note. As for elliptic T, i.e. $0 \notin \operatorname{spec}(\sigma_T)$, this generalizes the fundamental observation that its spectrum consists of isolated eigenvalues of finite multiplicity.

Proof. Let $\lambda_0 \in \operatorname{spec}(T)\setminus\operatorname{spec}(\sigma_T)$. We can assume that

$$|\lambda_0| > \sup\{\|\sigma_T(y,\zeta)\| \mid y \in Y \text{ and } \zeta \in T^*Y_y\}\,. \tag{18.8}$$

Then the symbol $\lambda_0 \operatorname{id} - \sigma_T(y,\zeta)$ is invertible and $\lambda_0 \operatorname{Id} - T$ is an elliptic pseudo-differential operator of order 0. Now either the Fredholm operator $\lambda_0 \operatorname{Id} - T$ is invertible (which is impossible since $\lambda_0 \in \operatorname{spec}(T)$) or it has a non-trivial null space $\ker(\lambda_0 \operatorname{Id} - T)$ of finite dimension, i.e. λ_0 is eigenvalue of T of finite multiplicity. We show that it is isolated. Let $\mathcal{N}_0$ denote the orthogonal projection onto

$\ker \lambda_0 \operatorname{Id} - T$. Then the operator $\lambda_0 \operatorname{Id} - T - \mathcal{N}_0$ is a self-adjoint Fredholm operator which is injective, hence surjective; and there must exist a whole real neighbourhood $|\lambda - \lambda_0| < \delta$ with

$$0 < \delta < |\lambda_0| - \sup\{\|\sigma_T(y,\zeta)\| \mid y \in Y \text{ and } \zeta \in T^*Y_y\}$$

such that $\lambda \operatorname{Id} - T - \mathcal{N}_0$ is invertible. If $0 < |\lambda - \lambda_0| < \delta$, it follows that $\lambda \notin \operatorname{spec}(T)$. For if $\lambda v - Tv = 0$ for such a $\lambda \neq \lambda_0$, then the section v is orthogonal to the range of $\mathcal{N}_0$ and so $\lambda v - Tv - \mathcal{N}_0 v = 0$; hence $v = 0$ and $\lambda \operatorname{Id} - T$ is in fact invertible. □

Corollary 18.10. *Let R be a pseudo-differential operator of order 0 over a closed Riemannian manifold Y defining an elliptic boundary condition. Let r denote the principal symbol of R. Then there exists an interval $[-\varepsilon_0, \varepsilon_0]$ such that for any $(y,\zeta) \in T^*Y$, the spectrum of $(r^*r)(y,\zeta)$ intersects $[-\varepsilon_0, \varepsilon_0]$ only at 0.*

Proof. Choose $\varepsilon_0 > 0$ so that

$$\operatorname{spec}(R^*R) \cap [-\varepsilon_0, \varepsilon_0] \subset \{0\}. \tag{18.9}$$

Then apply Proposition 18.8. □

Now it is obvious that formula

$$\mathcal{N}_R := \frac{1}{2\pi i} \int_{\lambda=\varepsilon_0} (R^*R - \lambda)^{-1} d\lambda \tag{18.10}$$

defines the orthogonal projection onto the kernel of R. The argument from the proof of Proposition 14.2 furthermore shows that $\mathcal{N}_R$ is a pseudo-differential operator of order 0 with principal symbol $p(y,\zeta)$ which is the orthogonal projection onto the kernel of the principal symbol $r(y,\zeta)$ of R. That proves the first part of the theorem. To prove the rest of the theorem, recall from Lemma 18.4 that $\mathcal{I}_{R^*} = \operatorname{Id} - \mathcal{N}_R$ defines the orthogonal projection onto the range of the operator R^*. Now we can repeat our considerations with the operator R replaced by R^*. This ends the proof of Theorem 18.5.

Note. To prove the statement of Theorem 18.5 we only used assumption (1) from Definition 18.1. Assumption (2) has various specific additional effects. It guarantees the existence of a suitable *elliptic fan* which implies the solutions of the boundary problem to be regular and brings about the finite dimension of the kernel. It also brings about the finite codimension of the range of A_R. This will be explained below.

The next proposition is a consequence of Theorem 18.5. It shows that each realization of a Dirac operator with arbitrary global elliptic boundary condition can be written as a realization with boundary condition defined by a projection. So, at least from the point of view of abstract theory, we can restrict ourselves to boundary conditions given by pseudo-differential *projections* with symbols satisfying condition (2) of Definition 18.1:

Proposition 18.11. *Let R be an elliptic boundary condition for a Dirac operator A and let $\mathcal{I}_{R^*}$ denote the orthogonal projection onto the range of R^*, i.e.* $\mathrm{Id} - \mathcal{I}_{R^*}$ *equals the orthogonal projection $\mathcal{N}_R$ onto the kernel of R. Then $\mathcal{I}_{R^*}$ defines an elliptic boundary condition for A and we have*

$$A_R = A_{\mathcal{I}_{R^*}} = A_{\mathrm{Id} - \mathcal{N}_R} . \tag{18.11}$$

Proof. It is easy to check conditions (1) and (2) of Definition 18.1 for the new boundary condition $\mathcal{I}_{R^*}$. Actually the range of a projection in a Hilbert space is always closed, hence also the range of the extensions $\mathcal{I}_{R^*}^{(s)}$ on the *Hilbertizable* spaces $H^s(Y;V)$. Assumption (2) on the symbol follows directly from the definition of the projection $\mathcal{I}_{R^*}$. Then the equality (18.11) of the realizations is a straightforward corollary of Theorem 18.5. □

Remarks 18.12. (a) Note that the principal symbol of the projection $\mathcal{I}_{R^*}$ does not necessarily coincide with the principal symbol p_+ of the Calderón projector $\mathcal{P}_+$. However, for any arbitrary elliptic boundary condition R, the principal symbol of the orthogonal projection $\mathcal{I}_{\mathcal{P}_+ R^*}$ is equal to p_+.

Before proving that, we sketch an example (to be further elaborated in Chapter 21): Consider a Dirac operator which decomposes near the boundary into the form

$$A = \begin{pmatrix} i & 0 \\ 0 & -i \end{pmatrix} \left(\partial_u + \begin{pmatrix} 0 & C_- \\ C_+ & 0 \end{pmatrix} \right)$$

with regard to a splitting $S^+|_Y \oplus S^-|_Y$ of the given Clifford bundle S over the boundary Y, and assume that the eigenvalues of the principal symbol

$$c(y, \zeta) = \begin{pmatrix} 0 & c_- \\ c_+ & 0 \end{pmatrix}$$

restricted to the cotangent sphere bundle are all contained in the set $\{1, -1\}$ (e.g. for $c_- = (c_+)^* = (c_+)^{-1}$). Then we get

$$p_+(y,\zeta) = \frac{1}{2}\begin{pmatrix} 1 & c_- \\ c_+ & 1 \end{pmatrix}.$$

Consider the elliptic boundary condition defined by the projection onto the first factor

$$R = R^* = \mathcal{I}_{R^*} = \begin{pmatrix} 1 & 0 \\ 0 & 0 \end{pmatrix}$$

which clearly has a principal symbol $r = \begin{pmatrix} 1 & 0 \\ 0 & 0 \end{pmatrix} \neq p_+$. On the level of symbols we have

$$R\mathcal{P}_+ \approx \frac{1}{2}\begin{pmatrix} 1 & c_- \\ 0 & 0 \end{pmatrix} \qquad \text{and} \qquad \mathcal{P}_+ R^* \approx \frac{1}{2}\begin{pmatrix} 1 & 0 \\ c_+ & 0 \end{pmatrix},$$

hence the range of the principal symbol of $R\mathcal{P}_+$ is the whole first factor and so coincides with range r. Therefore the orthogonal projections onto the range of R and onto the range of $R\mathcal{P}_+$ have the same principal symbol; and we shall see in Proposition 18.14a below that this always holds. Moreover

$$\operatorname{range} p_+ r^* = \{\tfrac{1}{2}\begin{pmatrix} s_+ \\ c_+ s_+ \end{pmatrix} \mid s_+ \in \dots\}$$

and

$$p_+\left(\frac{1}{2}\begin{pmatrix} s_+ \\ c_+ s_+ \end{pmatrix}\right) = \frac{1}{4}\begin{pmatrix} s_+ + s_+ \\ c_+ s_+ + c_+ s_+ . \end{pmatrix}$$

Hence the principal symbol of the projection $\mathcal{I}_{\mathcal{P}_+ R^*}$ onto the range of $\mathcal{P}_+ R^*$ is in our example equal to p_+; and, not in that example alone, but always, as we shall show in Proposition 18.14b below.

(b) So far we distinguished four ways of posing elliptic boundary problems for Dirac operators. We conclude:

(i) *Global elliptic boundary conditions* are defined by pseudo-differential operators of order 0 satisfying the assumptions of Definition 18.1. From the point of view of operator theory, they are the natural extensions of the restricted classical concept of local elliptic boundary conditions.

(ii) Elliptic boundary conditions defined via pseudo-differential *orthogonal projections* generate the same subset of elliptic boundary problem realizations in the space of unbounded operators in $L^2(Y;V)$ as in the first case; but calculations are much easier to carry out.

(iii) Our main interest is devoted to a more narrow class of elliptic boundary conditions, namely those defined by pseudo-differential orthogonal projections with the same principal symbol as the Calderón projector. (In the terminology explained in Chapter 15, that means one chooses a boundary condition in the *Grassmannian* of $\mathcal{P}_+$). We shall call this type of boundary conditions *generalized Atiyah-Patodi-Singer conditions.* Most of our geometric motivations and applications are related to this set of boundary conditions which is never empty ($\mathcal{P}_+$ itself always defines a suitable global elliptic boundary condition) and in fact of a remarkably rich structure as seen in Chapter 15 above and to be exploited later on.

(iv) Finally we have the (usually much smaller or even empty) space of *local elliptic boundary conditions* in the classical setting. Indeed, when the boundary of the underlying manifold is odd-dimensional, then that space is empty for Dirac operators, but not necessarily for elliptic complexes, systems, and coupling problems. It has some special features which we shall show in Chapter 20 and apply in Chapters 21, 24, and 25 below.

For the boundary integrals and for the realization of boundary value problems, we shall establish the Fredholm properties. To do that we need a version of Theorem 18.5 for the operator $R\mathcal{P}_+$. It is crucial to show that the range of that operator is closed. This can be derived from the existence of a left-parametrix for the boundary problem and will be discussed in Chapter 19. Here we present another proof based on a construction (the *elliptic fan*) of Birman & Solomyak [1982].

Lemma 18.13. *For any elliptic boundary condition R, the operator*

$$\Delta' := (\mathrm{Id} - \mathcal{P}_+) + \mathcal{P}_+ R^* R \mathcal{P}_+ \tag{18.12}$$

is an elliptic self-adjoint operator of order 0.

Proof. Since $(\mathrm{id} - p_+) + p_+ r^* r p_+$ is a self-adjoint transformation, its surjectivity follows from injectivity, and hence it suffices to show that its kernel is 0 at any $(y,\zeta) \in T^*Y \setminus \{0\}$. Let $v \in S_y^+$. Then

$$(\mathrm{id} - p_+(y,\zeta))\, v + p_+(y,\zeta) r^*(y,\zeta) r(y,\zeta) p_+(y,\zeta) v = 0$$

implies

$$(\mathrm{id} - p_+(y,\zeta))\, v = 0 \qquad \text{and} \qquad p_+(y,\zeta) r^*(y,\zeta) r(y,\zeta) p_+(y,\zeta) v = 0\,;$$

hence

$$v = p_+(y,\zeta) v \qquad \text{and} \qquad r(y,\zeta) p_+(y,\zeta) v = 0\,. \tag{18.13}$$

It follows from assumption (2) of Definition 18.1 that $v = p_+(y,\zeta) v = 0$. □

Since Δ' is an elliptic operator over a closed (= compact and without boundary) manifold, the kernel of Δ' consists of finitely many linearly independent smooth sections of S^+; more precisely

$$\ker \Delta' = \{ s \in L^2(Y; S^+|_Y) \mid Rs = 0 \text{ and } \mathcal{P}_+ s = s \} \subset C^\infty(Y; S^+|_Y)\,, \tag{18.14}$$

since $\mathcal{P}_+ R^* R s = 0$ implies $\langle R^* R s ; s \rangle = 0$ for $s \in \mathrm{range}(\mathcal{P}_+)$, hence $Rs = 0$. (In our notation we do not always distinguish between pseudo-differential operators of order zero defined on C^∞-spaces and their unique bounded extensions to L^2.)

Let K denote the orthogonal projection onto the kernel of the operator Δ'. As an operator of finite-dimensional range, K is a smoothing operator (i.e. a pseudo-differential operator of order $-\infty$); hence the operator

$$\Delta := \Delta' + K \tag{18.15}$$

has vanishing kernel and (because Δ is self-adjoint) also vanishing cokernel. Thus it is an invertible elliptic operator of order 0. Its inverse Δ^{-1} exists and is an elliptic pseudo-differential operator of order 0.

Proposition 18.14. *(a) For any elliptic boundary condition*

$$R : C^\infty(Y; S^+|_Y) \longrightarrow C^\infty(Y; V)\,,$$

the orthogonal projection $\mathcal{I}_{R\mathcal{P}_+}$ onto the range of the operator $R\mathcal{P}_+$ is a pseudo-differential operator and is given by the formula

$$\mathcal{I}_{R\mathcal{P}_+} := R\mathcal{P}_+ \Delta^{-1} \mathcal{P}_+ R^*\,, \tag{18.16}$$

where $\Delta = \Delta' + K$ *as in (18.15). In particular, the principal symbol of* $\mathcal{I}_{R\mathcal{P}_+}$ *is equal to the principal symbol of the projection* $\mathcal{I}_R$ *onto the range of* R.
(b) Similarly, the orthogonal projection $\mathcal{I}_{\mathcal{P}_+R^*}$ *onto the range of* $\mathcal{P}_+R^*$ *is a pseudo-differential operator given by*

$$\mathcal{I}_{\mathcal{P}_+R^*} := \Delta^{-1}\mathcal{P}_+R^*R\mathcal{P}_+ \,. \tag{18.17}$$

In particular, the principal symbol of $\mathcal{I}_{\mathcal{P}_+R^*}$ *is equal to* p_+.

Proof. First we refine the arguments of the preceding proof. With $A_1 := \mathrm{Id} - \mathcal{P}_+$ and $A_2 := R\mathcal{P}_+$ we clearly have, not only on the symbol level, but also on the operator level:

$$A_1A_2^* = 0 \quad \text{and} \quad A_2A_1^* = 0\,; \tag{18.18}$$

hence

$$\operatorname{coker} A_1^* = \ker A_1 \supset \operatorname{range} A_2^* \,.$$

Therefore the spaces range A_1^* and range A_2^* are mutually orthogonal and $A_1^*w_1 + A_2^*w_2 = 0$ implies $A_1^*w_1 = A_2^*w_2 = 0$ for all w_1 and w_2 and also for $w_i := A_is$, $i = 1,\ 2$ and $s \in \ker \Delta'$; hence

$$A_i^*A_iK = 0, \quad i = 1,\ 2\,, \tag{18.19}$$

where K again denotes the orthogonal projection onto $\ker \Delta'$. Since K is self-adjoint, so is $A_i^*A_iK$ and we also obtain $KA_i^*A_i = 0$ from (18.19). This provides the following equalities

$$\begin{aligned}\mathcal{P}_+R^*R\mathcal{P}_+\Delta &= \mathcal{P}_+R^*R\mathcal{P}_+(\Delta' + K)\\ &= \mathcal{P}_+R^*R\mathcal{P}_+R^*R\mathcal{P}_+ = \Delta\mathcal{P}_+R^*R\mathcal{P}_+\,,\end{aligned}$$

which imply

$$\mathcal{P}_+R^*R\mathcal{P}_+\Delta^{-1} = \Delta^{-1}\mathcal{P}_+R^*R\mathcal{P}_+ \,. \tag{18.20}$$

Now we can show that the operator $\mathcal{I}_{R\mathcal{P}_+}$, defined in (18.16), is identical to the orthogonal projection onto the range of the operator $R\mathcal{P}_+$. By definition (18.16), its range is contained in range $R\mathcal{P}_+$. The opposite inclusion comes from the following identities. Use (18.20):

$$\begin{aligned}R\mathcal{P}_+ &= R\mathcal{P}_+\Delta\Delta^{-1} = R\mathcal{P}_+((\mathrm{Id} - \mathcal{P}_+) + \mathcal{P}_+^*R^*R\mathcal{P}_+ + K)\Delta^{-1}\\ &= R\mathcal{P}_+(\mathcal{P}_+R^*R\mathcal{P}_+)\Delta^{-1} = (R\mathcal{P}_+\Delta^{-1}\mathcal{P}_+R^*)R\mathcal{P}_+ = \mathcal{I}_{R\mathcal{P}_+}R\mathcal{P}_+.\end{aligned} \tag{18.21}$$

Thus $\mathcal{I}_{R\mathcal{P}_+}$ is a self-adjoint pseudo-differential operator of order 0 with

$$\operatorname{range} \mathcal{I}_{R\mathcal{P}_+} = \operatorname{range} R\mathcal{P}_+ \ ,$$

and it must only be shown that $\mathcal{I}_{R\mathcal{P}_+}$ is a projection. Use equation (18.21):

$$\begin{aligned} \mathcal{I}^2_{R\mathcal{P}_+} &= \mathcal{I}_{R\mathcal{P}_+}(R\mathcal{P}_+\Delta^{-1}\mathcal{P}_+R^*) \\ &= (\mathcal{I}_{R\mathcal{P}_+}R\mathcal{P}_+)\Delta^{-1}\mathcal{P}_+R^* = R\mathcal{P}_+\Delta^{-1}\mathcal{P}_+R^* = \mathcal{I}_{R\mathcal{P}_+} \ . \end{aligned} \tag{18.22}$$

It follows that $q_+(y,\zeta)$, the principal symbol of the projection $\mathcal{I}_{R\mathcal{P}_+}$, is the orthogonal projection onto the range of $r(y,\zeta)$, where r denotes the principal symbol of the operator R. Hence $q_+ = q$, where q denotes the principal symbol of the orthogonal projection $\mathcal{I}_R$ onto the range of R. This ends the proof of (a). A calculation similar to the one presented above leads to the corresponding statement (b), namely
(18.23)

$$\mathcal{P}_+R^* = \Delta^{-1}\Delta\mathcal{P}_+R^* = (\Delta^{-1}\mathcal{P}_+R^*R\mathcal{P}_+)\mathcal{P}_+R^* = \mathcal{I}_{\mathcal{P}_+R^*}\mathcal{P}_+R^*,$$

which shows

$$\operatorname{range} \mathcal{P}_+R^* \subset \operatorname{range} \mathcal{I}_{\mathcal{P}_+R^*}$$

(the opposite inclusion is again obvious from rewriting $\mathcal{I}_{\mathcal{P}_+R^*} = \mathcal{P}_+R^* \cdots$ with (18.20)) and

$$\mathcal{I}_{\mathcal{P}_+R^*}{}^2 = (\mathcal{I}_{\mathcal{P}_+R^*}\mathcal{P}_+R^*)R\mathcal{P}_+\Delta^{-1} = \mathcal{I}_{\mathcal{P}_+R^*} \ .$$

From the definition and from (18.23), we finally get for the principal symbol of $\mathcal{I}_{\mathcal{P}_+R^*}$

$$\sigma_{\mathcal{I}_{\mathcal{P}_+R^*}} = \begin{cases} \text{id} & \text{on } \operatorname{range} p_+r^* \\ 0 & \text{on } \ker p_+ \ . \end{cases} \tag{18.24}$$

Whereas the space $\operatorname{range} \mathcal{P}_+R^*$ may be difficult to understand, it is obvious that $\operatorname{range} p_+r^*$ is the complement of $\operatorname{coker} p_+r^* = \ker rp_+ = \ker p_+$ because r is injective on $\operatorname{range} p_+$. This proves that $\sigma_{\mathcal{I}_{\mathcal{P}_+R^*}}$, the projection onto $\operatorname{range} p_+r^*$, is in fact equal to p_+. □

The following corollary follows immediately from the preceding proposition. It establishes the Fredholm properties of the *boundary integral* $R\mathcal{P}_+$ and constitutes one of the main results of this chapter and of the elementary analysis of (global) elliptic boundary problems for Dirac operators on the whole:

Corollary 18.15. *Let $\mathcal{P}_+$ denote the Calderón projector of a Dirac operator A over a manifold X with boundary Y and let R be an elliptic boundary condition. Then the operator*

$$R\mathcal{P}_+ : H_+ \longrightarrow \operatorname{range} R$$

is a Fredholm operator, where $H_+ = \operatorname{range}\mathcal{P}_+$ denotes the space of Cauchy data of A along Y.

Proof. It follows from (18.14) that the kernel of the operator $R\mathcal{P}_+$ is equal to the kernel of the operator Δ' which is an elliptic operator over a closed manifold. Hence $\ker R\mathcal{P}_+$ is finite-dimensional and contains only smooth sections.

Moreover, by Proposition 18.14a, the principal symbols of $\mathcal{I}_{R\mathcal{P}_+}$ and of $\mathcal{I}_R$ are equal, hence $\mathcal{I}_{R\mathcal{P}_+} - \mathcal{I}_R$ is compact. Therefore $\operatorname{range}(R\mathcal{P}_+)$ is a subspace of finite codimension in $\operatorname{range} R$ and closed since $\operatorname{range} R$ is closed. □

The linkage between the index theory of elliptic boundary problems and the theory of pseudo-differential Grassmannians is provided by the following proposition, a reformulation of the results of the preceding Proposition 18.14 and Corollary 18.15.

Proposition 18.16. *Let $R : C^\infty(Y; S|_Y) \to C^\infty(Y; V)$ be an elliptic boundary condition for a (total or chiral) Dirac operator A over a compact manifold X with boundary Y. Let p_+ denote the principal symbol of the Calderón projector $\mathcal{P}_+$ onto the space $H_+(A)$ of the Cauchy data of A.*
(a) We have

$$\mathcal{I}_{\mathcal{P}_+ R^*} \in \mathrm{Gr}_{p_+} \qquad \textit{and} \qquad \mathcal{I}_{R\mathcal{P}_+} \in \mathrm{Gr}_q\,,$$

where q denotes the principal symbol of the orthogonal projection $\mathcal{I}_R$ onto the range of R.
(b) The operator

$$T := R\mathcal{P}_+ : H_+(A) \to \operatorname{range}(R)$$

is a Fredholm operator and

$$\operatorname{index} T = \mathbf{i}(R, \mathcal{P}_+) = \underbrace{\mathbf{i}(\mathcal{I}_{\mathcal{P}_+ R^*}, \mathcal{P}_+)}_{\dim\ker T} + \underbrace{\mathbf{i}(\mathcal{I}_R, \mathcal{I}_{R\mathcal{P}_+})}_{-\dim\operatorname{coker} T}\,. \tag{18.25}$$

Proof. All results, except the formula (18.25), were obtained in Proposition 18.14 and Corollary 18.15. The index formula is a consequence of the following decomposition of the operator T:

$$\underbrace{H_+(A) \overset{\mathcal{I}_{\mathcal{P}_+R^*}}{\longrightarrow} \operatorname{range}(\mathcal{P}_+R^*) = (\ker R\mathcal{P}_+)^\perp}_{\text{index of this is } \mathbf{i}(\mathcal{I}_{\mathcal{P}_+R^*},\mathcal{P}_+)>0}$$

$$\underset{\cong}{\overset{R}{\longrightarrow}} \underbrace{\operatorname{range}(R\mathcal{P}_+) \hookrightarrow \operatorname{range}(R)}_{\mathbf{i}(\mathcal{I}_R,\mathcal{I}_{R\mathcal{P}_+})<0} .$$

□

Remarks 18.17. (a) It follows from (18.14) that the kernel of $R\mathcal{P}_+$ is equal to the kernel of the operator Δ' which is an elliptic operator over the closed manifold Y. Hence $\ker(R\mathcal{P}_+)$ is finite-dimensional and contains only smooth sections of the bundle $S^+|_Y$. Later on (from Proposition 20.3 and Theorem 20.8) we shall deduce that also $\operatorname{range}(R\mathcal{P}_+)$ is of finite codimension in $\operatorname{range}(R)$ and even of finite codimension in $C^\infty(Y;V)$ in the local elliptic case (see Lemma 20.11).
(b) Proposition 18.11 implies that there are four different *boundary integrals*

$$R\mathcal{P}_+,\ \mathcal{I}_{R^*}\mathcal{P}_+,\ R\,\mathrm{P}_\geq,\ \text{and } \mathcal{I}_{R^*}\,\mathrm{P}_\geq$$

naturally associated with every global elliptic boundary condition R. Note that the Calderón projector $\mathcal{P}_+$ depends on the Dirac operator A over the whole manifold with boundary, whereas the spectral projection $\mathrm{P}_\geq$ depends solely on the tangential part B of A over the boundary. From Corollary 18.15, one can derive that all these boundary integrals are Fredholm operators in the respective spaces. We shall see later on that some of their indices coincide and others may differ. More precisely, the index may change when replacing $\mathcal{P}_+$ by $\mathrm{P}_\geq$, but not when replacing R by the projection $\mathcal{I}_{R^*}$ onto the range of R^*.
(c) It is worth noticing that there exists a left inverse of the operator $R\mathcal{P}_+$ modulo operators of lower order, or, put differently: for any global elliptic boundary condition

$$R : C^\infty(Y;S^+|_Y) \longrightarrow C^\infty(Y,V),$$

there exists a pseudo-differential operator

$$T : C^\infty(Y;V) \longrightarrow C^\infty(Y;S^+|_Y)$$

such that $T R \mathcal{P}_+ - \mathcal{P}_+$ is a smoothing operator. This is a direct consequence of Proposition 18.14. Take for T the operator $\Delta^{-1}\mathcal{P}_+ R^*$ with Δ as defined in (18.15). Then the operator $T R \mathcal{P}_+$ equals the orthogonal projection $\mathcal{I}_{\mathcal{P}_+ R^*}$ onto the range of $\mathcal{P}_+ R^*$ which has principal symbol p_+. Moreover, if the boundary condition R is a projection, then we obtain that also $R\mathcal{P}_+ T - R$ is a smoothing operator for T, defined as above. This also follows from Proposition 18.14, since

$$R\mathcal{P}_+ T = R\mathcal{P}_+ \Delta^{-1} \mathcal{P}_+ R^* = \mathcal{I}_{R\mathcal{P}_+} ,$$

and the principal symbols of $\mathcal{I}_{R\mathcal{P}_+}$ and $\mathcal{I}_R = R$ coincide.

As in the theory of elliptic operators over closed manifolds, the existence of a left-parametrix provides a recursive argument for the regularity of solutions also in the case of elliptic boundary conditions. Alternatively, the regularity of the solutions of elliptic boundary problems can be derived from the regularity of elliptic problems over closed manifolds by reduction to the boundary. Both approaches will be explained in the following chapter.

19. Regularity of Solutions of Elliptic Boundary Problems

We derive norm estimates for boundary integrals and show the solutions of elliptic boundary problems for Dirac operators to have global regularity and all realizations to have a closed range.

The preceding chapter established some of the true meaning of the ellipticity condition of boundary problems in Definition 18.1, namely that given a Dirac operator $A : C^\infty(X; S^+) \to C^\infty(X; S^-)$ over a compact manifold X with boundary Y and given an elliptic boundary condition $R : C^\infty(Y; S^+) \to C^\infty(Y; V)$, then

(1) $R\mathcal{P}_+ : \operatorname{range}(\mathcal{P}_+) \to \operatorname{range}(R)$ is a Fredholm operator,
(2) $R\mathcal{P}_+$ has a left inverse T modulo smoothing operators,
(3) and any realization A_R can be rewritten as $A_{\mathcal{I}_{R^*}}$, where $Id - \mathcal{I}_{R^*}$ denotes the orthogonal projection $\mathcal{N}_R$ onto the kernel of R.

As before we suppress for typographical reasons the superscript $^+$ in the naming of the Dirac operator.

Before exploiting these results for the index theory of elliptic boundary problems, we shall discuss the other part of the meaning of the ellipticity condition: the regularity of the solutions of elliptic boundary problems. One major result of the theory of elliptic operators over closed manifolds is that the kernel only contains smooth solutions. However, the situation is radically different on manifolds with boundary, where the solution spaces $\ker_+(A, s) \subset H^s(X; S^+)$ may depend on s, see Chapter 12 above. Nevertheless, the solutions of homogeneous elliptic boundary problems are still smooth.

Theorem 19.1. *Let $v \in H^s(X; S^+)$ be a solution of the boundary problem (A, R), i.e. $Av = 0$ and $R(\gamma_0 v) = 0$. Then v is a smooth section of S^+ over X.*

Proof. We have $Av = 0$ and $R(\gamma_0 v) = 0$ which shows that $h := \gamma_0 v$ is an element of

$$\ker R\mathcal{P}_+ \quad \text{and} \quad \ker((\mathrm{Id} - \mathcal{P}_+) + \mathcal{P}_+ R^* R\mathcal{P}_+) .$$

This last operator is an elliptic operator (see Lemma 18.13), hence h is a smooth section of $S^+|_Y$ and $v = K_+(h)$ is a smooth section of S^+ over X. $\square$

For later use notice:

Corollary 19.2. *The various null spaces of an elliptic boundary problem R for a Dirac operator A can be naturally identified:*

$$\ker A_R = \ker \begin{pmatrix} A \\ R\gamma_0 \end{pmatrix} \cong \ker\{R\mathcal{P}_+ : H_+(A) \to \operatorname{range}(R)\}\,.$$

Proof. Let $A_R^{(s)}$ denote the operator A with the domain

$$\operatorname{dom} A_R^{(s)} := \{v \in H^s(X; S^+) \mid R(v|_Y) = 0\} \tag{19.1}$$

for any real $s > \frac{1}{2}$. Here $R(v|_Y) = 0$ means that v belongs to the kernel of the composite continuous map

$$H^s(X;S^+) \xrightarrow{\gamma_0} H^{s-\frac{1}{2}}(Y; S^+|_Y) \xrightarrow{R^{(s-\frac{1}{2})}} R^{(s-\frac{1}{2})}(Y;V)\,,$$

where γ_0 denotes the restriction to the boundary and $R^{(s-\frac{1}{2})}$ the extension of R to $H^{s-\frac{1}{2}}(Y;S^+|_Y)$. (We constantly write A_R instead of $A_R^{(1)}$.) Clearly

$$\ker A_R^{(s)} = \ker \begin{pmatrix} A^{(s)} \\ \\ R^{(s-\frac{1}{2})}\gamma_0 \end{pmatrix};$$

hence, by Theorem 19.1, $\ker A_R^{(s)}$ is independent of s and in particular

$$\ker A_R = \ker \begin{pmatrix} A \\ R \end{pmatrix}. \tag{19.2}$$

Now let $v \in \ker A_R$. Then, by the preceding theorem, $v \in C^\infty(X;S^+)$; hence $v|_Y$ belongs to the Cauchy data space $H_+(A) \subset C^\infty(Y;S^+|_Y)$, so $\mathcal{P}_+(v|_Y) = v|_Y$ and $R\mathcal{P}_+(v|_Y) = R(v|_Y) = 0$. This proves that the trace operator

$$\gamma_0 : C^\infty(X;S^+) \to C^\infty(Y;S^+|_Y)$$

maps $\ker A_R$ into $\ker R\mathcal{P}_+|_{H_+}$. Because the boundary data determine the solution of $Av = 0$ uniquely (Theorem 12.4a which gives $K_+ : H_+(A,s) \cong \ker_+(A,s)$), γ_0 provides an isomorphism

$$\ker A_R \cong \ker(R\mathcal{P}_+|_{H_+}) \,. \tag{19.3}$$

□

It is a little more difficult to prove a general regularity theorem for inhomogeneous elliptic boundary problems. We choose the following formulation: For given $f \in H^s(X; S^-)$, $g \in H^{s+\frac{1}{2}}(Y; V)$, and $s > -\frac{1}{2}$, find a solution of

$$\begin{cases} Av &= f \\ R(\gamma_0 v) &= g \,. \end{cases}$$

Assume that $v \in H^t$ is a solution of that problem. Notice that, by Theorem 13.8, the trace $\gamma_0 v \in H^{t-\frac{1}{2}}$ is well-defined. Then one wants to show that actually v is an element of the space $H^{s+1}(X; S^+)$.

There are different ways of proving that result. We choose one which uses the existence of a left-parametrix for the operator

$$\begin{pmatrix} A \\ \\ R\gamma_0 \end{pmatrix} : C^\infty(X; S^+) \longrightarrow \begin{matrix} C^\infty(X; S^-) \\ \oplus \\ C^\infty(Y; V) \end{matrix} \,.$$

We begin with a lemma.

Lemma 19.3. *Let* $R : C^\infty(Y; S^+|_Y) \to C^\infty(Y; V)$ *be an elliptic boundary condition for a Dirac operator* $A : C^\infty(X; S^+) \to C^\infty(X; S^-)$*. Then the pseudo-differential operator*

$$\begin{pmatrix} R \\ \\ \operatorname{Id} - \mathcal{P}_+ \end{pmatrix} : C^\infty(Y; S^+|_Y) \longrightarrow \begin{matrix} C^\infty(Y; V) \\ \oplus \\ C^\infty(Y; S^+|_Y) \end{matrix} \tag{19.4}$$

has an injective principal symbol and there exists an operator

$$Q = (Q_1, Q_2) : \begin{matrix} C^\infty(Y; V) \\ \oplus \\ C^\infty(Y; S^+|_Y) \end{matrix} \longrightarrow C^\infty(Y; S^+) \tag{19.5}$$

which is a left-parametrix for the operator $\begin{pmatrix} R \\ \operatorname{Id} - \mathcal{P}_+ \end{pmatrix}$, *i.e.*

$$Q \circ \begin{pmatrix} R \\ \operatorname{Id} - \mathcal{P}_+ \end{pmatrix} = \operatorname{Id} + k \,, \tag{19.6}$$

where k is a smoothing operator.

Proof. The principal symbol $\begin{pmatrix} r(y,\zeta) \\ \mathrm{id} - \pi_+(y,\zeta) \end{pmatrix}$ of $\begin{pmatrix} R \\ \mathrm{Id} - \mathcal{P}_+ \end{pmatrix}$ is obviously injective for any $(y,\zeta) \in T^*Y \setminus \{0\}$, since one can decompose the fibre S_y^+ into $\operatorname{range} p_+(y,\zeta) \oplus \operatorname{range}(\mathrm{id} - p_+(y,\zeta))$. Then $r(y,\zeta)$ is an injection on the first factor because of the regularity condition (2) of Definition 18.1, and $\mathrm{id} - p_+(y,\zeta)$ is an injection on the second factor.

Choose a pseudo-differential operator $Q = (Q_1, Q_2)$ of order 0 with principal symbol (q_1, q_2) providing a left inverse of $\begin{pmatrix} r \\ \mathrm{id} - p_+ \end{pmatrix}$. This implies (19.6) by the standard argument. □

Corollary 19.4. *For any real s, the extension*

$$\begin{pmatrix} R^{(s)} \\ \\ \mathrm{Id} - \mathcal{P}_+^{(s)} \end{pmatrix} : H^s(Y; S^+|_Y) \longrightarrow \begin{matrix} H^s(Y;V) \\ \oplus \\ H^s(Y; S^+|_Y) \end{matrix} \tag{19.7}$$

of the operator $\begin{pmatrix} R \\ \mathrm{Id} - \mathcal{P}_+ \end{pmatrix}$ has finite-dimensional kernel and closed range as well as the operator

$$R\mathcal{P}_+ : \operatorname{range}(\mathcal{P}_+) \longrightarrow C^\infty(Y;V) \tag{19.8}$$

and its extensions.

Proof. For convenience we write the operator in (19.7) as $T : H_1 \to H_2$ and its left-parametrix $\widehat{T} : H_2 \to H_1$ so that $\widehat{T}T - \mathrm{Id}$ is compact. Then we show that T has closed range by applying the following standard chain of arguments.

Clearly $\ker T \subset \ker \widehat{T}T$. Hence it follows from $\widehat{T}T - \mathrm{Id}$ compact that $\dim\ker T \le \dim\ker \widehat{T}T < \infty$. Since $(\widehat{T}T - \mathrm{Id})^* = T^*\widehat{T}^* - \mathrm{Id}$ is likewise compact, it follows from $\operatorname{range}(T^*) \supset \operatorname{range}(T^*\widehat{T}^*)$ that $\dim\operatorname{coker} T^* \le \dim\operatorname{coker} T^*\widehat{T}^* < \infty$, so $\operatorname{range}(T^*)$ is closed, hence also $\operatorname{range}(T)$ is closed, e.g. by Proposition 18.6.

Now it is immediate that

$$\begin{aligned} f \in \ker \begin{pmatrix} R \\ \mathrm{Id} - \mathcal{P}_+ \end{pmatrix} &\iff Rf = 0 \text{ and } (\mathrm{Id} - P_+)f = 0 \\ &\iff Rf = 0 \text{ and } P_+ f = f \\ &\iff f \in \ker R\mathcal{P}_+ \,. \end{aligned} \tag{19.9}$$

Moreover, also the range of the corresponding extension of $R\mathcal{P}_+$ is closed, since it can be written as

$$\text{(19.10)} \quad \operatorname{range} R^{(s)}\mathcal{P}_+^{(s)} = \left\{ \begin{pmatrix} w \\ v \end{pmatrix} \in \operatorname{range} \begin{pmatrix} R^{(s)} \\ \operatorname{Id} - \mathcal{P}_+^{(s)} \end{pmatrix} \mid v = 0 \right\}.$$

□

Note. Corollary 19.4 gives another proof of the existence of a pseudo-differential projection onto the range of the operator $R\mathcal{P}_+$ (earlier established in Proposition 18.14a).

It is a remarkable fact that any left-parametrix for the boundary integral $\begin{pmatrix} R \\ \operatorname{Id} - \mathcal{P}_+ \end{pmatrix}$ provides an explicit left-parametrix for the whole problem $\begin{pmatrix} A \\ R\gamma_0 \end{pmatrix}$. (To ease the reading we suppress the superscripts $^{(s)}$ of the various operator extensions when the context excludes ambiguity).

Theorem 19.5. *Let* $Q = (Q_1, Q_2)$ *be a left-parametrix of the operator* $\begin{pmatrix} R \\ \operatorname{Id} - \mathcal{P}_+ \end{pmatrix}$. *Then the operator*

$$\text{(19.11)} \qquad \mathcal{S} := \Big((\operatorname{Id} + K_+ Q_2 (\operatorname{Id} - \mathcal{P}_+)\gamma_0) r^+ \widetilde{A}^{-1} e^+ \;;\; K_+ Q_1 \Big)$$

is a left-parametrix for the operator $\begin{pmatrix} A \\ R\gamma_0 \end{pmatrix}$.

Proof. Check

$$\begin{aligned}
&\mathcal{S} \circ \begin{pmatrix} A \\ R\gamma_0 \end{pmatrix} \\
&= \Big((\operatorname{Id} + K_+ Q_2 (\operatorname{Id} - \mathcal{P}_+)\gamma_0) r^+ \widetilde{A}^{-1} e^+ \;;\; K_+ Q_1 \Big) \circ \begin{pmatrix} A \\ R\gamma_0 \end{pmatrix} \\
&= (\operatorname{Id} + K_+ Q_2 (\operatorname{Id} - \mathcal{P}_+)\gamma_0)\, r^+ \widetilde{A}^{-1} e^+ A + K_+ Q_1 R \gamma_0 \\
&\overset{\text{Lemma 12.7}}{=} \operatorname{Id} - K_+\gamma_0 + K_+ Q_2 (\operatorname{Id} - \mathcal{P}_+)\gamma_0 (\operatorname{Id} - K_+\gamma_0) \\
&\qquad\qquad + K_+ Q_1 R \gamma_0 \qquad\qquad \text{(19.12)} \\
&= \operatorname{Id} - K_+\gamma_0 + K_+ \left(Q_2 (\operatorname{Id} - \mathcal{P}_+) + Q_1 R \right) \gamma_0 \\
&\qquad\qquad - K_+ Q_2 (\operatorname{Id} - \mathcal{P}_+)\gamma_0 K_+ \gamma_0 \\
&\overset{\text{(19.6) and } \mathcal{P}_+ = \gamma_0 K_+}{=} \operatorname{Id} - K_+\gamma_0 + K_+ (\operatorname{Id} + k)\gamma_0 \\
&\qquad\qquad - K_+ Q_2 (\operatorname{Id} - \mathcal{P}_+)\mathcal{P}_+ \gamma_0 \\
&= \operatorname{Id} + K_+ k \gamma_0 .
\end{aligned}$$

□

When checking the continuity for the various maps involved, notice that for $s > \frac{1}{2}$

$$(19.13)\qquad \begin{pmatrix} A \\ \\ R\gamma_0 \end{pmatrix} : H^s(X;S^+) \longrightarrow \begin{matrix} H^{s-1}(X;S^-) \\ \oplus \\ H^{s-\frac{1}{2}}(Y;V) \end{matrix} ,$$

while

$$(19.14)\qquad \mathcal{S} : \begin{matrix} H^{s-1}(X;S^-) \\ \oplus \\ H^{s-\frac{1}{2}}(Y;V) \end{matrix} \longrightarrow H^s(X;S^+) .$$

Now we derive a *lifting jack* for regularity (see Figure 19.1):

Theorem 19.6. *Let A be a Dirac operator over a compact (even-dimensional) manifold X with boundary Y and let R be an elliptic boundary condition for A. Let $v \in H^t(X;S^+)$ be given, and assume that $Av \in H^s(X;S^-)$ and $R(v|_Y) \in H^{s+\frac{1}{2}}(Y;V)$ for given s, t real, and $s > -\frac{1}{2}$. Then v belongs to $H^{s+1}(X;S^+)$ and we have a constant $C = C(s,t)$ independent of v such that*

$$(19.15)\qquad \|v\|_{s+1} \leq C\left(\|R(v|_Y)\|_{s+\frac{1}{2}} + \|v\|_t + \|Av\|_s\right).$$

Proof. Assume $\frac{1}{2} < t \leq s+1$. Then

$$(19.16)\quad H^{s+1}(X;S^+) \ni \mathcal{S}\begin{pmatrix} A \\ R\gamma_0 \end{pmatrix} v = \mathcal{S}\begin{pmatrix} Av \\ R\gamma_0 v \end{pmatrix} = v + K_+ k\gamma_0 v .$$

The element $K_+ k\gamma_0 v$ is a smooth section of S^+; hence

$$(19.17)\qquad v = \mathcal{S}\begin{pmatrix} A \\ R\gamma_0 \end{pmatrix} v - K_+ k\gamma_0 v$$

is an element of H^{s+1}. □

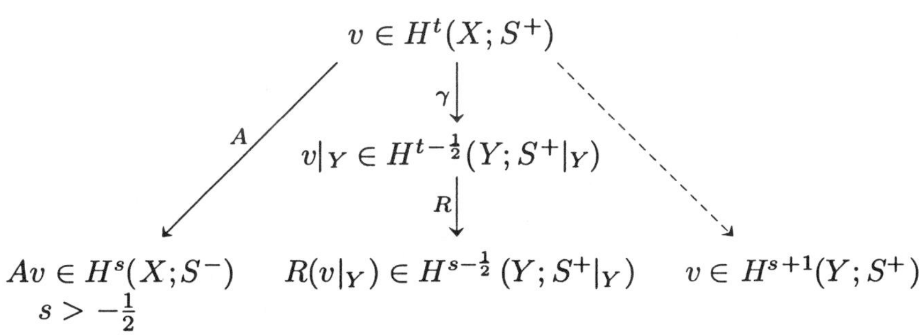

Fig. 19.1 *Lifting jack* by Theorem 19.6 ($s > -1/2$ and t arbitrary)

Corollary 19.7. *Let* $v \in L^2(X;S^+)$, $Av \in L^2(X;S^-)$, *and* $R\gamma_0(v) = 0$. *Then* v *belongs to* $H^1(X;S^+)$.

Remark 19.8. On the contrary, $\begin{pmatrix} A \\ R\gamma_0 \end{pmatrix}$ does not in general have a right-parametrix. Further details on left and right ellipticity and parametrices can be found in Grubb [1977], from which the proof of Theorem 19.5 is taken. Fortunately we only need Lemma 18.13 and Proposition 18.14 to show that the kernel and cokernel of the operators A_R and $\begin{pmatrix} A \\ R\gamma_0 \end{pmatrix}$ consist of smooth sections.

We shall close this section with a discussion of the closedness of the range of the operators $\begin{pmatrix} A \\ R\gamma_0 \end{pmatrix}$ and $A_R = A_R^{(1)}$. That the respective ranges are closed is a consequence of the following elementary lemma.

Lemma 19.9. *Let* $T : H_1 \longrightarrow H_2$ *denote a bounded operator between Hilbert spaces* H_1 *and* H_2. *Assume that* T *has a left-parametrix, i.e. there exists a bounded operator* $S : H_2 \longrightarrow H_1$ *such that*

$$ST = \mathrm{Id}_{H_1} + k\,,$$

where $k : H_1 \longrightarrow H_1$ *is a compact operator. Then* T *has closed range.*

Proof. According to Proposition 9.4, it must be shown that $T|_{(\ker T)^\perp}$ is invertible (i.e. the inverse is bounded). This is equivalent to the existence of a constant $c > 0$ such that

$$\|f\|_{H_1} \leq c\|Tf\|_{H_2} \qquad \text{for any } f \in (\ker T)^\perp\,. \tag{19.18}$$

Assume that such a constant does not exist. Then there exists a sequence $\{f_j\} \subset (\ker T)^\perp$ such that

$$\|f_j\| = 1 \text{ and } \|Tf_j\| \leq \tfrac{1}{j}\,. \tag{19.19}$$

Since k is compact, the sequence $\{kf_j\}$ has a convergent subsequence and so also does the sequence $\{f_j = STf_j - kf_j\}$, which differs from $\{-kf_j\}$ only by the term STf_j which converges to 0.

Let f_0 denote the limit of this subsequence. Then f_0 is an element of $(\ker S)^\perp$ with $\|f_0\| = 1$. On the other hand $Tf_0 = 0$; hence f_0 has to be equal 0 which is a contradiction to (19.19). Hence the lemma is proved. □

Applying Lemma 19.3 and Theorem 19.5 yields at once:

Corollary 19.10. *For any $s > \frac{1}{2}$, the range of the operator*

$$\begin{pmatrix} A \\ \\ R\gamma_0 \end{pmatrix} : H^s(X;S^+) \longrightarrow \begin{matrix} H^{s-1}(X;S^-) \\ \oplus \\ H^{s-\frac{1}{2}}(Y;V) \end{matrix}$$

is closed.

A direct consequence is the following proposition:

Proposition 19.11. *The range of the operator A_R is closed.*

Proof. Let $\{x_n\}$ be a sequence of elements of $\operatorname{dom} A_R$ such that

$$Ax_n \longrightarrow y \in L^2(X;S^-).$$

By definition $R(\gamma_0 x_n) = 0$ for any n, hence

$$\left\{ \begin{pmatrix} Ax_n \\ 0 \end{pmatrix} \right\}$$

is a sequence of elements belonging to the range of the operator

$$\begin{pmatrix} A \\ \\ R\gamma_0 \end{pmatrix} : H^1(X;S^+) \longrightarrow \begin{matrix} L^2(X;S^-) \\ \oplus \\ L^2(Y;V) \end{matrix} .$$

Then by Corollary 19.10 there exists $w \in H^1(X;S^+)$ such that

$$\begin{pmatrix} Aw \\ R(\gamma_0 w) \end{pmatrix} = \begin{pmatrix} y \\ 0 \end{pmatrix},$$

which ends the proof of the proposition. □

20. Fredholm Property of the Operator A_R

We obtain the closedness of the realization A_R, an explicit formula for the adjoint boundary value problem, the standard Fredholm properties of A_R, and various relations between its index and the indices of the related mapping pair and boundary integral in the general (global elliptic) case as well as in the local elliptic case.

To keep the treatment transparent we use chiral notation and write A^- for A^* in this chapter, whether we deal with either the total or chiral Dirac operator. Recall the concept of the realization A_R^+ of a Dirac operator

$$A^+ : C^\infty(X; S^+) \to C^\infty(X; S^-)$$

over a smooth manifold X of even dimension with boundary Y under an elliptic boundary condition

$$R : C^\infty(Y; S^+|_Y) \to C^\infty(Y; V)$$

and the major results obtained in Chapters 18 and 19:
1. By $A^+{}_R$, we denote the operator A^+ with the domain

$$\text{(20.1)} \qquad \operatorname{dom} A^+{}_R := \{u \in H^1(X; S^+) \mid R(u|_Y) = 0\}\,.$$

Here $R(u|_Y) = 0$ means that u belongs to the kernel of the composite continuous map

$$H^1(X; S^+) \xrightarrow{\gamma_0} L^2(Y; S^+|_Y) \xrightarrow{R^{(0)}} L^2(Y; V)\,,$$

where γ_0 denotes the restriction to the boundary, and $R^{(0)}$ is the well-defined continuous extension of R to $H^0(Y; S^+|_Y) = L^2(Y; S^+|_Y)$. Note that γ_0 is even continuous $H^1(X) \to H^{1/2}(Y)$ (Theorem 11.4 and Corollary 11.8). Whether we call the operator $A^+{}_R$, unbounded in L^2, the realization of A^+ *or* of R, depends on our focus.
2. In the same way the operator $A_R^{+(s)}$ is defined with the domain

$$\text{(20.2)} \qquad \operatorname{dom} A_R^{+\,(s)} := \{u \in H^s(X; S^+) \mid R(u|_Y) = 0\}$$

for any real $s > \frac{1}{2}$. It maps $\operatorname{dom} A_R^{+\,(s)}$ into $H^{s-1}(X; S^-)$ and shares all important properties of the operator $A_R^+ = A_R^{+\,(1)}$.
3. Proposition 18.11 showed that the orthogonal projection $\mathcal{I}_{R^*}$ of $C^\infty(Y; S^+|_Y)$ onto the range of R^* is a pseudo-differential operator and defines an elliptic boundary condition for A^+ with

$$A_R^+ = A_{\mathcal{I}_{R^*}}^+ . \tag{20.3}$$

Note that $Id - \mathcal{I}_{R^*}$ is the orthogonal projection of $C^\infty(Y; S^+|_Y)$ onto the kernel of R.
4. In Corollary 18.15 we have seen that the operator

$$R\mathcal{P}_+(A^+) : \operatorname{range} \mathcal{P}_+(A^+) \to \operatorname{range} R \tag{20.4}$$

is a Fredholm operator. Here $\mathcal{P}_+(A^+)$ denotes the Calderón projector of A^+, so that $\operatorname{range} \mathcal{P}_+(A^+)$ becomes the Cauchy data space $H_+(A^+)$. Roughly speaking, this is one half of the meaning of an elliptic boundary condition.
5. In Theorem 19.1, resp. Corollary 19.2, we achieved that the kernel of the pair

$$(A^{+\,(t)}, R^{(t-\frac{1}{2})}) : H^t(X; S^+) \to H^{t-1}(X; S^-) \oplus H^{t-\frac{1}{2}}(Y; V) \tag{20.5}$$

is well-defined for any real t and consists of smooth sections. If $t > \frac{1}{2}$, it coincides with $\ker A_R^{+\,(t)}$ as noticed in equation (19.2) above. Again, roughly speaking, this regularity property constitutes the second half of the meaning of an elliptic boundary condition.

However, these two aspects of elliptic boundary conditions, Fredholm property and regularity, are intimately interrelated. We saw that in the preceding chapters and shall now elaborate further.

We shall show that $A_R^+ : \operatorname{dom} A_R^+ \to L^2(X; S^-)$ is a Fredholm operator. First we prove that A_R^+ is closed. (Proposition 19.11 established that the range of A_R^+ is closed, which is quite a different statement).

Lemma 20.1. *The operator*

$$A_R^+ : \operatorname{dom} A_R^+ \to L^2(X; S^-)$$

is a closed operator from $L^2(X; S^+)$ to $L^2(X; S^-)$.

Proof. Let $u \in L^2(X; S^+)$, $w \in L^2(X; S^-)$, and let $\{u_n\}$ be a sequence of elements of $\operatorname{dom} A_R^+$ such that

$$u_n \to u \text{ in } L^2(X; S^+) \text{ and } A^+ u_n \to w \text{ in } L^2(X; S^-) . \tag{20.6}$$

We have to show that u belongs to $\operatorname{dom} A_R^+$ and that $A^+u = w$.

We use the *left-parametrix* $r^+\tilde{A}^{-1}e^+$ for A^+ introduced in Lemma 12.7:

$$(r^+\tilde{A}^{-1}e^+)A^+u_n = u_n - K_+\gamma_0 u_n\,, \tag{20.7}$$

where

$$K_+ : H^{-\frac{1}{2}}(Y; S^+|_Y) \longrightarrow \ker_+(A^+, 0)\,.$$

Then

$$\begin{aligned}\|(u_n - u_m) - K_+\gamma_0(u_n - u_m)\|_{H^1} &= \|(r^+\tilde{A}^{-1}e^+)A^+(u_n - u_m)\|_{H^1}\\ &\le C\|A^+(u_n - u_m)\|_{L^2}\,,\end{aligned}$$

which shows that $\{v_n := u_n - K_+\gamma_0 u_n\}$ is convergent in $H^1(X; S^+)$ to the limit $v := \lim_{n\to\infty} v_n$. Therefore $\{K_+\gamma_0 u_n = u_n - v_n\}$ has in particular a limit in $L^2(X; S^+)$ and, by Theorem 12.4, $\{\mathcal{P}_+\gamma_0 u_n = \gamma_0 K_+\gamma_0 u_n\}$ is convergent in $H^{-\frac{1}{2}}(Y; S^+|_Y)$.

Now $A^+v = A^+u$ in H^{-1}, hence $u - v \in \ker_+(A^+, 0)$ and the trace and Cauchy datum $\gamma_0(u-v) \in H_+(A^+, 0) \subset H^{-\frac{1}{2}}(Y; S^+|_Y)$ are well-defined. This shows that $\gamma_0(u)$ exists:

$$\gamma_0(u) = \gamma_0(v) - \gamma_0(u - v) \in H^{-\frac{1}{2}}(Y; S^+|_Y)\,, \tag{20.8}$$

and moreover,

$$\begin{aligned}R\gamma_0(u) &= R\left(\overbrace{\gamma_0(v)}^{\in H^{-1}} + \overbrace{\gamma_0(u - v)}^{\in H_+(A^+,0)}\right)\\ &= \lim_{n\to\infty} R\left(\gamma_0(v_n) + \gamma_0(u_n - v_n)\right) = \lim_{n\to\infty} R\left(\gamma_0(u_n)\right) = 0\,.\end{aligned}$$

Now we have $u \in L^2$ by assumption;

$$\begin{aligned}A^+u &= A^+(v + u - v) = A^+v + 0 = A^+v\\ &= \lim_{n\to\infty} A^+(u_n - K_+\gamma_0 u_n) = \lim_{n\to\infty} A^+(u_n) = w \in L^2\,;\end{aligned}$$

and $R(\gamma_0(u)) = 0$. It follows from Corollary 19.7 that $u \in H^1(X; S^+)$ and so u belongs to the domain of A_R. □

Remarks 20.2. (a) In Corollary 20.5 below, we shall show that the operator A_R^+ is actually the L^2-closure of the operator $A_R^{+(\infty)}$: $C^\infty(X;S^+;R) \to C^\infty(X;S^-)$, where $C^\infty(X;S^+;R)$ denotes the space of C^∞-sections u which satisfy $R(u|_Y) = 0$.
(b) Instead of the preceding explicit sequence argument in the proof of Lemma 20.1, we could also argue in a distributional way to prove that $A^+u \in L^2$: Since A^+ extends to a bounded operator from the whole of L^2 to the distributional Sobolev space $H^{-1}(X;S^-)$, we have $A^+u = w$ in a distributional sense. More precisely, we obtain

$$\langle A^+u;\varphi\rangle \;=\; \langle A^+\lim u_n;\varphi\rangle \;=\; \langle \lim A^+u_n;\varphi\rangle \;=\; \langle w;\varphi\rangle$$

for all test sections $\varphi \in C^\infty(X;S^-)$ and hence for all $\varphi \in L^2(X;S^-)$. So A^+u belongs to $L^2(X;S^-)$ and equals w.
(c) There are various other ways to prove that the realization A_R^+ of a boundary value problem is closed. The proof of Lemma 20.1 shows, however, that it is not necessary to apply the whole machinery of *a priori* inequalities and subtle analysis on the collar N of the boundary Y. Whether one uses our sequential or our distributional argument, the closedness of A_R^+ is a direct consequence of the simple *lifting jack* (Corollary 19.7) provided by the Calderón projector and the related explicit potentials and parametrices.

Now we can easily find the adjoint operator of a given realization A_R^+. It is an especially nice feature of the theory of (global) elliptic boundary value problems for Dirac operators that its class of realizations is closed under the taking of adjoints.

Proposition 20.3. *Let R be an elliptic boundary condition for a given Dirac operator A^+. We assume that A^+ takes the form $G(y)(\partial_t + B)$ close to Y with unitary G. Let $\mathcal{I}_{R^*}$ denote the L^2-orthogonal projection of $C^\infty(Y;S^+|_Y)$ onto the range of R^*; hence* $\ker R = \ker \mathcal{I}_{R^*}$ *and $A_R^+ = A_{\mathcal{I}_{R^*}}^+$. Then the L^2-orthogonal projection*

$$(20.9)\qquad Q := G(Id - \mathcal{I}_{R^*})G^* : C^\infty(Y;S^-|_Y) \to C^\infty(Y;S^-|_Y)$$

is an elliptic boundary condition for the operator $(A^+)^* = A^-$. Moreover

$$(20.10)\qquad (A_R^+)^* = (A_{\mathcal{I}_{R^*}}^+)^* = A_Q^- .$$

Remarks 20.4. Before proving Proposition 20.3, we recall the three different meanings of *taking adjoints*:
(a) For the pseudo-differential operator

$$R : C^\infty(Y; S^+|_Y) \to C^\infty(Y; V) \tag{20.11}$$

of order 0 acting between sections in Hermitian bundles over a compact Riemannian manifold without boundary, we denote the *formal adjoint* by R^*; it is the uniquely defined pseudo-differential operator of order 0

$$R^* : C^\infty(Y; V) \to C^\infty(Y; S^+|_Y)$$

with

$$\langle Rf; g\rangle_{L^2(Y;V)} = \langle f; R^*g\rangle_{L^2(Y;S^+|_Y)} \tag{20.12}$$

for all $f \in C^\infty(Y; S^+|_Y)$ and $g \in C^\infty(Y; V)$. Moreover, the continuous extension of the formal adjoint R^* to a bounded operator $(R^*)^{(0)} : L^2(Y; V) \to L^2(Y; S^+|_Y)$ is the *adjoint operator* of the extension $R^{(0)}$ in the true Hilbert space meaning.
(b) The Dirac operator

$$A^+ : C^\infty(X; S^+) \to C^\infty(X; S^-) \tag{20.13}$$

is a differential operator of order 1 acting between sections in Hermitian bundles over a compact Riemannian manifold *with* boundary. Here the *formal adjoint* is a differential operator of order 1

$$(A^+)^* : C^\infty(X; S^-) \to C^\infty(X; S^+) \tag{20.14}$$

uniquely defined by the condition

$$\langle A^+u; v\rangle_{L^2(X;S^-)} = \langle u; (A^+)^*v\rangle_{L^2(X;S^+)} \tag{20.15}$$

for all $u \in C^\infty(X; S^+)$, $v \in C^\infty(X; S^-)$ with $\operatorname{supp} u$, $\operatorname{supp} v$ contained in $X \setminus Y$. We have seen in Chapter 7 that $(A^+)^* = A^-$ is an operator with an immediate geometric meaning like A^+. Moreover, we have shown Green's formula

$$\langle A^+u; v\rangle_{L^2(X;S^-)} = \langle u; A^-v\rangle_{L^2(X;S^+)} - \langle Gu|_Y; v|_Y\rangle_{L^2(Y;S^-|_Y)} \tag{20.16}$$

for spinors with support not necessarily contained in the interior of X. So in contrast to the situation over closed manifolds (i.e. compact

and without boundary), where formal adjoint operators become adjoint when extended to suitable Hilbert spaces, we must here impose boundary conditions to make formal adjoint operators adjoint.
(c) *The adjoint operator* $(A_R^+)^*$ of the realization

$$A_R^+ : \operatorname{dom} A_R^+ \to L^2(X; S^-) \tag{20.17}$$

will be taken in the sense of the theory of unbounded linear operators in the Hilbert space, i.e. $v \in \operatorname{dom}(A_R^+)^*$, if and only if there exists a $w \in L^2(X; S^+)$ such that

$$\langle A^+ u; v\rangle_{L^2(X;S^-)} \;=\; \langle u; w\rangle_{L^2(X;S^+)} \text{ for all } u \in \operatorname{dom} A_R^+ . \tag{20.18}$$

Clearly w is uniquely determined, since $\operatorname{dom} A_R^+$ is dense in $L^2(X; S^+)$. Note that condition (20.18) is equivalent to the existence of a positive real constant C depending on v such that

$$|\langle A^+ u; v\rangle| \le C\|u\| \text{ for all } u \in \operatorname{dom} A_R^+ . \tag{20.19}$$

In fact, $C = \|A^- v\|$ since

$$|\langle A^+ u; v\rangle| = |\langle u, A^- v\rangle| \le \|A^- v\|\,\|u\| , \tag{20.20}$$

see also e.g. Reed & Simon [1972; Definitions VIII. 1.4/5].

Proof of Proposition 20.3. Let us consider a fixed $v \in \operatorname{dom} A_Q^-$. Close to Y the operator $A^- = (A^+)^*$ takes the form

$$(-\partial_t + B^*)G(y)^* = -G(y)^*(\partial_t - G(y)B^*G(y)^*) . \tag{20.21}$$

Then Green's formula yields, for any $u \in \operatorname{dom} A^+_{\mathcal{I}_{R^*}}$,

$$\langle A^- v; u\rangle \;=\; \langle v; A^+ u\rangle + \int_Y (G^* v|_Y; u|_Y) = \;\langle v; A^+ u\rangle . \tag{20.22}$$

The integral over Y vanishes since $\mathcal{I}_{R^*}(u|_Y)$ vanishes, so $u|_Y \in (\operatorname{range} \mathcal{I}_{R^*})^\perp$; and $(Id - \mathcal{I}_{R^*})G^* v|_Y$ vanishes, so $G^* v|_Y \in \operatorname{range} \mathcal{I}_{R^*}$. From (20.22), we get

$$|\langle A^+ u; v\rangle| = |\langle u; A^- v\rangle| \le \|A^- v\|\,\|u\| \tag{20.23}$$

which shows that $v \in \operatorname{dom}(A_R^+)^*$, cf. (20.19) in Remark 20.4c above.

For the opposite inclusion we consider a fixed $v \in \operatorname{dom}(A_R^+)^*$, so $v \in L^2(X; S^-)$ and $A^- v \in L^2(X; S^+)$; hence it follows from Corollary 13.7 that $v|_Y$ is a well-defined element of $H^{-\frac{1}{2}}(X; S^-)$. We want to show that v belongs to $\operatorname{dom} A_Q^-$.

It suffices to show that $(Id - \mathcal{I}_{R^*})G^* v|_Y$ vanishes as a distribution. So for any $f \in C^\infty(Y; S^+|_Y)$ we determine

$$\begin{aligned} \langle (Id - \mathcal{I}_{R^*})G^*(v|_Y); f\rangle &= \langle G^*(v|_Y); (Id - \mathcal{I}_{R^*})f\rangle = \langle G^*(v|_Y); u|_Y\rangle \\ \text{(20.24)} \qquad & \quad \text{for any } u \in C^\infty(X; S^+) \text{ with } u|_Y = (Id - \mathcal{I}_{R^*})f \\ &= \langle G^*(v|_Y); (\varphi u)|_Y\rangle = \langle A^- v; \varphi u\rangle - \langle v; A^+(\varphi u)\rangle \,. \end{aligned}$$

Here φ is any arbitrarily smooth function on X with $\varphi \equiv 1$ in an open neighbourhood of Y. By a suitable choice of φ, we can make the L^2-norm of φ arbitrarily small, see Figure 20.1. This yields the vanishing of

$$|\langle A^- v; \varphi u\rangle| \le \|A^- v\| \cdot \|\varphi u\| \,. \tag{20.25}$$

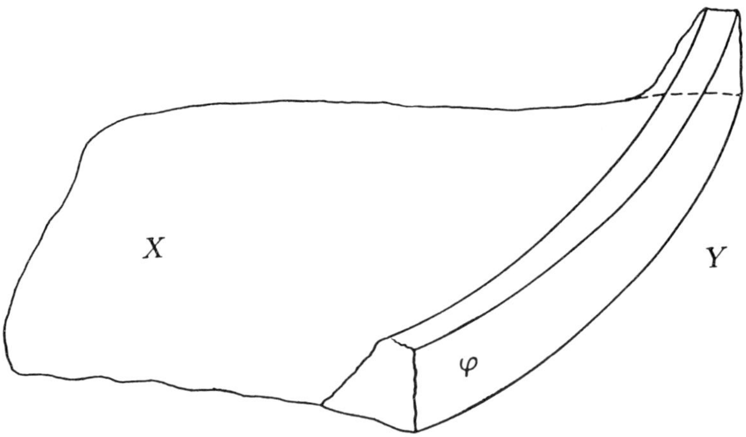

Fig. 20.1 A suitable choice of φ

Since the Dirac operator, as a differential operator of positive order, is not bounded in L^2, we cannot apply the same argument to prove that also the second term in the final difference of (20.24) can be made arbitrarily small. Instead we exploit that $v \in \operatorname{dom}(A_R^+)^*$ which yields

$$|\langle v; A^+(\varphi u)\rangle| \le C \|\varphi u\| \,. \tag{20.26}$$

From (20.24), (20.25), and (20.26) we find $Q(v|_Y) = 0$, as desired.

It remains to prove that Q satisfies the conditions (1) and (2) of Definition 18.1. First (1): for any real s, the Green forms G and $G^* = G^{-1}$ extend to invertible bounded operators between the related Sobolev spaces $H^s(Y; S^+|_Y)$ and $H^s(Y; S^-|_Y)$. Since $\mathcal{I}_{R^*}$ and $Id - \mathcal{I}_{R^*}$ are projections they have closed range, and hence also the extension of Q over $H^s(Y; S^-|_Y)$ has closed range.

Now we must check the ellipticity condition (2), that is, we have to check the boundary condition Q on the symbol level. Therefore we have to show that the principal symbol of Q

$$\sigma_0(Q) = \sigma_0\left(G(Id - \mathcal{I}_{R^*})G^{-1}\right) = G\,\sigma_0(Id - \mathcal{I}_{R^*})\,G^{-1} \tag{20.27}$$

provides an isomorphism

$$(S_y^-)_\eta^> \quad \xrightarrow{\cong} \quad \operatorname{range}\sigma_0(Q)(y,\eta) \tag{20.28}$$

for any $y \in Y$ and $\eta \in T^*Y_y$ with $\eta \neq 0$. Here $S^{-,>}$ denotes the vector bundle over $T^*Y \setminus 0$ generated in (y,η) by the eigenvectors of the principal symbol of the *tangential part* $-GB^*G^{-1}$ of

(20.29)

$$(A^+)^* = A^- = (-\partial_t + B^*)G^* = -G^*(\partial_t - GB^*G^*) \qquad \text{close to } Y,$$

namely

$$\sigma_1(-GB^*G^{-1})(y,\eta) = -G(y)\sigma_1(B^*)(y,\eta)G^{-1}(y) : S_y^- \to S_y^-,$$

where we only take the eigenvectors which have eigenvalues with positive real part. (Since $A^\pm$ are elliptic, there are no purely imaginary eigenvalues and especially no vanishing eigenvalues neither of $\sigma_1(B)(y,\eta)$, nor of $\sigma_1(-GB^*G^{-1})(y,\eta)$ for $\eta \neq 0$). Clearly the vector space $S_{y,\eta}^{-,>}$ is generated by vectors of the form $G(y)e$, where $e \in S_y^+$ is an eigenvector of $\sigma_1(B)(y,\eta)$ with negative real part eigenvalue.

Recall from Corollary 12.6 that the closures of the infinite-dimensional spaces of Cauchy data $H_+(A^+)$ and $G^{-1}(H_+(A^-))$ are orthogonal and complementary in $L^2(Y; S^+|_Y)$. Similarly, but more easily, we now obtain that the finite-dimensional vector spaces $S_{y,\eta}^{+,>}$ and $G^{-1}(S_{y,\eta}^{-,>})$ are orthogonal and complementary in the space S_y^+. Moreover, $\operatorname{range}\sigma_0(\mathcal{I}_{R^*})(y,\eta)$ and $\operatorname{range}\sigma_0(Id - \mathcal{I}_{R^*})(y,\eta)$ are orthogonal and complementary in S_y^+ by definition. Therefore, if $\mathcal{I}_{R^*}$ is an elliptic boundary condition for A^+, its symbol provides an isomorphism from $S_{y,\eta}^{+,>}$ onto $\operatorname{range}\sigma_0(\mathcal{I}_{R^*})(y,\eta)$ and hence its complementary map

(20.30)

$$G^{-1}(y)(S_{y,\eta}^{-,>}) \qquad \xrightarrow{\sigma_0(Id-\mathcal{I}_{R^*})(y,\eta)} \qquad \operatorname{range}\sigma_0(Id - \mathcal{I}_{R^*})(y,\eta)$$

must also be an isomorphism and so (20.28). □

We can use the arguments of the preceding proposition to reprove the closedness of the operator $A_R^+ = A_R^{+\,(1)}$, which was obtained in Lemma 20.1. In fact, we have:

Corollary 20.5. *Let $A_R^{+\,(\infty)}$ denote the operator A^+ with the domain*

$$\{u \in C^\infty(X; S^+) \mid R(u|_Y) = 0\}\,. \tag{20.31}$$

Then A_R^+ is the closure of $A_R^{+\,(\infty)}$ as an operator from a subset of $L^2(X; S^+)$ into $L^2(X; S^-)$.

Proof. A slight modification of the arguments, presented in the proof of Proposition 20.3, gives the following list of inclusions to be proved:

(i) $$\operatorname{dom} A_Q^- \subset \operatorname{dom}(A_R^{+\,(\infty)})^*$$

(ii) $$\operatorname{dom}(A_R^{+\,(\infty)})^* \subset \operatorname{dom} A_Q^-$$

(iii) $$\operatorname{dom} A_R^+ \subset \operatorname{dom}(A_Q^-)^*$$

(iv) $$\operatorname{dom}(A_Q^-)^* \subset \operatorname{dom} A_R^+\,,$$

where $Q = G(Id - \mathcal{I}_{R^*})G^*$ and $\mathcal{I}_{R^*}$ an orthogonal projection with $A_R^+ = A_{\mathcal{I}_{R^*}}^+$.

Let T denote the operator $A_R^{+\,(\infty)}$. Then $\operatorname{dom} T \subset \operatorname{dom} A_R^+$ which implies (i), namely

$$\operatorname{dom} T^* \supset \operatorname{dom}(A_R^+)^* = \operatorname{dom} A_Q^-\,. \tag{20.32}$$

In order to prove (ii) we show for any $v \in \operatorname{dom} T^*$ that $v|_Y$ is well-defined in $H^{-\frac{1}{2}}(Y; S^-|_Y)$ and $Q(v|_Y) = 0$. We can repeat all the arguments of the second part of the proof of Proposition 20.3, where only C^∞-test functions were involved. (i) and (ii) show that

$$T^* = (A_{R,1}^+)^* = A_Q^-\,. \tag{20.33}$$

In order to prove (iii) we fix a $u \in \operatorname{dom} A_R^+$ and apply once again Green's formula which yields the estimate

$$|\langle A^- v; u\rangle| \le C_u \|v\| \tag{20.34}$$

for all $v \in \operatorname{dom} A_Q^-$, since $G^*(v|_Y)$ and $u|_Y$ are orthogonal by definition.

To prove (iv) we show that $u|_Y$ is well-defined in $H^{-\frac{1}{2}}(Y; S^+|_Y)$, and $\mathcal{I}_{R^*}(u|_Y) = 0$ in a distributional sense for any $u \in \mathrm{dom}(A_Q^-)^*$. We make the following modifications of the earlier proof:

(20.35)

$$\begin{aligned}\langle \mathcal{I}_{R^*}(u|_Y); f\rangle &= \langle u|_Y; \mathcal{I}_{R^*} f\rangle = \langle Gu|_Y; G\mathcal{I}_{R^*} f\rangle = \langle Gu|_Y; w|_Y\rangle \\ &\qquad \text{with } w \in C^\infty(X; S^-) \text{ and } w|_Y = G\mathcal{I}_{R^*} f \\ &= \langle Gu|_Y; (\varphi w)|_Y\rangle = \langle u; A^-\varphi w\rangle - \langle A^+u; \varphi w\rangle,\end{aligned}$$

which shows the vanishing of $\mathcal{I}_{R^*}(u|_Y)$ for suitable C^∞-functions φ with $\varphi \equiv 1$ close to Y and $\|\varphi\|_{L^2}$ arbitrarily small.

(iii) and (iv) yield $(A_Q^-)^* = A_R^+$, so $A_R^+ = (A_R^{+(\infty)})^{**}$ which in the Hilbert space means that A_R^+ is the closure of $A_R^{+(\infty)}$. □

The space of closed extensions of a Dirac operator A^+ (or of suitable restrictions) does not only consist of the space

$$\mathcal{R} := \{A_R^+ \mid R \text{ elliptic boundary problem}\}$$

of all realizations of elliptic boundary problems, but contains also two other prominent closed operators, $A^+_{\min}$ and $A^+_{\max}$ (see also Figure 20.2):

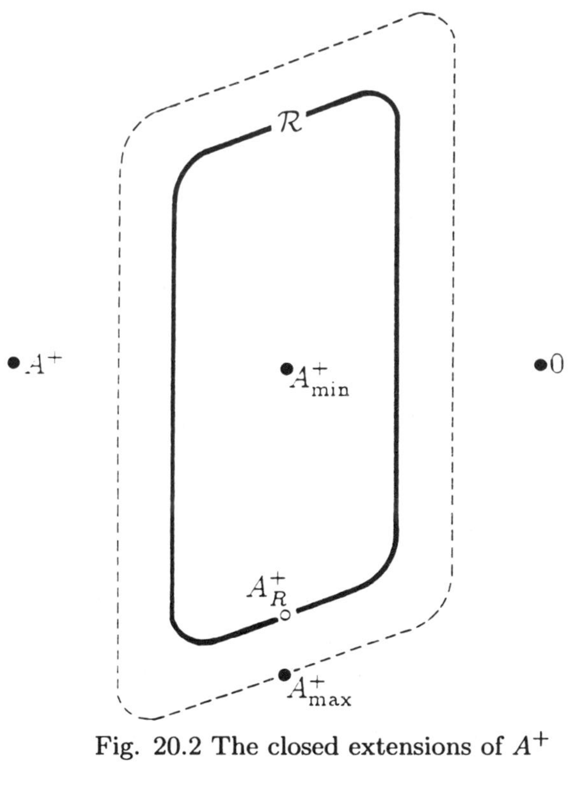

Fig. 20.2 The closed extensions of A^+

Definition 20.6. Let $A^+_{\min}$ and $A^+_{\max}$ denote the unbounded operators in L^2 which act like A^+ and are determined by

$$\operatorname{dom} A^+_{\min} := \{v \in L^2(X;S^+) \mid A^+v \in L^2(X;S^-) \text{ and } \gamma_0 v = 0\}\,,$$
$$\operatorname{dom} A^+_{\max} := \{v \in L^2(X;S^+) \mid A^+v \in L^2(X;S^-)\}.$$

Similarly we define $A^-_{\min}$ and $A^-_{\max}$.

Proposition 20.7. *The operators $A^+_{\min}$ and $A^+_{\max}$ are closed operators from subspaces of $L^2(X;S^+)$ into $L^2(X;S^-)$ and in fact*

$$\operatorname{dom} A^+_{\min} = \{v \in H^1(X;S^+) \mid \gamma_0 v = 0\}\,.$$

Moreover

$$(A^+_{\min})^* = A^-_{\max} \quad \text{and} \quad (A^+_{\max})^* = A^-_{\min}\,.$$

Note. We have

$$A^+_{\min} \subset A^+_R \subset A^+_{\max} \qquad \text{for all } A^+_R \in \mathcal{R}$$

in the space of closed unbounded operators from $L^2(X;S^+)$ into $L^2(X;S^-)$ (see Figure 20.2).

Proof. To prove that $A^+_{\min}$ is closed, let $v_n \to v$ in $L^2(X;S^+)$, $A^+v_n \to w$ in $L^2(X;S^-)$, and $\gamma_0 v_n = 0$. Then

$$v_n = (\operatorname{Id} - K_+\gamma_0)v_n = (r^+\tilde{A}^{-1}e^+)A^+v_n\,;$$

hence, by Corollary 19.7, $v_n \in H^1$ and

$$\|v_n\|_{H^1} = \|(\operatorname{Id} - K_+\gamma_0)v_n\|_{H^1} \le C\|A^+v_n\|_{L^2}\,. \tag{20.36}$$

As in the proof of Lemma 20.1, we get that v_n is convergent in H^1 to a limit v' with $\gamma_0 v' = 0$, but then it is convergent in L^2 to the same limit; hence $v = v'$ and $A^+v = \lim A^+v_n = w$.

To prove that $A^+_{\max}$ is closed, we begin as above, but do not assume $\gamma_0 v_n = 0$. Nevertheless, when we replace v_n by $v_n' := (\operatorname{Id} - K_+\gamma_0)v_n$, (20.36) remains valid and provides once again a limit $v' \in H^1$ with

$$A^+v' = \lim_{n\to\infty} A^+(v_n - K_+(\gamma_0 v_n)) = \lim A^+v_n = w\,.$$

Moreover, $K_+(\gamma_0 v_n) = v_n - v'_n \to v - v'$ in L^2; hence $v - v' \in \ker_+(A^+, 0)$ and $A^+ v = A^+(v' + (v - v')) = w$.

Similarly we prove $(A^+_{\min})^* = A^-_{\max}$: Let $v_0, w_0 \in L^2$ and

$$\langle A^+ v; w_0\rangle = \langle v; v_0\rangle$$

for any $v \in \operatorname{dom} A^+_{\min}$. We have to show $v_0 = A^- w_0$. Choose a sequence $w_n \to w_0$ in L^2, consisting of smooth sections, and get

$$\langle A^+ v; w_n\rangle = \langle v; A^- w_n\rangle$$

by Green's formula. $C_0^\infty \subset \operatorname{dom} A^+_{\min}$, hence $\operatorname{dom} A^+_{\min}$ is dense in L^2. Therefore the sequence $v_n := A^- w_n$ converges in L^2. Let $v' := \lim v_n$; hence $w'_n := r^+ \widetilde{A^-}^{-1} e^+ v_n = (\mathrm{Id} - K_+\gamma_0) w_n$ is convergent in H^1 with $w' := \lim w_n \in H^1$ and $A^- w' = \lim A^- w'_n = \lim A^- w_n$, and we obtain

$$\langle A^+ v; w_0\rangle = \lim\langle v; A^- w_n\rangle = \langle v; A^- w'\rangle\,,$$

for any $v \in \operatorname{dom} A^+_{\min}$, hence $v_0 = A^- w'$ in L^2. Since $A^+_{\min}$ is closed, we get $A^+_{\min} = (A^+_{\min})^{**} = (A^-_{\max})^*$, and there is nothing left to prove. □

Now we derive a formula for the index of the realization of an elliptic boundary value problem $R : C^\infty(Y; S^+|_Y) \to C^\infty(Y; V)$ for a Dirac operator $A^+ : C^\infty(X; S^+) \to C^\infty(X; S^-)$. Recall that the boundary integral

$$R\mathcal{P}_+(A^+) : H_+(A^+) \to \operatorname{range}(R)$$

is a Fredholm operator by Corollary 18.15. The following theorem provides a new proof of that fact and extends it to the realization.

Theorem 20.8. *The operator*

$$A^+_R = A^+_{\mathcal{I}_{R^*}} : \operatorname{dom} A^+_R \to L^2(X; S^-)$$

is a Fredholm operator with

$$\begin{aligned}\operatorname{index} A^+_R &= \dim\ker A^+_R - \dim\ker(A^+_R)^* \\ &= \operatorname{index}\{R\mathcal{P}_+(A^+) : H_+(A^+) \to \operatorname{range}(R)\} \\ &=: \mathbf{i}(R, \mathcal{P}_+(A^+)).\end{aligned}$$

Proof. Let T denote the operator $R\mathcal{P}_+(A^+)$ acting from $H_+(A^+)$ to range(R). We have already shown that

(1) range(A_R^+) is closed (by Proposition 19.11);
(2) $\ker A_R^+ \cong \ker T$ (by Corollary 19.2);
(3) and $(A_R^+)^* = A^-_{G(\mathrm{Id}-\mathcal{I}_{R^*})G^{-1}}$ (by Proposition 20.3);

hence we only have to identify $\operatorname{coker} A_R$ with $\operatorname{coker} T$:
(20.37)

$$\begin{aligned}\operatorname{coker} A_R &= \ker A^-_{G(\mathrm{Id}-\mathcal{I}_{R^*})G^{-1}} = \ker G(\mathrm{Id}-\mathcal{I}_{R^*})G^{-1}\mathcal{P}_+(A^-)\\ &\cong \ker(\mathrm{Id}-\mathcal{I}_{R^*})(\mathrm{Id}-\mathcal{P}_+(A^+))\\ &= \{u \mid \mathcal{I}_{R^*}u = u \text{ and } \mathcal{P}_+(A^+)u = 0\}\\ &= \ker\{\mathcal{P}_+(A^+) : \mathrm{range}(R^*) \to H_+(A^+)\}\\ &= \ker\{\mathcal{P}_+(A^+)R^* : \mathrm{range}(R) \to H_+(A^+)\}\,.\end{aligned}$$

For the last equation we once more exploit the fact that the boundary condition R (and hence R^*) has closed range by definition. That implies that $R^* : \mathrm{range}(R) \longrightarrow \mathrm{range}(R^*)$ is an isomorphism. Clearly the last space in (20.37) is isomorphic to $\operatorname{coker} T$. □

Note. In the following we shall return to the notation $\mathcal{P}_+$ for $\mathcal{P}_+(A^+)$.

From Theorem 20.8 we get that the index of the Calderón boundary problem $A^+_{\mathcal{P}_+}$ vanishes, since the corresponding boundary integral is just the identity. This can also be seen from Corollary 12.6 and Proposition 20.3 which give $(A^+_{\mathcal{P}_+})^* = A^-_{\mathcal{P}_-}$. By definition of the Calderón projector, we have $\ker A^+_{\mathcal{P}_+} = \ker A^-_{\mathcal{P}_-} = \{0\}$. For the Atiyah-Patodi-Singer boundary condition (= the spectral projection $\mathrm{P}_{\geq}(A^+)$ onto the linear span of the eigensections of the tangential part of A^+ corresponding to non-negative eigenvalues), the situation is quite different. First, we have

$$(A^+_{\mathrm{P}_{\geq}(A^+)})^* = A^-_{G(\mathrm{Id}-\mathrm{P}_{\geq}(A^+))G^*} = A^-_{\mathrm{P}_{<}(A^-)}$$

with an inevitable asymmetry in the boundary condition for a nontrivial kernel of the tangential component B. Moreover, the kernel of the realization $A^+_{\mathrm{P}_{\geq}}$ does not necessarily vanish, and the indices of $A^+_{\mathcal{P}_+}$ and $A^+_{\mathrm{P}_{\geq}}$ coincide if and only if $\mathcal{P}_+$ and $\mathrm{P}_{\geq}$ belong to the same connected component of the Grassmannian $\mathrm{Gr}_{\mathcal{P}_+}$. We shall investigate this more closely in Chapters 21, 25, and 26.

Next we explain special features of the index of elliptic boundary problems of local type. Let A be a Dirac operator or a system of Dirac operators which admit a *local elliptic* condition (cf. Remark 18.2c). Then $R : C^\infty(Y; S^+|_Y) \to C^\infty(Y; W)$ is a pseudo-differential operator of order 0 such that

$$r \circ p_+(y,\zeta) : \operatorname{range} p_+(y,\zeta) \longrightarrow W_y$$

is an isomorphism. In particular the principal symbol of R is surjective.

Lemma 20.9. *Let $S : C^\infty(Y; V) \to C^\infty(Y; W)$ be a pseudo-differential operator of order 0 such that the principal symbol $s(y,\zeta) : V_y \to W_y$ of S is surjective for any (y,ζ). Then S has finite-dimensional cokernel and in particular the range of S is closed.*

Proof. The operator $SS^* : C^\infty(Y; W) \to C^\infty(Y; W)$ is an elliptic operator since its principal symbol ss^* is an isomorphism. Therefore $\ker SS^* = \ker S^*$ is finite-dimensional. □

We reformulate Corollary 19.10:

Proposition 20.10. *For any real $s > \frac{1}{2}$, the operator*

$$\begin{pmatrix} A \\ R \end{pmatrix} : H^s(X; S^+) \to \begin{matrix} H^{s-1}(X; S^-) \\ \oplus \\ H^{s-\frac{1}{2}}(Y; W) \end{matrix}$$

has a closed range.

It remains to compute the cokernel of $\begin{pmatrix} A \\ R \end{pmatrix}$. We have the following elementary lemma which sharpens Corollary 18.15, but is valid only for local elliptic boundary conditions:

Lemma 20.11. *For local elliptic boundary conditions the operator*

$$R\mathcal{P}_+ : H_+(A) \to C^\infty(Y; W)$$

is a Fredholm operator.

Proof. Once again we may consider everything on the level of Sobolev spaces. It is enough to show that $\dim\ker R\mathcal{P}_+ < \infty$ and $\dim\operatorname{coker} R\mathcal{P}_+ < \infty$. We have already proved the first estimate. Moreover

$$\begin{aligned} \operatorname{coker} R\mathcal{P}_+ &= \ker(R\mathcal{P}_+)^* = \ker \mathcal{P}_+ R^* \\ &= \ker(R\mathcal{P}_+)\mathcal{P}_+ R^* = \ker R\mathcal{P}_+ R^* \,, \end{aligned}$$

and the operator $R\mathcal{P}_+ R^*$ is elliptic. □

Theorem 20.12. *For local elliptic boundary conditions one has*

$$\operatorname{index}\begin{pmatrix} A \\ R \end{pmatrix} = \operatorname{index}\{R\mathcal{P}_+(A) : H_+(A) \to C^\infty(Y;W)\}\,.$$

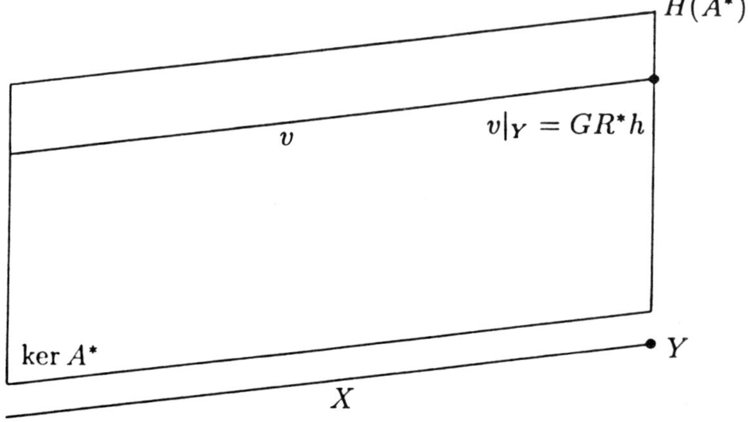

Fig. 20.3 The parametrization of coker(A,R) by GR^*h

Proof. With regard to the results obtained earlier, it only remains to identify the cokernel of (A, R) with $\operatorname{coker} R\mathcal{P}_+(A) = \ker \mathcal{P}_+(A)R^*$ (see Figure 20.3):

$$\begin{aligned}\operatorname{coker}(A,R) = \{(v,h) \in C^\infty(X;S^-) \oplus C^\infty(Y;W) \mid \langle As; v\rangle \\ + \int_Y \langle R(s|_Y); h\rangle\, dy = 0 \text{ for any } s \in C^\infty(X;S^+)\},\end{aligned}$$

and by Green's formula

$$\begin{aligned} 0 &= \langle As; v\rangle + \int_Y \langle R(s|_Y); v|_Y\rangle\, dy \\ &= \langle s; A^*v\rangle + \int_Y \langle s|_Y; R^*h - G^{-1}(v|_Y)\rangle dy\,. \end{aligned}$$

Once again this implies $A^*v = 0$ and

$$\begin{aligned} \operatorname{coker}(A,R) &\cong \{(q,h) \in C^\infty(Y;S^-|_Y) \oplus C^\infty(Y;W) \\ &\qquad \text{with } P_+(A^*)q = q\tau \text{ and } R^*h = q\} \\ &\cong \{(r,h) \in C^\infty(Y;S^+|_Y) \oplus C^\infty(Y;W) \\ &\qquad \text{with } \mathcal{P}_+ r = 0 \text{ and } r = R^*h\} \\ &= \operatorname{coker}(\mathcal{P}_+(A))\,. \end{aligned}$$

□

For later use we give the following reformulation of Theorem 20.12 which determines the relation between index A_R and index(A, R) (notice that we suppress the subscripts $_+$ of the Calderón projector and the Cauchy data space):

Theorem 20.13. *Under the preceding conditions*

$$\begin{aligned} \operatorname{index}(A,R) &= \operatorname{index}\{R\mathcal{P}(A) : H(A) \to C^\infty(Y;W)\} \\ &= \operatorname{index} A_R - \dim(C^\infty(Y;W)/\operatorname{range}(R)) . \end{aligned} \tag{20.38}$$

Note. Decompose the kernel of the operator

$$\mathcal{P}(A)R^* : C^\infty(Y;V) \xrightarrow{R^*} C^\infty(Y;E|_Y) \xrightarrow{\mathcal{P}(A)} H(A)$$

into two finite-dimensional subspaces

$$\ker \mathcal{P}(A)R^* = W_1 \oplus W_2$$

with $W_1 := \ker R^*$ and W_2 the orthogonal complement of W_1 in $\ker \mathcal{P}(A)R^*$ (see Figure 20.4). Since $\ker(A,R) = \ker A_R$ and $\operatorname{coker}(A,R) \cong \operatorname{coker}(R\mathcal{P}) = W_1 \oplus W_2$ are already established, Theorem 20.13 shows that

$$\operatorname{coker} A_R \cong W_2 . \tag{20.39}$$

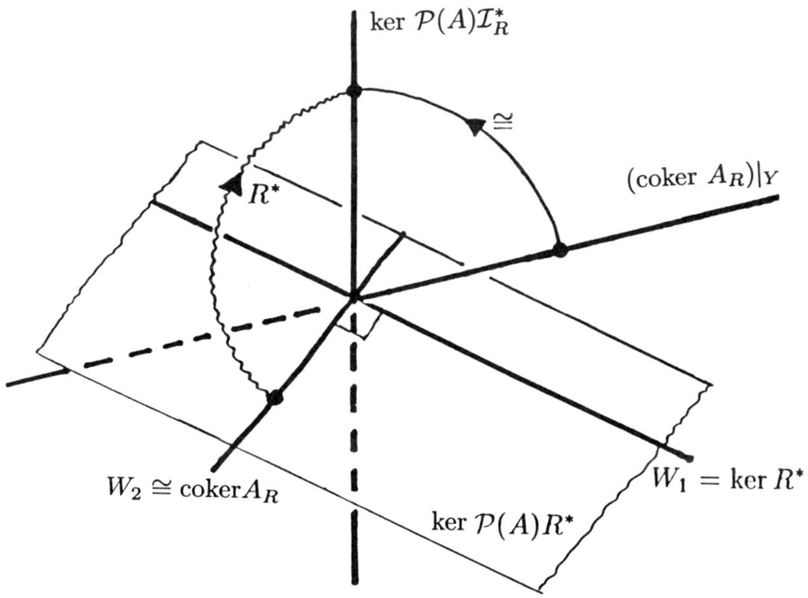

Fig. 20.4 The decomposition of ker$\mathcal{P}(A)R^*$ and the isomorphism (coker A_R)$|_Y \cong W_2$

Corollary 20.14. *For local elliptic boundary conditions we have* $\operatorname{index} A_R = \operatorname{index}(A, R)$, *if and only if* R *is a surjective operator.*

Remark 20.15. We have seen in Chapter 15 above that a compact perturbation of R can change the connected component of the corresponding projection $\mathcal{I}_{R^*}$ and we shall see in Section 22F below that in general the index of elliptic boundary problems is not a topological invariant (see Theorem 15.12 and Remark 22.25). However, when we restrict ourselves to the class of local elliptic conditions, the index becomes a topological invariant and we have a corresponding Atiyah-Bott index formula (see Atiyah & Bott [1964] and also Hörmander [1985; Section 20]). We do not discuss the Atiyah-Bott index formula here and refer the reader to the works of Atiyah, Bott, and Hörmander mentioned before.

21. Exchanges on the Boundary: Agranovič-Dynin Type Formulas and the Cobordism Theorem for Dirac Operators

We prove two variants of the Agranovič-Dynin formula and the cobordism theorem.

In the preceding chapter we obtained the following index formula for the realization A_R of an elliptic boundary condition $R = R^2$ (cf. Theorem 20.8):

$$\text{index}\, A_R = \text{index}\,\{R\mathcal{P}(A) : H(A) \to \text{range}\, R\}\ , \tag{21.1}$$

where $\mathcal{P}(A)$ denotes the Calderón projector onto the Cauchy data space $H(A)$ corresponding to the Dirac operator A.

The obvious question is whether this integer is a topological invariant. Contrary to the case of *local* elliptic boundary problems, the answer is negative. The index is not a homotopy invariant of the principal symbols of the operators A and R. It is not difficult to deliver a specific example.

Example 21.1. Let $X := D^2$ and take the Cauchy-Riemann operator

$$\bar{\partial} = \frac{\partial}{\partial x} + i\frac{\partial}{\partial y} = e^{i\varphi}\left(\frac{\partial}{\partial r} + \frac{i}{r}\frac{\partial}{\partial \varphi}\right)\ ,$$

where we take the standard polar coordinates (r, φ) close to the boundary. In this special case the choice of the orientation of the transversal coordinate makes no real difference, but only a sign change for the whole expression, since r is in both components here. The Calderón projector $\mathcal{P}(\bar{\partial})$ is the orthogonal projection of $C^\infty(S^1; \mathbf{C})$ (orthogonal with respect to $L^2(S^1; \mathbf{C})$) onto the space $\mathcal{L}_{\mathbf{C}}\{e_k(\varphi) = e^{ik\varphi}\}_{k\geq 0}$ of the Cauchy data of all holomorphic functions on the disk D^2. According to the results of Chapter 12, the projection $\mathcal{P}(\bar{\partial})$ is a pseudo-differential operator of order 0 with the principal symbol

$$p_+(x;\xi) = \begin{cases} 1 & \text{for } \xi > 0 \\ 0 & \text{for } \xi < 0\,. \end{cases}$$

Actually, in this particular situation the Calderón projector $\mathcal{P}(\bar\partial)$ coincides with the spectral projection $\mathrm{P}(i\partial/\,r\partial\varphi)$.

Now let us consider projections P_k given by the formula

$$P_k := e^{ik\varphi}\mathcal{P}(\bar\partial)e^{-ik\varphi} \qquad \text{for } k \in \mathbf{Z}.$$

It follows from Theorem 20.8, see equation (21.1), that:

$$\text{(21.2)} \qquad \operatorname{index} \bar\partial_{P_k} = \operatorname{index} P_k\mathcal{P}(\bar\partial) = \operatorname{index} P_k P_0 = k\,,$$

since

$$\ker P_k P_0 = \mathcal{L}_{\mathbf{C}}\{e^{im\varphi} \mid 0 \le m < k\},\ \operatorname{range} P_k = \mathcal{L}_{\mathbf{C}}\{e^{im\varphi} \mid m \ge k\},$$
$$\text{and } \operatorname{range} P_k P_0 = \mathcal{L}_{\mathbf{C}}\{e^{im\varphi} \mid m \ge \max\{0,k\}\}.$$

Therefore

	$\dim\ker P_kP_0$:	$\dim\operatorname{coker} P_kP_0$:
$k \ge 0$:	k	0
$k < 0$:	0	$-k$

which proves (21.2). Nevertheless the principal symbol for any P_k is p_+ and hence is unchanged.

As an immediate corollary of Theorem 20.13, let us mention a simple variant of the Agranovič-Dynin formula for local elliptic boundary conditions:

Theorem 21.2. *Let $A : C^\infty(X;S) \to C^\infty(X;S')$ be (possibly a system of) Dirac operators over a compact manifold X with boundary Y, and for $j = 1,2$ let $R_j : C^\infty(Y;S|_Y) \to C^\infty(Y;V_j)$ be a local elliptic condition. Then*

$$\text{(21.3)} \qquad \operatorname{index}(A,R_2) - \operatorname{index}(A,R_1) = \operatorname{index} R_2\mathcal{P}(A)R_1\,.$$

Proof. We apply Theorem 20.12:

$$\begin{aligned}
\operatorname{index}(A,R_2) - \operatorname{index}(A,R_1)&\\
&= \operatorname{index}\{R_2\mathcal{P}(A) : H(A) \to C^\infty(Y;V_2)\}\\
&\qquad - \operatorname{index}\{R_1\mathcal{P}(A) : H(A) \to C^\infty(Y;V_1)\}\\
&= \operatorname{index}\{R_2\mathcal{P}(A) : H(A) \to C^\infty(Y;V_2)\}\\
&\qquad + \operatorname{index}\{\mathcal{P}(A)R_1^* : C^\infty(Y;V_1) \to H(A)\}\\
&= \operatorname{index} R_2\mathcal{P}(A)R_1^*\,.
\end{aligned}$$

□

With regard to Theorem 20.13 and Corollary 20.14, the preceding theorem has the following direct consequence which is crucial for the proof of the cobordism theorem for Dirac operators.

Proposition 21.3. *To the previous assumptions, add that, for $j = 1, 2$, the local elliptic boundary operator R_j is an orthogonal projection onto a subbundle V_j of $S|_Y$, i.e. $R_j = R_j^2 = R_j^*$. Then*

$$\begin{aligned} &\operatorname{index} A_{R_2} - \operatorname{index} A_{R_1} \\ &\qquad = \operatorname{index}(A, R_2) - \operatorname{index}(A, R_1) \\ &\qquad = \operatorname{index}\{R_2 \mathcal{P}(A) R_1 : C^\infty(Y; V_1) \to C^\infty(Y; V_2)\}\,. \end{aligned} \tag{21.4}$$

Note. Elliptic boundary conditions of the type assumed in the preceding theorem exist for Dirac operators only in the odd-dimensional case, but for systems in the even-dimensional case as well.

Proof. Theorem 20.13 and Corollary 20.14 tell us that

$$\operatorname{index} A_{R_j} = \operatorname{index}(A, R_j) = \mathbf{i}(R_j, \mathcal{P}(A))\,.$$

Therefore we have

$$\begin{aligned} &\operatorname{index} A_{R_2} - \operatorname{index} A_{R_1} \\ &\qquad = \mathbf{i}(R_2, \mathcal{P}(A)) - \mathbf{i}(R_1, \mathcal{P}(A)) \\ &\qquad = \mathbf{i}(R_2, \mathcal{P}(A)) + \mathbf{i}(\mathcal{P}(A), R_1) \\ &\qquad = \operatorname{index}\{R_2 \mathcal{P}(A) R_1 : C^\infty(Y; V_1) \to C^\infty(Y; V_2)\}\,. \end{aligned} \tag{21.5}$$

□

Treating elliptic boundary problems as projections yields a corresponding variant of the Agranovič-Dynin formula for global boundary conditions of generalized Atiyah-Patodi-Singer type, i.e. belonging to the Grassmannian Gr_{p_+} (whereas in the preceding theorem the projections R_1, R_2 do not in general belong to Gr_{p_+}):

Proposition 21.4. *Let A be a Dirac operator over a compact manifold X with boundary Y, and let R_1, $R_2 \in \mathrm{Gr}_{p_+}$, where p_+ denotes the principal symbol of the Calderón projector $\mathcal{P}(A)$ (= the principal symbol of the spectral projection $\mathrm{P}_\geq$ of the tangential part B of A). Then*

$$\operatorname{index} A_{R_2} - \operatorname{index} A_{R_1} = \mathbf{i}(R_2, R_1)\,. \tag{21.6}$$

Proof. By Theorem 20.8 we have

$$\operatorname{index} A_{R_j} = \mathbf{i}(R_j, \mathcal{P}(A))\,. \tag{21.7}$$

Then formula (21.6) is a consequence of Proposition 15.15. □

The main purpose of this chapter, however, is to give a proof of the famous *cobordism theorem* for Dirac operators based on Proposition 21.3 (applied to an odd-dimensional case).

Let Y be an odd-dimensional compact Riemannian manifold with boundary Z. We assume that the metric of Y is a product near the boundary. Let S be a bundle of $C\ell(Y)$-modules over Y with compatible connection. Let

$$B : C^\infty(Y;S) \to C^\infty(Y,S)$$

denote the corresponding Dirac operator. We have a decomposition of B close to the boundary Z of the form

$$B = \Gamma(\partial_u + A)\,, \tag{21.8}$$

where Γ is a unitary bundle automorphism with $\Gamma^2 = -Id_{S|_Z}$, and hence Γ defines a decomposition of $S|_Z$ into the direct sum $S^+ \oplus S^-$ of the subbundles of the $\pm i$-eigenvalues of $\{\Gamma_z\}_{z\in Z}$.

Theorem 21.5. *(a) The operator B takes the following form close to Z with respect to the decomposition $S|_Z \cong S^+ \oplus S^-$:*

$$B = \begin{pmatrix} i & 0 \\ 0 & -i \end{pmatrix} \left(\partial_u + \begin{pmatrix} 0 & A_- \\ A_+ & 0 \end{pmatrix} \right). \tag{21.9}$$

(b) Let $P_\pm : S \to S^\pm$ denote the orthogonal projections of S onto $S^\pm$. Then the operators $\mathcal{B}_\pm$ defined by

$$\begin{cases} \operatorname{dom} \mathcal{B}_\pm := \{ s \in H^1(Y;S) \mid P_\pm(s|_Y) = 0 \} \\ \mathcal{B}_\pm := B \end{cases} \tag{21.10}$$

are well-defined local elliptic boundary problems, and we have the following equations:

(I) $\mathcal{B}_+^* = \mathcal{B}_-$,

(II) $\operatorname{index} \mathcal{B}_\mp - \operatorname{index} \mathcal{B}_\pm = \operatorname{index} A_\pm$,

(III) $\operatorname{index} \mathcal{B}_\pm = 0$.

Note. Equations (II) and (III) contain the famous cobordism theorem for the index of operators of Dirac type. The cobordism theorem provided the main step in the original proof of the index theorem (see Palais [1965; Chapter XVII]) and has various generalizations.

Corollary 21.6. (Cobordism theorem). *Let* $A_+ : C^\infty(Z;S^+) \to C^\infty(Z;S^-)$ *denote the restriction of a Dirac operator on a closed even-dimensional manifold* Z *to the spinors of positive chirality. Then the index of* A_+ *vanishes if the couple* (Z,S^+) *is a "boundary". This means that there exists a manifold* X *with boundary* Z *and a bundle* S *of Clifford modules over* X *such that* S *restricted to* Z *is equal to* $S^+ \oplus S^-$.

Proof of Theorem 21.5. *(a)* The special form of the decomposition of B follows from Chapter 7.
(b) As in Theorem 7.3, we have $A_- = A_+^*$. This shows that B is formally self-adjoint. That the unbounded operators $\mathcal{B}_+^*$ and $\mathcal{B}_-$ are mutually adjoint follows immediately from the definition, see Remark 20.4c.

Equation (III) is an elementary consequence of Green's formula. We show that

$$\ker(\mathcal{B}_+) = \{0\} = \ker(\mathcal{B}_-)\,. \tag{21.11}$$

Let s be an element of the kernel of $\mathcal{B}_+$, i.e. $Bs = 0$ and $s_+ = 0$, where we write $s|_Z = (s_+, s_-)$, and $s_\pm$ denotes the projection $P_\pm s$. Then

$$\begin{aligned} 0 = \langle Bs; s\rangle &= \langle s; Bs\rangle - \int_Z \langle \Gamma(y)s(y); s(y)\rangle\, dvol(y) \\ &= -\int_Z \langle \Gamma(y)s(y); s(y)\rangle\, dvol(y) = i\int_Z \langle s_-(y); s_-(y)\rangle\, dvol(y) \\ &= \|s_-\|_{L^2(S|_Z)}\,. \end{aligned}$$

It follows that

$$\ker(\mathcal{B}_+) = \{s \mid Bs = 0 \text{ and } s|_Z = (s_+, s_-) = 0\} = \{0\} \tag{21.12}$$

and then $\operatorname{index} \mathcal{B}_\pm = 0$ because of (I).

Now we prove (II). We recall the Agranovič-Dynin formula (Proposition 21.3):

$$\operatorname{index} B_{R_1} - \operatorname{index} B_{R_2} = \operatorname{index} R_1 \mathcal{P}(B) R_2\,,$$

where the boundary conditions R_i are given by projections. Hence we have

$$\operatorname{index} \mathcal{B}_- - \operatorname{index} \mathcal{B}_+ = \operatorname{index} P_- \mathcal{P}(B) P_+. \tag{21.13}$$

We show that the principal symbol of the 0-order pseudo-differential operator $P_-\mathcal{P}(B)P_+$ is equal (up to a factor $\frac{1}{2}$) to the principal symbol of the first-order differential operator A_+, when restricted to the cotangent sphere bundle $S(Z)$.

Let us denote by a the principal symbol of the (total) tangential operator A and let $a_\pm$ denote the principal symbols of the operators $A_\pm$. They satisfy the following equalities on $S(Z)$:

$$(21.14) \qquad a_-a_+ = Id_{S^+} \quad \text{and} \quad a_+a_- = Id_{S^-}.$$

Hence the eigenvalues of

$$a(z,\zeta) = \begin{pmatrix} 0 & a_- \\ a_+ & 0 \end{pmatrix}$$

are all contained in the set $\{1, -1\}$. We have shown in Proposition 14.2 that the principal symbol p_+ of the Calderón projection $\mathcal{P}(B)$ is equal to the principal symbol of the (positive) spectral projection of the operator A. So $p_+(z,\zeta) : S_z \to S_z$ is the orthogonal projection onto the subspace of S_z spanned by the eigenvectors of $a(z,\zeta)$ corresponding to the positive eigenvalues. Here $z \in Z$ and $\zeta \in T^*(Z)_z$ is of length 1. Then we have the equality

$$p_+(z,\zeta) = \frac{1}{2}\left(Id + a(z,\zeta)\right) = \frac{1}{2}\begin{pmatrix} 1 & a_- \\ a_+ & 1 \end{pmatrix},$$

and hence

$$\begin{aligned} \sigma_0(P_-\mathcal{P}(B)P_+)(z,\zeta) &= P_- \cdot p_+(z,\zeta) \cdot P_+ \\ &= \frac{1}{2}\begin{pmatrix} 0 & 0 \\ 0 & 1 \end{pmatrix}\begin{pmatrix} 1 & a_- \\ a_+ & 1 \end{pmatrix}\begin{pmatrix} 1 & 0 \\ 0 & 0 \end{pmatrix} = \frac{1}{2}a_+ . \end{aligned}$$

□

22. The Index Theorem for Atiyah-Patodi-Singer Problems

We present a proof of the index theorem for Atiyah-Patodi-Singer boundary problems.

22A. Preliminary Remarks

Let $A^+ : C^\infty(X;S^+) \to C^\infty(X;S^-)$ be a Dirac operator on an even-dimensional smooth compact Riemannian manifold X with boundary Y. Actually the odd-dimensional case is not different and we shall make the necessary comments after the proof of the index theorem. We assume that the Riemannian structure on X and the Hermitian structure on S are products in a collar neighbourhood $N \cong [0,1] \times Y$ of the boundary Y in X, that is, the metrics do not depend on the normal coordinate u when restricted to $Y_u := \{u\} \times Y$ so that $S|_{Y_u}$ can be identified with $S|_Y$.

Recall that A^+ then has the following form on N

$$A^+ = G(\partial_u + B)\,, \tag{22.1}$$

where $G : S^+|_Y \to S^-|_Y$ is a unitary bundle isomorphism, namely the Clifford multiplication by the inward normal vector, and $B : C^\infty(Y;S^+|_Y) \to C^\infty(Y;S^+|_Y)$ denotes the corresponding Dirac operator on Y. Note that G and B do not depend on the normal coordinate on N. Moreover, the *tangential operator* B is self-adjoint and elliptic of first order. Let $\{\lambda_k;\varphi_k\}_{k\in\mathbf{Z}}$ be a spectral decomposition of $L^2(Y;S^+|_Y)$ generated by B with $\lambda_k \geq 0$ for $k \geq 0$. We know from Proposition 14.2 that the orthogonal projection $\mathrm{P}_\geq := \mathrm{P}_{\geq 0}$ onto $\mathcal{L}_{\mathbf{C}}\{\varphi_k\}_{k\geq 0}$ is a pseudo-differential operator and we have a well-defined Fredholm operator

$$\mathcal{A} := A^+_{\mathrm{P}_\geq} : \operatorname{dom} A^+_{\mathrm{P}_\geq} \to L^2(X;S^-)\,. \tag{22.2}$$

It follows from the elementary index theory presented in Chapter 20 (see Theorem 20.8) that

$$\operatorname{index}\mathcal{A} = \mathbf{i}(\mathrm{P}_\geq, \mathcal{P}_+)\,, \tag{22.3}$$

where $\mathcal{P}_+$ denotes the Calderón projector of A^+ and

$$\mathbf{i}(\mathrm{P}_{\geq}, \mathcal{P}_+) = \operatorname{index}\{\mathrm{P}_{\geq}\, \mathcal{P}_+ : \operatorname{range} \mathcal{P}_+ \to \operatorname{range} \mathrm{P}_{\geq}\}$$

the virtual codimension of $\mathrm{P}_{\geq}$ in $\mathcal{P}_+$.

Formula (22.3) does not involve the geometric data suitable for applications to concrete problems. Deeper insight is provided by the geometric Atiyah-Patodi-Singer index theorem for the operator $\mathcal{A}$ (Theorem 22.18 below). Following Atiyah, Patodi & Singer [1975], we shall present a proof of this theorem. As in the heat equation proof of the Atiyah-Singer index theorem (accounted for e.g. in Gilkey [1984] and Berline, Getzler & Vergne [1992]), the base of the Atiyah-Patodi-Singer index theorem is the study of suitable *heat operators* and the related *heat kernels* and *traces*. A novel feature is of course to keep control of the boundary conditions. This is possible due to our assumption of the product structure in the collar neighbourhood of the boundary providing for explicit computations of the traces of the heat kernels involved.

We begin with some formulas for the adjoint boundary problem and the related natural problems of second order:

Lemma 22.1. *Let $\mathcal{A}$ denote the closed operator from $L^2(X;S^+)$ to $L^2(X;S^-)$ which acts like A^+ and is determined by* $\operatorname{dom}\mathcal{A} = \{s \in H^1(X;S^+) \mid \mathrm{P}_{\geq}(s|_Y) = 0\}$. *Then*
(a) The adjoint operator $\mathcal{A}^*$ acts like A^- and is determined by

$$\operatorname{dom}\mathcal{A}^* = \{s \in H^1(X;S^-) \mid \mathrm{P}_{<}\, G^*(s|_Y) = 0\},$$

where $\mathrm{P}_{<} := \mathrm{P}_{<0}$ denotes the spectral projection of the tangential part B of A^+ corresponding to the eigenvalues of B in the interval $(-\infty, 0)$.
(b) The operator $\mathcal{A}^*\mathcal{A}$ acts like A^-A^+ and is determined by

$$\operatorname{dom}\mathcal{A}^*\mathcal{A} = \{s \in H^2(X;S^+) \mid \mathrm{P}_{\geq}(s|_Y) = 0 \textit{ and } \partial_u \mathrm{P}_{<}(s|_{Y_u})_{u=0} = -B(\mathrm{P}_{<}(s|_Y))\}.$$

(c) The operator $\mathcal{A}\mathcal{A}^*$ acts like A^+A^- and is determined by

$$\operatorname{dom}\mathcal{A}\mathcal{A}^* = \{s \in H^2(X;S^-) \mid \mathrm{P}_{<}(G^* s|_Y) = 0 \textit{ and } \partial_u \mathrm{P}_{\geq}(G^* s|_{Y_u})_{u=0} = B(\mathrm{P}_{\geq}(G^* s|_Y))\}.$$

(d) The null space $\ker \mathcal{A}^*\mathcal{A}$ *coincides with that of $\mathcal{A}$, while the null space of $\mathcal{A}\mathcal{A}^*$ coincides with that of $\mathcal{A}^*$.*

Note. Here $\mathrm{P}_{\geq}(s|_Y) = 0$ means that s belongs to the kernel of the composite continuous map

$$H^1(X; S^+) \xrightarrow{\gamma_0} L^2(Y; S^+|_Y) \xrightarrow{\mathrm{P}_{\geq}} L^2(Y; S^+|_Y),$$

where γ_0 denotes the restriction to the boundary. Note that γ_0 is even continuous $H^1(X) \to H^{1/2}(Y)$ (Theorem 11.4).

Proof. Recall that $\mathrm{P}_< := \mathrm{P}_{<0} = \mathrm{Id} - \mathrm{P}_{\geq}$ and

$$(A^+)^* = A^- = -G^*(\partial_u - GBG^*) \tag{22.4}$$

on the collar. Then *(a)* is a reformulation of Proposition 20.3. As for *(b)* we have that

$$\operatorname{dom} \mathcal{A}^*\mathcal{A} = \{s \in H^2(X; S^+) \mid s \in \operatorname{dom} \mathcal{A} \text{ and } A^+ s \in \operatorname{dom} \mathcal{A}^*\};$$

and now *(b)* is an elementary consequence of the commutativity of the tangential operator B with its spectral projections $\mathrm{P}_{\geq}$ and $\mathrm{P}_<$. The computation of the domain in *(c)* follows in a similar way. Finally *(d)* is inherent in the concept of an adjoint operator, but can also be seen from the explicit form of the two boundary problems which actually fit appropriately. □

The preceding lemma allows us to use the traces of the kernels of the corresponding heat operators to calculate the index of the boundary problem $\mathcal{A}$. We prove an Atiyah-Bott trace formula for boundary problems in the same way as in the case of closed manifolds (see e.g. Gilkey [1984; Lemma 1.6.5] or Berline, Getzler & Vergne [1992; Theorem 3.50]).

Let $\lambda \in \mathbf{R}$ and let n_λ and n^*_λ denote the dimensions of the λ-eigenspaces V_λ of the operator $\mathcal{A}^*\mathcal{A}$ and V^*_λ of the operator $\mathcal{A}\mathcal{A}^*$. They are finite, since the null space of elliptic boundary problems is finite-dimensional (as shown in Corollary 19.4). Recall that the eigenspaces consist of smooth sections by Theorem 19.1. From Lemma 22.1 we get

$$\mathcal{A}(\mathcal{A}^*\mathcal{A}) = (\mathcal{A}\mathcal{A}^*)\mathcal{A} \quad \text{and} \quad \mathcal{A}^*(\mathcal{A}\mathcal{A}^*) = (\mathcal{A}^*\mathcal{A})\mathcal{A}^*$$

so that $\mathcal{A}$ and $\mathcal{A}^*$ interchange the eigenspaces V_λ and V^*_λ, and moreover induce an isomorphism

$$V_\lambda \xrightarrow{\mathcal{A}} V^*_\lambda \xrightarrow{\mathcal{A}^*} V_\lambda \tag{22.5}$$

between them if $\lambda \neq 0$. Thus $n_\lambda - n^*_\lambda = 0$ for $\lambda \neq 0$, and we compute for $t > 0$

$$\begin{aligned}\operatorname{tr} e^{-t\mathcal{A}^*\mathcal{A}} - \operatorname{tr} e^{-t\mathcal{A}\mathcal{A}^*} &= \sum_\lambda (n_\lambda - n^*_\lambda) e^{-t\lambda} = n_0 - n^*_0 \\ &= \dim\ker \mathcal{A}^*\mathcal{A} - \dim\ker \mathcal{A}\mathcal{A}^* \\ &\overset{\text{Lemma 22.1d}}{=} \dim\ker \mathcal{A} - \dim\ker \mathcal{A}^* \overset{\text{Thm. 20.8}}{=} \operatorname{index} \mathcal{A}.\end{aligned}$$

This proves

Lemma 22.2. *The following equality holds for any* $t > 0$

$$\operatorname{index} \mathcal{A} = \operatorname{tr} e^{-t\mathcal{A}^*\mathcal{A}} - \operatorname{tr} e^{-t\mathcal{A}\mathcal{A}^*}. \tag{22.6}$$

Formula (22.6) is the base of the proof of the Atiyah-Patodi-Singer index theorem. The proof splits into three parts:

(1) We construct the heat kernels $\mathcal{E}_c$ and $\mathcal{E}_{c*}$ of the operators $e^{-t\overline{\mathcal{A}}^*\overline{\mathcal{A}}}$ and $e^{-t\overline{\mathcal{A}}\overline{\mathcal{A}}^*}$, where $\overline{\mathcal{A}}$ denotes the corresponding boundary problem on the infinite cylinder $[0,\infty) \times Y$. We have also given the corresponding heat kernels $\mathcal{E}_d$ and $\mathcal{E}_{d*}$ for the operators $\widetilde{A}^*\widetilde{A}$ and $\widetilde{A}\widetilde{A}^*$ on the closed double $\widetilde{X}$ of X (see Gilkey [1984], Roe [1988], or Berline, Getzler & Vergne [1992] for the construction).

(2) We use Duhamel's principle to construct the kernels $\mathcal{E}$ and $\mathcal{E}_*$ for the heat operators $e^{-t\mathcal{A}^*\mathcal{A}}$ and $e^{-t\mathcal{A}\mathcal{A}^*}$.

(3) We study $\lim_{t\to 0_+}(\operatorname{tr} e^{-t\mathcal{A}^*\mathcal{A}} - \operatorname{tr} e^{-t\mathcal{A}\mathcal{A}^*})$ to obtain the final index formula.

22B. Heat Kernels on the Cylinder

We construct the kernels $\mathcal{E}_c$ and $\mathcal{E}_{c*}$ of the operators $e^{-t\overline{\mathcal{A}}^*\overline{\mathcal{A}}}$ and $e^{-t\overline{\mathcal{A}}\overline{\mathcal{A}}^*}$ on the semi-infinite cylinder $[0,\infty) \times Y$, where $\overline{\mathcal{A}}$ acts like $\overline{A}$ and $\overline{A}$ is given by formula (22.1), but with $u \in [0,\infty)$. For simplicity consider the *normalized* Dirac operator

$$D := \partial_u + B \tag{22.7}$$

on the cylinder acting on sections of $\rho^*(S^+|_Y)$ ($=S^+|_Y$ lifted to $\mathbf{R}_+ \times Y$ under the natural projection ρ). In this section the assumption $G = \mathrm{Id}$ is no restriction, since

$$\overline{A}^*\overline{A} = -\partial_u^2 + B^2$$

for arbitrary unitary G.

Clearly D is elliptic and its formal adjoint is

$$D^* := -\partial_u + B\,. \tag{22.8}$$

We shall show

Proposition 22.3. *Let $\mathcal{D}$ and $\mathcal{D}^*$ denote the closure of the operators in $L^2(\mathbf{R}_+ \times Y; \rho^*(S^\pm|_Y))$ defined by D and D^* with domains given by smooth functions subject to the boundary conditions*

$$\mathrm{P}_{\geq}(f|_{\{0\}\times Y}) = 0 \tag{22.9}$$

and

$$\mathrm{P}_{<}(f|_{\{0\}\times Y}) = 0 \tag{22.10}$$

respectively. Then $\mathcal{D}$ and $\mathcal{D}^*$ are adjoints of each other, and

$$\operatorname{dom}\mathcal{D} = \{f \in H^1(\mathbf{R}_+ \times Y; \rho^*(S^+|_Y)) \mid \mathrm{P}_{\geq}(f|_{\{0\}\times Y}) = 0\} \tag{22.11}$$

and

$$\operatorname{dom}\mathcal{D}^* = \{f \in H^1(\mathbf{R}_+ \times Y; \rho^*(S^-|_Y)) \mid \mathrm{P}_{<}(f|_{\{0\}\times Y}) = 0\}\,. \tag{22.12}$$

Here $\mathrm{P}_{\geq}(f|_{\{0\}\times Y}) = 0$ means for $f \in H^1(\mathbf{R}_+ \times Y; \rho^*(S^+|_Y))$ that f belongs to the kernel of the composite continuous map

$$H^1(\mathbf{R}_+ \times Y; \rho^*(S^+|_Y)) \xrightarrow{\gamma_0} L^2(Y; S^+|_Y) \xrightarrow{\mathrm{P}_{\geq}} L^2(Y; S^+|_Y)\,,$$

where γ_0 denotes the restriction to the boundary.

Note. The preceding proposition resembles Lemma 22.1a but has a different message: Whereas Lemma 22.1a and its originator, Proposition 20.3, are about elliptic boundary problems over a *compact* manifold (so that we can apply Fredholm operator theory), we are now dealing with *open* manifolds which require another set of arguments. We present these arguments in a series of lemmata about evolution equations which establish a parametrix for $\mathcal{D}$ (and similarly for $\mathcal{D}^*$) and are of independent interest for the later computation of the heat kernels. We need some principal facts about the *Laplace transformation.* Nice proofs and additional information can be found in Carslaw & Jaeger [1906/1988] where one of the oldest, but most elementary and satisfactory explanations, is given. For more advanced readers we suggest a modern presentation such as that of Doetsch [1970].

The following distinctions must be made:

- Let $C^\infty_{\text{comp}}(\mathbf{R}_+ \times Y; \rho^* S^+|_Y)$ denote the C^∞-sections with compact support (i.e. vanishing for $u \geq C$);
- let $C^\infty(\mathbf{R}_+ \times Y; \rho^* S^+|_Y; \mathrm{P}_\geq)$ denote the space of all C^∞-sections satisfying the boundary condition (22.9) (for $u = 0$) with subspace $C^\infty_{\text{comp}}(\mathbf{R}_+ \times Y; \rho^* S^+|_Y; \mathrm{P}_\geq)$;
- and let $C^\infty_0(\mathbf{R}_+ \times Y; \rho^*(S^+|_Y))$ denote the space of C^∞-sections with (compact) support totally contained in the interior of the cylinder $\mathbf{R}_+ \times Y$ (i.e. vanishing for $u \geq C$ and for $u \leq c$).

For later use notice that the inclusions

$$
\begin{aligned}
(22.13) \quad C^\infty_0(\mathbf{R}_+ \times Y; \rho^*(S^+|_Y)) &\subset C^\infty_{\text{comp}}(\mathbf{R}_+ \times Y; \rho^*(S^+|_Y); \mathrm{P}_\geq) \\
&\subset C^\infty_{\text{comp}}(\mathbf{R}_+ \times Y; \rho^*(S^+|_Y)) \subset L^2(\mathbf{R}_+ \times Y; \rho^*(S^+|_Y))
\end{aligned}
$$

are dense.

We shall show that any equation $Df = g$ with

$$g \in C^\infty_{\text{comp}}(\mathbf{R}_+ \times Y; \rho^* S^+|_Y)$$

has a unique solution $f \in C^\infty(\mathbf{R}_+ \times Y; \rho^* S^+|_Y; \mathrm{P}_\geq)$. More precisely:

Proposition 22.4. *For any normalized Dirac operator $D = \partial_u + B$ over the cylinder $\mathbf{R}_+ \times Y$ there exists a linear operator*

$$Q : C^\infty_{\text{comp}}(\mathbf{R}_+ \times Y; \rho^* S^+|_Y) \longrightarrow C^\infty(\mathbf{R}_+ \times Y; \rho^* S^+|_Y; \mathrm{P}_\geq)$$

such that

(i) *$DQg = g$ for all $g \in C^\infty_{\text{comp}}(\mathbf{R}_+ \times Y; \rho^* S^+|_Y)$.*

(ii) *$QDf = f$ for all $f \in C^\infty(\mathbf{R}_+ \times Y; \rho^* S^+|_Y; \mathrm{P}_\geq)$.*

(iii) *Q extends to a continuous map $H^{s-1} \to H^s_{\text{loc}}$ for all integers $s \geq 1$. Here H^s_{loc} denotes the space of sections which are locally in the Sobolev space H^s.*

(iv) *The kernel $Q(u, y; v, z)$ of Q is a section in a suitable homomorphism bundle (with u, v running in $\mathbf{R}_+$ and y, z in Y) which is C^∞ for $u \neq v$.*

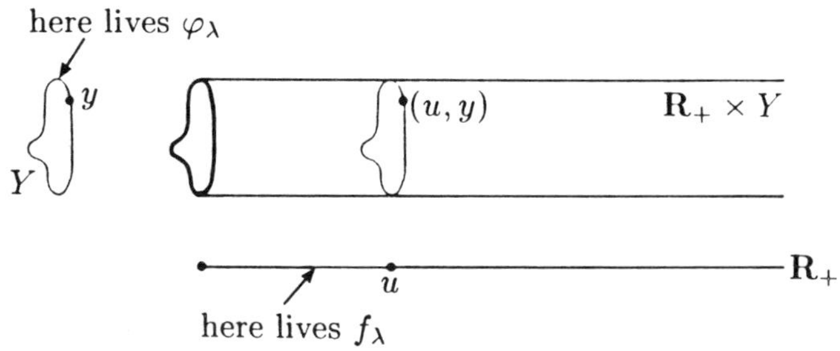

Fig. 22.1 Separation of variables over $\mathbf{R}_+ \times Y$

Proof. To solve

(22.14) $\quad Df = g$

with $f \in C^\infty(\mathbf{R}_+ \times Y; \rho^* S^+|_Y; \mathrm{P}_\geq)$ for $g \in C^\infty_{\text{comp}}(\mathbf{R}_+ \times Y; \rho^* S^+|_Y)$

we make the usual separation of variables (products of sections from $L^2(\mathbf{R}_+)$ and from $L^2(Y; S^+|_Y)$ span $L^2(\mathbf{R}_+ \times Y; \rho^* S^+|_Y)$, see Figure 22.1) and expand f and g in terms of the previously fixed orthonormal basis $\{\varphi_\lambda\}_{\lambda \in \text{spec}(B)}$ of eigenfunctions of B (here and in the following we repeat $\lambda \in \text{spec}(B)$ corresponding to its multiplicity):

$$f(u, y) = \sum_\lambda f_\lambda(u) \varphi_\lambda(y) \quad \text{and} \quad g(u, y) = \sum_\lambda g_\lambda(u) \varphi_\lambda(y) \,.$$

Then the partial differential equation (22.14) reduces to a family of ordinary differential equations

$$\text{(22.15)} \qquad \left(\frac{d}{du} + \lambda \right) f_\lambda = g \,, \qquad \lambda \in \text{spec}(B) \,,$$

half of them with initial conditions

$$\text{(22.16)} \qquad f_\lambda(0) = 0 \qquad \text{for } \lambda \geq 0 \,.$$

The textbook solution of (22.15) is

$$f_\lambda(u) = f_\lambda(0) e^{-\lambda u} + \int_0^u e^{\lambda(v-u)} g_\lambda(v) \, dv \,.$$

With (22.16) and an appropriate choice of

$$\text{(22.17)} \qquad f_\lambda(0) := - \int_0^\infty e^{\lambda v} g_\lambda(v) dv \qquad \text{for } \lambda < 0,$$

we get explicit solutions

$$f_\lambda(u) = \begin{cases} \int_0^u e^{\lambda(v-u)} g_\lambda(v)\,dv & \text{for } \lambda \geq 0 \\ \int_u^\infty e^{\lambda(v-u)} g_\lambda(v)\,dv & \text{for } \lambda < 0 \end{cases} \tag{22.18}$$

so that formally the operator

$$Q := \sum_\lambda Q_\lambda \qquad \text{with} \qquad Q_\lambda : g_\lambda(u)\varphi_\lambda(y) \mapsto f_\lambda(u)\varphi_\lambda(y)$$

satisfies (i) and (ii).

To show that this formal solution converges, we estimate the Sobolev norms. This will prove (iii) as well. To do that, rewrite the equations (22.15) in terms of the Laplace transform

$$\widetilde{f}(\xi) := \int_0^\infty e^{-iu\xi} f(u)du$$

and get

$$\widetilde{f_\lambda}(\xi) = \frac{\widetilde{g_\lambda}(\xi) + f_\lambda(0)}{i\xi + \lambda}. \tag{22.19}$$

We compute the $L^2(\mathbf{R})$-norms. Clearly

$$\lambda^2 \|\widetilde{f_\lambda}\|^2 \leq \|\widetilde{g_\lambda}\|^2 \qquad \text{for } \lambda \geq 0.$$

For $\lambda < 0$ we need more subtlety. Using elementary arithmetic, we first deduce from (22.19)

$$\|\widetilde{f_\lambda}\|^2 \leq 2\left\{\frac{\|g_\lambda\|^2}{\lambda^2} + |f_\lambda(0)|^2 \int_{-\infty}^\infty \frac{d\xi}{\xi^2 + \lambda^2}\right\}. \tag{22.20}$$

From

$$\int_{-\infty}^\infty \frac{d\xi}{\xi^2 + \lambda^2} = \frac{\pi}{|\lambda|} \tag{22.21}$$

and

$$\left\|e^{\lambda\cdot}\right\|^2 \cdot \|g_\lambda\|^2 \overset{\text{Parseval}}{=} \frac{1}{2|\lambda|} \cdot \frac{1}{2\pi} \|\widetilde{g_\lambda}\|^2 \text{ for } \lambda < 0,$$

we get

$$\lambda^2\|\widetilde{f}_\lambda\|^2 \leq 2\left\{\|\widetilde{g}_\lambda\|^2 + \frac{1}{4}\|\widetilde{g}_\lambda\|^2\right\} \qquad \text{for } \lambda < 0\,.$$

Apply Parseval's formula once more and get

$$(22.23) \qquad |\lambda|\,\|f_\lambda\| \leq 2\,\|g_\lambda\| \qquad \text{for all } \lambda \in \operatorname{spec}(B)\,.$$

Since the f_λ satisfy (22.15) this gives

$$(22.24) \qquad \left\|\frac{df_\lambda}{du}\right\| \leq 3\|g_\lambda\|\,.$$

Recall that a suitable norm $\|\cdot\|_1$ for the Sobolev space $H^1(\mathbf{R}_+ \times Y;\rho^*S^+|_Y)$ may be defined by

$$\|f\|_1^2 := \|f\|^2 + \left\|\frac{\partial f}{\partial u}\right\|^2 + \|Bf\|^2$$

since B is a first-order elliptic operator. So we set

$$\|f\|_1^2 = \sum_\lambda\left\{(1+\lambda^2)\|f_\lambda\|^2 + \left\|\frac{\partial f_\lambda}{\partial u}\right\|^2\right\}$$

and deduce continuity of

$$Q : H^0(\mathbf{R}_+ \times Y;\rho^*S^+|_Y) \to H^1_{\text{loc}}(\mathbf{R}_+ \times Y;\rho^*S^+|_Y)$$

by (22.23) and (22.24) which establishes (iii) for $k = 1$. What we have is actually a true continuous operator $Q' := \sum_{\lambda\neq 0} Q_\lambda : H^0 \to H^1$, and for $\lambda = 0$, we use

$$f_0(u) = \int_0^\infty g_0(v)\,dv$$

instead of (22.23).

To establish (iii) for general k, multiply (22.23) by powers of λ and differentiate (22.15) with respect to u. This leads to

$$|\lambda|^p\left\|\frac{\partial^q f_\lambda}{\partial u^q}\right\| \;\leq\; 3|\lambda|^{p+q-r-1}\,\|g_\lambda\| + \sum_{r=1}^{q-1}|\lambda|^{p+q-r-1}\left\|\frac{\partial^r g_\lambda}{\partial u^r}\right\|\,.$$

We may repeat the arguments given above, but instead of the norm $\|\cdot\|_1$, we now introduce a norm $\|\cdot\|_k$ for the Sobolev space H^k by taking the square root of the sum of the L^2-norms of all $B^p(\partial/\partial u)^q f$ for $p+q \le k$.

Proving (iv) is an exercise in *heat equation* arguments. From (22.18), one deduces that Q can be written as a convolution in the u-variable with the two-sided evolution operator

$$H(t) := \begin{cases} e^{-t|B|}\,\mathrm{P}_{\geq} & t \geq 0 \\ -e^{t|B|}\,\mathrm{P}_{<} & t < 0\,, \end{cases} \tag{22.25}$$

where $|B| := B\,\mathrm{P}_{\geq} - B\,\mathrm{P}_{<}$. More explicitly, let $g \in C^\infty_{\mathrm{comp}}(\mathbf{R}_+ \times Y; \rho^*(S^+|_Y))$ and let $\mathcal{H}(t,y,z)$ denote the kernel of the operator $H(t)$; then

$$Qg(u,y) = \int_{\mathbf{R}_+\times Y} \mathcal{H}(u-v,y,z)g(v,z)dvdz\,,$$

hence $Q(u,y;v,z) = \mathcal{H}(u-v,y,z)$. So we verify (iv) by proving the smoothness of $\mathcal{H}(t,y,z)$ for $t \neq 0$. By (22.25), we only need to show that the kernel $\mathcal{E}(t,y,z)$ of the heat operator $e^{-t|B|}$ is C^∞ on $\mathbf{R}_+ \times Y \times Y$, which actually is a well-known consequence of the spectral theorem and widely used in heat kernel expansions. Alternatively, the smoothness of $\mathcal{E}$ can be deduced as follows: For $0 < t_0 < t_1 < \infty$, we get from (22.18)

$$\begin{aligned} &\int_{t_0}^{t_1}\int_{Y\times Y}\left|B^p\left(\frac{\partial}{\partial t}\right)^q \mathcal{E}(t,y,z)\right|^2 dydzdt \\ &= \int_{t_0}^{t_1}\left(\sum_\lambda \lambda^{2p+2q}e^{-2t|\lambda|}\right)dt < (t_1-t_0)\sum_\lambda \lambda^{2p+2q}e^{-2t_0|\lambda|} < \infty \end{aligned}$$

for arbitrary integers p,q. The last estimate follows from

$$\sum \lambda^{2k}e^{-2t_0|\lambda|} < C\sum \lambda^{-2N}$$

which converges for large N. It follows that $\mathcal{E}$ belongs to all Sobolev spaces H^k for arbitrary $k = p+q$ and so by the Sobolev lemma is C^∞. □

Remark 22.5. If we replace D by D^* and the boundary condition (22.9) by (22.10), the only difference in the proof above occurs for $\lambda = 0$, in which case we shift between (22.16) and (22.17) and set

$$f_0(u) := -\int_u^\infty g_0(v)\,dv\,.$$

Therefore, D^* also has a fundamental solution R with properties (i*, ii*, iii*, iv*) like Q in (i, ii, iii, iv).

Proof of Proposition 22.3. As usual it needs some consideration to take the closure and the adjoint of an unbounded operator. Let us explain the problem by distinguishing certain levels:
(I) As mentioned above, the operators $D = \partial_u + B$ and $D^* = -\partial_u + B$ are *formal adjoints* of each other, namely
(22.26)
$$\langle Df; g\rangle = \langle f; D^*g\rangle \qquad \text{for all } f,\ g \in C_0^\infty(\mathbf{R}_+ \times Y; \rho^*(S^+|_Y)).$$

(II) Next notice that (22.26) remains valid for all

$$f \in C^\infty_{\mathrm{comp}}(\mathbf{R}_+ \times Y; \rho^*(S^+|_Y); \mathrm{P}_{\geq})$$

and

$$g \in C^\infty_{\mathrm{comp}}(\mathbf{R}_+ \times Y; \rho^*(S^+|_Y); \mathrm{P}_{<}).$$

Expand f and g with respect to a complete orthonormal system of eigenfunctions of B. Note that $\mathrm{P}_{\geq}(f|_{\{0\}\times Y}) = 0$ and $\mathrm{P}_{<}(g|_{\{0\}\times Y}) = 0$. Then

$$\begin{aligned}
\langle Df; g\rangle &= \left\langle \sum_\lambda \left(\frac{df_\lambda}{du}(u) + \lambda f_\lambda(u)\right)\varphi_\lambda(y)\ ;\ \sum_\mu g_\mu(u)\varphi_\mu(y)\right\rangle \\
&= \sum_\lambda \int_0^\infty f'_\lambda(u)\overline{g_\lambda(u)}\,du + \sum_\lambda \lambda \int_0^\infty f_\lambda(u)\overline{g_\lambda(u)}\,du \\
&= \sum_\lambda - \int_0^\infty f_\lambda(u)\overline{g'_\lambda(u)}\,du + \sum_\lambda \int_0^\infty f_\lambda(u)\overline{\lambda g_\lambda(u)}\,du \\
&\qquad + \sum_\lambda \underbrace{f_\lambda(0)\overline{g_\lambda(0)}}_{=0} \\
&= \langle f; D^*g\rangle .
\end{aligned}$$

(III) Now we have to prove that the L^2-closures $\mathcal{D}$ and $\mathcal{D}^*$ of the unbounded operators

$$D \quad \text{with } \mathrm{dom}_D := C^\infty(\mathbf{R}_+ \times Y; \rho^*(S^+|_Y); \mathrm{P}_{\geq})$$

and

$$D^* \quad \text{with } \mathrm{dom}_{D^*} := C^\infty(\mathbf{R}_+ \times Y; \rho^*(S^+|_Y); \mathrm{P}_{<})$$

are adjoints of each other. Then, decompose the Hilbert space $L^2(\mathbf{R}_+ \times Y; \rho^*(S^+|_Y))$ into two parts $H' \oplus H''$, where H'' involves the zero-eigenvalue of B and H' all the non-zero eigenvalues. Then $\mathcal{D}$ decomposes accordingly, since by definition $\mathcal{D}(H') \subset H'$, $\mathcal{D}(H'') \subset H''$, and the projection of

$$\operatorname{dom}\mathcal{D} = \{f \mid \exists\{f_\nu\},\ f_\nu \in \operatorname{dom}_D, f_\nu \to f, \text{ and } Df_\nu \to g := \mathcal{D}f\}$$

on H' along H'' does not lead out of $\operatorname{dom}\mathcal{D}$ (cf. Kato [1976; Section II.5.6]). Similarly $\mathcal{D}^*$ decomposes. We show that the parts $\mathcal{D}'$ and $\mathcal{D}''$ of $\mathcal{D}$ and the parts $\mathcal{D}^{*\prime}$ and $\mathcal{D}^{*\prime\prime}$ of $\mathcal{D}^*$ in H', H'' respectively, are mutually adjoint. On H'' one has $D = \partial/\partial u$, $D^* = -\partial/\partial u$ and the adjointness of $\mathcal{D}''$ and $\mathcal{D}^{*\prime\prime}$ is clear. On H' the fundamental solution of Proposition 22.3 gives a bounded inverse Q' for $\mathcal{D}'$, and similarly we get a bounded inverse R' for $D^{*\prime}$. Then $R' = (Q')^*$ follows by continuity from the fact that $\langle Df; g\rangle = \langle f; D^*g\rangle$ for $f, g \in C^\infty_{\text{comp}}$ and satisfying (22.9) and (22.10) respectively. Since adjoints commute with inverses (proved e.g. in Kato [1976; Theorem III.5.30]), the adjointness of $\mathcal{D}'$ and $\mathcal{D}^{*\prime}$ follows.

It remains to prove that

$$\operatorname{dom}\mathcal{D} = \ker(\mathrm{P}_{\geq} \circ \gamma_0)\,.$$

Clearly we have

$$\operatorname{dom}\mathcal{D} \subset \ker(\mathrm{P}_{\geq} \circ \gamma_0) \text{ and } \operatorname{dom}\mathcal{D}^* \subset \ker(\mathrm{P}_{<} \circ \gamma_0)\,.$$

Since $\mathrm{P}_{\geq}$ and $\mathrm{P}_{<}$ are complementary orthogonal projections and $\mathcal{D} = ad(\mathcal{D}^*)$, we conclude

$$\operatorname{dom}\mathcal{D} = \operatorname{dom} ad(\mathcal{D}^*) \supset \ker(\mathrm{P}_{\geq} \circ \gamma_0)\,. \qquad \square$$

Definition 22.6. With any (self-adjoint) Dirac operator

$$B : C^\infty(Y; S^+|_Y) \to C^\infty(Y; S^+|_Y)$$

over a closed manifold Y, we associate various operators over the cylinder $\mathbf{R}_+ \times Y$ and various kernels:
(a) The self-adjoint (closed) *cylindrical Dirac Laplacians*

$$\Delta_{\mathrm{c}} := \mathcal{D}^*\mathcal{D} \quad \text{and} \quad \Delta_{\mathrm{c}*} := \mathcal{D}\mathcal{D}^*$$

both act like $-\partial^2/\partial u^2 + B^2$ and are distinguished by

$$\operatorname{dom}\Delta_{\mathrm{c}} := \Big\{ f \in H^2(\mathbf{R}_+ \times Y; \rho^* S^+|_Y) \mid \mathrm{P}_{\geq}(f|_{\{0\}\times Y}) = 0 \quad \text{and} \quad \mathrm{P}_{<}\Big\{ \Big(\frac{\partial f}{\partial u} + Bf \Big)_{u=0} \Big\} = 0 \Big\}$$

and

$$\operatorname{dom}\Delta_{\mathrm{c}*} := \Big\{ f \in H^2(\mathbf{R}_+ \times Y; \rho^* S^+|_Y) \mid \mathrm{P}_{<}(f|_{\{0\}\times Y}) = 0 \quad \text{and} \quad \mathrm{P}_{\geq}\Big\{ \Big(\frac{\partial f}{\partial u} - Bf \Big)_{u=0} \Big\} = 0 \Big\}.$$

(b) The *heat operators* $e^{-t\Delta_{\mathrm{c}}}$ and $e^{-t\Delta_{\mathrm{c}*}}$ are associated with Δ_{c} and $\Delta_{\mathrm{c}*}$, and by the spectral theorem, they are well-defined bounded (in fact smoothing) operators on $L^2(\mathbf{R}_+ \times Y; \rho^*(S^+|_Y))$ for $t > 0$ (see also Figure 22.2).

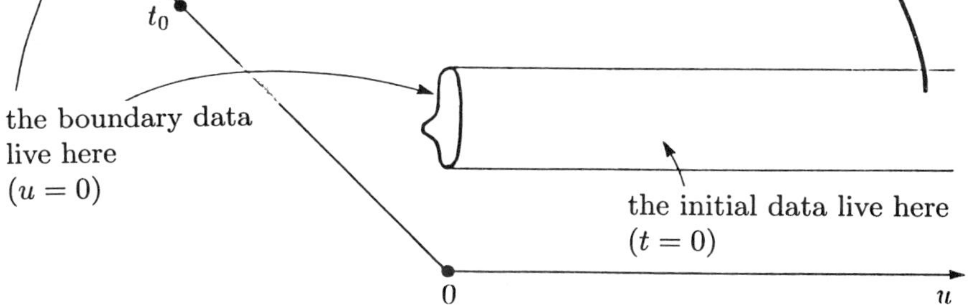

Fig. 22.2 The action of the cylindrical heat operator $e^{-t\Delta_{\mathrm{c}}}$

(c) For $j \in \{c, c*\}$, let $\mathcal{E}_j(t; u, y; v, z)$ denote the *cylindrical heat kernel*, i.e. the kernel of $e^{-t\Delta_j}$; let $\mathcal{K}(t; u, y)$ denote the *symmetric (cylindrical) heat kernel*, i.e. the kernel of $e^{-t\Delta_{\mathrm{c}}} - e^{-t\Delta_{\mathrm{c}*}}$ at the point $(u, y; u, y)$

of the diagonal of $(\mathbf{R}_+ \times Y) \times (\mathbf{R}_+ \times Y)$; and let $\mathcal{K}(t)$ denote the *symmetric integral*

$$\mathcal{K}(t) := \int_0^\infty \int_Y \mathcal{K}(t; u, y)\, dy du\,.$$

Remarks 22.7. (a) With regard to the polarization of the spinors in $L^2(\mathbf{R}_+ \times Y; \rho^*(S^+|_Y))$ introduced by the orthogonal spectral projections $\mathrm{P}_\geq$ and $\mathrm{P}_<$, the operator Δ_{c} ($\Delta_{\mathrm{c}*}$ respectively) is equal to the Dirac Laplacian $-\partial^2/\partial u^2 + B^2$ with Dirichlet boundary condition on the non-negative (negative) polarized spinors and a modified Neumann boundary condition on the negative (non-negative) polarized spinors.

(b) Recall that on $\mathbf{R}^n$ the scalar heat kernel is given by the exact Poisson formula

$$e(t; x, x') = (4\pi t)^{-n/2} e^{-d^2(x,x')/4t}\,, \tag{22.27}$$

where $d(x, x')$ denotes the Euclidean distance. The scalar heat kernel is a smooth function on $(0, \infty) \times \mathbf{R}^n \times \mathbf{R}^n$, which is a fundamental solution of the heat equation $\partial_t + \Delta_x$, i.e.

(1) it solves the equation

$$\left(\frac{\partial}{\partial t} - \frac{\partial^2}{\partial {x_1}^2} - \cdots - \frac{\partial^2}{\partial {x_n}^2}\right) e(t; x, x') = 0$$

for all $x' \in \mathbf{R}^n$ under the initial condition that $\lim_{t\to 0} e(t; x, x')$ is the Dirac distribution $\delta(x, x')$;

(2) alternatively stated, it defines an integral operator

$$(E_t s)(x) := \int_{\mathbf{R}^n} e(t; x, x')\, s(x')\, dx'$$

which solves the heat equation for any given initial distribution s;

(3) stated yet differently, it is the kernel of the heat evolution operator $e^{-t\Delta_x}$, i.e. $E_t s$ is approximated by $\sum_k (-t)^k/k!\, \Delta^k s$;

(4) this yields a weak form of Weyl's theorem

$$\operatorname{tr} e^{-t\Delta} = \sum_\lambda e^{-t\lambda} \sim C\, t^{-n/2} \qquad \text{as } t \to 0\,. \tag{22.28}$$

(c) Let P be any elliptic self-adjoint differential operator of second order with a positive definite principal symbol acting on sections of a Hermitian vector bundle E over a closed m-dimensional Riemannian manifold M. It is well-known that the existence of a heat kernel (homomorphism) $e(t;x,x')$ of the operator e^{-tP} can be established with properties corresponding to the properties $(1, \dots, 4)$ listed above; that it is usually impossible to find an exact expression for $e(t;x,x')$; that it can, however, be estimated for bounded positive t with certain constants $c_1, c_2 > 0$ by

$$(22.29) \qquad |e(t;x,x')| \leq c_1 t^{-m/2} \cdot e^{-c_2 \frac{d^2(x,x')}{t}}$$

where $d(x,x')$ denotes the geodesic distance. This inequality can be derived from approximate solutions as explained e.g. in Gilkey [1984; Lemma 1.7.4], Roe [1988; Chapter 5], and Berline, Getzler & Vergne [1992; Theorem 2.30]. Note that then the weak form of Weyl's theorem generalizes to

$$(22.30) \qquad \operatorname{tr} e^{-tP} = \sum_{\lambda \in \operatorname{spec} P} e^{-t\lambda} \sim C\, t^{-m/2} \qquad \text{as } t \to 0\,.$$

We note a minor generalization of (22.29) which will be useful in cutting and pasting heat kernels and discussing the η-invariant below. Let $Q : C^\infty(M;E) \to C^\infty(M;E)$ be an auxiliary differential operator of order 1. Then Qe^{-tP} is a smoothing operator with kernel $Q_x e(t;x,x')$, where Q_x denotes the operator Q acting on the x-variable. Instead of (22.29), we get

$$(22.31) \qquad |Q_x e(t;x,x')| \leq c_1 t^{-(m+1)/2} \cdot e^{-c_2 \frac{d^2(x,x')}{t}}$$

for sufficiently small positive t and suitable positive constants c_1, c_2.

The following proposition extends the validity of (22.29) for $M := Y$, $m := n-1$, and $P := B^2$ to an estimate concerning the cylindrical heat equation. (Below in Section 22C we shall apply (22.29) and (22.31) to the heat equation over the closed double with $M := \widetilde{X}$, $m := n$, and $P := \widetilde{A^-}\widetilde{A^+}$). It asserts that after imposing boundary conditions, the contribution outside the diagonal ($u = v$) is still asymptotically negligible.

Proposition 22.8. *For any normalized Dirac operator $\partial_u + B$ over the cylinder $\mathbf{R}_+ \times Y$ with $\dim Y = n-1$, the kernels $\mathcal{E}_c(t;u,y;v,z)$ and*

$\mathcal{E}_{c*}(t;u,y;v,z)$ of the corresponding heat operators $e^{-t\Delta_c}$ and $e^{-t\Delta_{c*}}$ are exponentially small in t as $t \to 0$ for $u \neq v$. More precisely they are bound by

$$C \cdot t^{-\frac{n}{2}} \cdot e^{-\frac{(u-v)^2}{4t}} \tag{22.32}$$

for some constant $C > 0$ as $t \to 0$.

Proof. To solve the elliptic differential equation $\Delta_c f = 0$ with $f \in \operatorname{dom}\Delta_c$, one expands f in terms of the eigensections φ_λ of B, as in the proof of Proposition 22.4. Then, if $f(u,y) = \sum f_\lambda(u)\varphi_\lambda(y)$, for each λ one must study the ordinary differential equation

$$-\frac{d^2 f_\lambda}{du^2} + \lambda f_\lambda = 0 \qquad \text{for } u \geq 0$$

with the boundary conditions

$$(\text{Dir}{\geq}) \qquad f_\lambda(0) = 0 \qquad \text{if } \lambda \geq 0\,,$$

$$(mod\,\text{Neu}{<}) \qquad \left(\frac{df_\lambda}{du} + \lambda f_\lambda\right)_{u=0} = 0 \qquad \text{if } \lambda < 0\,.$$

For $\lambda \geq 0$, consider the function
(22.33)

$$e_\lambda(t;u,v) := \frac{e^{-\lambda^2 t}}{2\sqrt{\pi t}} \left\{ e^{-\frac{(u-v)^2}{4t}} - e^{-\frac{(u+v)^2}{4t}} \right\} \qquad \text{for } u,v \geq 0 \text{ and } t > 0.$$

It vanishes for $u = 0$; as $t \to 0$, it approaches 0 for $u \neq v$ and ∞ for $u = v$; finally, for fixed t and v, apply the operator

$$\frac{\partial}{\partial t} - \frac{\partial^2}{\partial u^2} + \lambda^2, \tag{22.34}$$

and then all terms cancel as one can easily check. Hence it is a fundamental solution for the heat operator (22.34) with the boundary condition (Dir≥). For the boundary condition ($mod\,$Neu<), standard Laplace transform methods (accounted for e.g. in Carslaw & Jaeger [1959; Section 14.2]) give for $\lambda < 0$

$$\begin{aligned} e_\lambda(t;u,v) := \frac{e^{-\lambda^2 t}}{2\sqrt{\pi t}} &\left\{ e^{-\frac{(u-v)^2}{4t}} + e^{-\frac{(u+v)^2}{4t}} \right\} \\ &+ \lambda e^{-\lambda(u+v)} \operatorname{erfc}\left(\frac{u+v}{2\sqrt{t}} - \lambda\sqrt{t} \right), \end{aligned} \tag{22.35}$$

where $u, v \geq 0$, $t > 0$ and erfc denotes the *complementary error function* defined by

$$\operatorname{erfc}(x) := \frac{2}{\sqrt{\pi}} \int_x^\infty e^{-\xi^2}\, d\xi\,.$$

Note that $\operatorname{erfc}(0) = 1$. The kernel $\mathcal{E}_c(t; u, y; v, z)$ of the heat operator $e^{-t\Delta_c}$ is therefore explicitly given in terms of the eigenfunctions, namely by multiplying $e_\lambda(t; u, v)$ by $\varphi_\lambda(y) \otimes \varphi_\lambda^*(z)$ and summing over all $\lambda \in \operatorname{spec} B$.

Clearly (22.33) is bounded by $e^{-\lambda^2 t}/\sqrt{\pi t} \cdot e^{-(u-v)^2/4t}$ for $u, v \geq 0$. Since

$$\int_x^\infty e^{-\xi^2}\, d\xi < e^{-x^2}$$

we have, for $\lambda < 0$,

$$\begin{aligned}
&\left|\lambda e^{-\lambda(u+v)} \operatorname{erfc}\left(\frac{u+v}{2\sqrt{t}} - \lambda\sqrt{t}\right)\right| \\
&\qquad < |\lambda| e^{-\lambda(u+v)} \frac{2}{\sqrt{\pi}} e^{-\frac{(u+v)^2}{4t}} e^{2(u+v)\lambda} e^{-\lambda^2 t} \\
&\qquad \leq \frac{2|\lambda|}{\sqrt{\pi}} e^{-\lambda^2 t} e^{-\frac{(u-v)^2}{4t}},
\end{aligned}$$

so that the fundamental solutions (22.33) and (22.35) both are bounded by

$$\left\{ \frac{e^{-\lambda^2 t}}{\sqrt{\pi t}} + \frac{2|\lambda| e^{-\lambda^2 t}}{\sqrt{\pi}} \right\} e^{-\frac{(u-v)^2}{4t}}\,. \tag{22.36}$$

Using the inequality $x \leq e^{x^2/2}$ for $x := |\lambda|\sqrt{t}$ and taking into account the elementary bound $2|\varphi_\lambda(y) \otimes \varphi_\lambda^*(z)| \leq |\varphi_\lambda(y)|^2 + |\varphi_\lambda(z)|^2$, it follows that $\mathcal{E}_c(t; u, y; v, z)$ is bounded by

$$\frac{3}{2\sqrt{\pi t}} \left\{ \sum_\lambda e^{-\lambda^2 t/2} \left(|\varphi_\lambda(y)|^2 + |\varphi_\lambda(z)|^2\right) \right\} e^{-\frac{(u-v)^2}{4t}}\,. \tag{22.37}$$

Since the kernel of e^{-tB^2} on the diagonal of $Y \times Y$ is bounded by $Ct^{-(n-1)/2}$ with $\dim Y = n - 1$ (see (22.29)), the proposition follows for $\mathcal{E}_c$.

For the operator Δ_{c*}, the boundary conditions for each λ are

$$\text{(Dir<)} \qquad f_\lambda(0) = 0 \quad \text{if } \lambda < 0\,,$$

$$(mod\text{Neu}\geq) \qquad \left(-\frac{df_\lambda}{du} + \lambda f_\lambda\right)_{u=0} = 0 \quad \text{if } \lambda \geq 0\,.$$

The fundamental solution $e_{*\lambda}(t;u,v)$ of $\frac{\partial}{\partial t} - \frac{\partial^2}{\partial u^2} + \lambda^2$ for (Dir<) is again given by (22.33), while for (modNeu $\geq$) we must use (22.35) with λ changed to $-\lambda$. One sees that then the bounds (22.36) and (22.37) remain valid and the proposition follows also for $\mathcal{E}_{c*}$. □

As announced in Definition 22.6c, we are primarily interested in the contribution from the diagonal and in the difference between $e^{-\Delta_c}$ and $e^{-t\Delta_{c*}}$ expressed by the symmetric (cylindrical) heat kernel $\mathcal{K}(t;u,y)$ and the symmetric integral $\mathcal{K}(t) := \int_0^\infty \int_Y \mathcal{K}(t;u,y)\,dydu$. For the cylindrical heat kernel, the following corollary provides the final formulas which we shall apply below in Section 22D.

Corollary 22.9. *Let $\partial_u + B$ be a normalized Dirac operator over the infinite cylinder $\mathbf{R}_+ \times Y$ and let $\mathcal{K}(t)$ denote the corresponding symmetric heat kernel integral. Then one has*
(a)

$$\mathcal{K}(t) = -\sum_\lambda \frac{\operatorname{sign}\lambda}{2}\operatorname{erfc}(|\lambda|\sqrt{t})\,. \tag{22.38}$$

(b) For $\Re(s)$ large, one has

$$\int_0^\infty (\mathcal{K}(t) + \frac{1}{2}h)t^{s-1}\,dt = -\frac{\Gamma(s+\frac{1}{2})}{2s\sqrt{\pi}}\eta_B(2s)\,. \tag{22.39}$$

(c) If $\mathcal{K}(t)$ has an asymptotic expansion

$$\mathcal{K}(t) \sim \sum_{k>-n} a_k t^{k/2} \qquad \textit{as } t \to 0\,, \tag{22.40}$$

then

$$\eta_B(2s) = -\frac{2s\sqrt{\pi}}{\Gamma(s+\frac{1}{2})}\left\{\frac{h}{2s} + \sum_{k=-n+1}^{N} \frac{a_k}{\frac{1}{2}k+s} + \theta_N(s)\right\}, \tag{22.41}$$

where $\theta_N(s)$ *is holomorphic for* $\Re(s) > \frac{1}{2}(N+1)$.

Recall that $h := \dim\ker B$ is the multiplicity of the 0-th eigenvalue, $\eta_B(s) := \sum_{\lambda_n \neq 0} \operatorname{sign}(\lambda_n)|\lambda_n|^{-s}$, and Γ is defined by $\Gamma(s) := \int_0^\infty t^{s-1}e^{-t}dt$ for $\Re(s) > 0$ and extended to a meromorphic function on $\mathbf{C}$ with isolated simple poles at $s = 0, -1, -2, \ldots$.

Proof. *(a)* For $t > 0$, $u \geq 0$, and $y \in Y$ one determines the kernel of $e^{-t\Delta_c} - e^{-t\Delta_{c*}}$ at the diagonal point $(u, y; u, y) \in (\mathbf{R}_+ \times Y) \times (\mathbf{R}_+ \times Y)$ by

$$
\begin{aligned}
\mathcal{K}(t;u,y) &= \sum_\lambda \{e_\lambda(t;u,u) - e_{*\lambda}(t;u,u)\}\, \varphi_\lambda(y) \otimes \varphi_\lambda^*(y) \\
&= \sum_{\lambda \geq 0} \left\{ \frac{-e^{-\lambda^2 t}}{2\sqrt{\pi t}} e^{-u^2/t} - \frac{e^{-\lambda^2 t}}{2\sqrt{\pi t}} e^{-u^2/t} \right. \\
&\qquad \left. + \lambda e^{2\lambda u} \operatorname{erfc}\left(\frac{u}{\sqrt{t}} + \lambda\sqrt{t}\right)\right\} |\varphi_\lambda(y)|^2 \,\mathrm{Id} \\
&\quad + \sum_{\lambda<0} \left\{ \frac{e^{-\lambda^2 t}}{2\sqrt{\pi t}} e^{-u^2/t} + \lambda e^{-2\lambda u}\operatorname{erfc}\left(\frac{u}{\sqrt{t}} - \lambda\sqrt{t}\right) \right. \\
&\qquad \left. + \frac{e^{-\lambda^2 t}}{2\sqrt{\pi t}} e^{-u^2/t}\right\} |\varphi_\lambda(y)|^2 \,\mathrm{Id} \\
&= \sum_\lambda \operatorname{sign}(\lambda) \left\{ -\frac{e^{-\lambda^2 t}e^{-u^2/t}}{\sqrt{\pi t}} \right. \\
&\qquad \left. + |\lambda| e^{2|\lambda|u}\operatorname{erfc}\left(\frac{u}{\sqrt{t}} + |\lambda|\sqrt{t}\right)\right\} |\varphi_\lambda(y)|^2 \,\mathrm{Id} \\
&= \sum_\lambda \operatorname{sign}(\lambda) \frac{\partial}{\partial u} \left\{ \frac{1}{2} e^{2|\lambda|u} \operatorname{erfc}\left(\frac{u}{\sqrt{t}} + |\lambda|\sqrt{t}\right)\right\} |\varphi_\lambda(y)|^2 ,
\end{aligned} \tag{22.42}
$$

where, for convenience, $\operatorname{sign}(0) := 1$. Integrating over $\mathbf{R}_+ \times Y$, one obtains

$$
\mathcal{K}(t) = \int_0^\infty \int_Y \mathcal{K}(t;u,y)\,dy\,du = \sum_\lambda \frac{\operatorname{sign}(\lambda)}{2} \operatorname{erfc}(|\lambda|\sqrt{t}). \tag{22.43}
$$

(b) Note that, as $t \to 0$ in (22.43), $\mathcal{K}(t) \to -\frac{1}{2}h$. Moreover, $\mathcal{K}(t) + \frac{1}{2}h \to 0$ exponentially as $t \to \infty$. Also (22.43) shows that

$$
|\mathcal{K}(t)| \leq \frac{1}{\sqrt{\pi}} \sum_\lambda e^{-\lambda^2 t} \overset{(22.30)}{<} C\, t^{-(n-1)/2} \qquad \text{as } t \to 0.
$$

Hence for $\Re(s)$ large

$$\int_0^\infty (\mathcal{K}(t) + \frac{1}{2}h)\, t^{s-1}\, dt$$

converges. Integrating by parts, one gets

$$\int_0^\infty (\mathcal{K}(t)+\frac{1}{2}h)\, t^{s-1}\, dt = -\frac{\Gamma(s+\frac{1}{2})}{2s\sqrt{\pi}} \sum_{\lambda\neq 0} \frac{\operatorname{sign}(\lambda)}{|\lambda|^{2s}} = -\frac{\Gamma(s+\frac{1}{2})}{2s\sqrt{\pi}} \eta_B(2s)$$

by definition of $\eta_B(s)$. This follows directly from the definition of the Γ-function

$$\Gamma(s+\frac{1}{2}) = \int_0^\infty \frac{1}{\sqrt{\theta}}\, \theta^s e^{-\theta}\, d\theta$$

by setting $\theta = \lambda^2 t$ and then summing. Moreover, one uses

$$\mathcal{K}'(t) = \frac{1}{2\sqrt{\pi t}} \sum_\lambda \lambda e^{-\lambda^2 t}$$

which is obtained by differentiating (22.43) with respect to t.
(c) This is an immediate consequence of (22.39). □

Remarks 22.10. (a) Equation (22.41) gives an expression for the continuation of $\eta_B(2s)$ to the whole s-plane. In particular, $\eta_B(s)$ is holomorphic near $s=0$ and its value at $s=0$ is given by

$$\eta_B(0) = -(2a_0 + h)\,. \tag{22.44}$$

(b) Finally, the asymptotic expansion (22.40) is unchanged if, instead of defining $\mathcal{K}(t)$ by integrating over the infinite cylinder $\mathbf{R}_+ \times Y$, we integrate only over a large finite cylinder $[0, R] \times Y$ for some $R > 0$. In fact, from (22.42) one gets that the difference to (22.43) is

$$\sum_\lambda \operatorname{sign}(\lambda)\, \frac{1}{2} e^{2|\lambda|R} \operatorname{erfc}(R/\sqrt{t} + |\lambda|\sqrt{t})$$

which is bounded by

$$(e^{-R^2/t}/\sqrt{\pi}) \sum_\lambda e^{-\lambda^2 t} \overset{(22.30)}{<} C\, e^{-R^2/t}\, t^{-n/2}$$

and hence is exponentially small.

22C. Duhamel's Principle. Heat Kernels on Manifolds with Boundary

This section investigates the kernels $\mathcal{E}$ and $\mathcal{E}_*$ of the heat operators $e^{-t\mathcal{A}^*\mathcal{A}}$ and $e^{-t\mathcal{A}\mathcal{A}^*}$. Recall that $\mathcal{A}$ denotes the realization of the Dirac operator A^+ over the compact manifold X with boundary Y defined by the Atiyah-Patodi-Singer boundary problem. Following the presentation in McKean & Singer [1967] (see also Douglas & Wojciechowski [1991] and Klimek & Wojciechowski [1992], [1993]), we prove the fundamental properties of the kernels which are needed in order to establish the index formula.

As in the case of closed manifolds, referred to in Remark 22.7c above, it is usually impossible to find an exact expression for $\mathcal{E}(t,x,x')$. The kernel can, however, be estimated for bounded positive t in a similar way as in the estimate (22.29); this inequality can be derived from approximate solutions obtained by patching the heat kernel $\mathcal{E}_{\mathrm{d}}$ and $\mathcal{E}_{\mathrm{c}}$ together. Here let $\mathcal{E}_{\mathrm{d}}$ denote the kernel of the heat operator $e^{-t\widetilde{A}^*\widetilde{A}}$ of the invertible extension $\widetilde{A}$ of A^+ over the closed double $\widetilde{X}$ of X. Its restriction to X is well-defined and satisfies all of our demands for a fundamental solution for the operator $\partial_t + \mathcal{A}^*\mathcal{A}$, except for the boundary conditions of Lemma 22.1 and Definition 22.6a. Let $\mathcal{E}_{\mathrm{c}}$ denote the kernel of the heat operator $e^{-t\mathcal{D}^*\mathcal{D}}$ of the normalized Dirac operator $\mathcal{D}$ over the cylinder $\mathbf{R}_+ \times Y$ investigated above. Restricted to a collar neighbourhood $N = I \times Y$ of Y in X, it satisfies the boundary conditions but does not extend automatically to the whole of X. (See Figure 22.3).

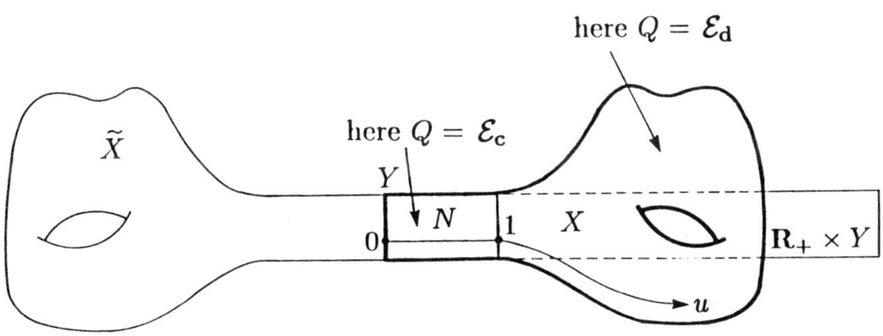

Fig. 22.3 The making of a *parametrix* for $e^{-t\mathcal{A}^*\mathcal{A}}$

We construct an approximate fundamental solution for the operator $\partial_t + \mathcal{A}^*\mathcal{A}$, sometimes called a *parametrix* of the operator $e^{-t\mathcal{A}^*\mathcal{A}}$. Let $\rho(a,b)$ be an increasing smooth function of the real variable u such that

$$\rho = \begin{cases} 0 & \text{for } u \leq a \\ 1 & \text{for } u \geq b\,. \end{cases} \tag{22.45}$$

We define four smooth functions $\varphi_1, \varphi_2, \psi_1, \psi_2$ of the normal variable u on $N := [0,1] \times Y$ which have obvious extensions to the whole of X, to the whole of $\widetilde{X}$, and to the whole of $\mathbf{R}_+ \times Y$ (φ_1, ψ_1 being extended by 0 and φ_2, ψ_2 being extended by 1, see Figure 22.4):

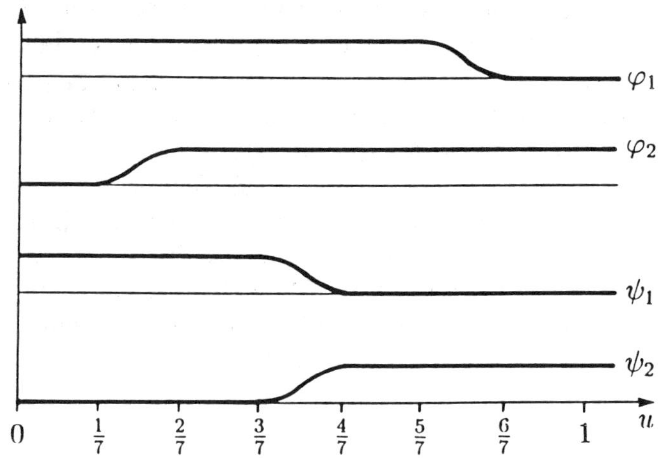

Fig. 22.4 Cut-off functions

$$\begin{aligned} \varphi_1 &:= 1 - \rho(\tfrac{5}{7}, \tfrac{6}{7})\,; \quad \varphi_2 := \rho(\tfrac{1}{7}, \tfrac{2}{7})\,; \\ \psi_1 &:= 1 - \rho(\tfrac{3}{7}, \tfrac{4}{7})\,; \quad \psi_2 := 1 - \psi_1\,. \end{aligned} \tag{22.46}$$

Then $\{\psi_1, \psi_2\}$ is a smooth partition of unity suitable to the open covering

$$\{\{u > 3/7\} \text{ and } \{u < 4/7\}\}$$

of X with $\varphi_j \equiv 1$ on $\operatorname{supp} \psi_j$ and

$$\text{(22.47)} \qquad \operatorname{dist}\left(\operatorname{supp}(\frac{\partial \varphi_j}{\partial u})\ ;\ \operatorname{supp}(\psi_j)\right) \geq \frac{1}{7}\,.$$

For $t > 0,\ x, x' \in X$ we define

$$\text{(22.48)} \qquad Q(t; x, x') := \sum_{j=1}^{2} \varphi_j(x) \mathcal{E}_j(t; x, x') \psi_j(x')\,,$$

where $\mathcal{E}_1 := \mathcal{E}_{\mathrm{c}}$, $\mathcal{E}_2 := \mathcal{E}_{\mathrm{d}}$, and the φ_j, ψ_j are considered as multiplication operators. Clearly Q is an approximate fundamental solution of the operator $\partial_t + \mathcal{A}^*\mathcal{A}$ over X which satisfies the initial and boundary conditions, but not the equation over the whole of X. We define a *parametrix* Q_* of the operator $e^{-t\mathcal{A}\mathcal{A}^*}$ in an analogous way.

Lemma 22.11. *Under the preceding conditions let*

$$C(t; x, x') := -(\frac{\partial}{\partial t} + \mathcal{A}^*\mathcal{A}) Q(\cdot; \cdot, x')|_{(t;x,x')}$$

denote the error term induced by the approximate fundamental solution Q. *It vanishes for all* $x \in X \setminus ([1/7, 6/7] \times Y)$ *and also for* $d(x, x') < 1/7$. *The following inequality holds:*

$$\text{(22.49)} \qquad |C(t; x, x')| \leq c_1 e^{-c_2 d^2(x,x')/t}$$

for some constants $c_1, c_2 > 0$ *as* $t \to 0$.

Note. The concept underlying the whole range of arguments between the preceding Lemma 22.11 and Theorem 22.14 below is referred to as *Duhamel's principle.* It goes back to Duhamel [1833], where the heat equation with *time dependent* boundary conditions is solved by taking the derivative of a superposition (convolution) of the solutions at successive times for the corresponding boundary conditions with *frozen* coefficients, see also Carslaw & Jaeger [1988; Section 1.14]. Similarly one speaks about Duhamel's principle when solving the inhomogeneous wave equation by replacing the *force term* on the right side by a succession of suitable initial *impulses* applied at various times to the solution of the homogeneous equation. The underlying idea is always the same, namely

- write the error term between the true solution and the approximate solution at time t as an integral over the derivative of a convolution of the true and the approximate solution,

- iterate and apply power series analysis to get *Levi's sum* for the fundamental solution,
- derive the needed estimates which ensure convergence and the vanishing of the error term for $t \to 0$,
- and estimate the pole of the heat kernel for $t \to 0$ up to terms of magnitude $t^{-n/2}$.

(See Levi [1909], Minakshisundaram & Pleijel [1949], and McKean & Singer [1967].)

Proof. Since both $\mathcal{E}$ and Q satisfy the initial condition $\mathcal{E} \to \mathrm{Id}$ and $Q \to \mathrm{Id}$ for $t \to 0$, we have on the operator level

$$\begin{aligned} \mathcal{E}(t) - Q(t) &= \mathcal{E}(t)Q(0) - \mathcal{E}(0)Q(t) \\ &= \int_0^t \frac{d}{ds}\left(\mathcal{E}(s)\,Q(t-s)\right) ds \\ &= \int_0^t \left(\mathcal{E}(s)\{-\mathcal{A}^*\mathcal{A} + \frac{d}{ds}\}Q(t-s)\right) ds \\ &= \int_0^t \left(\mathcal{E}(s)\{-(\mathcal{A}^*\mathcal{A} + \frac{d}{d(t-s)})\}Q(t-s)\right) ds \\ &= \int_0^t \mathcal{E}(s)\,C(t-s)\,ds\,. \end{aligned} \tag{22.50}$$

Hence on the kernel level we have the equality

$$\mathcal{E}(t;x,x') = Q(t;x,x') + \int_0^t \int_X \mathcal{E}(s;x,z)C(t-s;z,x')\,dzds\,. \tag{22.51}$$

Next we derive a formula for the kernel of $C(t)$. We fix an $x' \in X$. Then we have

(22.52)

$$\begin{aligned} -C(t;x,x') &= \left(\frac{d}{dt} + \mathcal{A}^*\mathcal{A}\right) Q(t;x,x') \\ &= \left(\frac{d}{dt} + \mathcal{A}^*\mathcal{A}\right)\left(\sum_{j=1}^2 \varphi_j(x)\mathcal{E}_j(t;x,x')\psi_j(x')\right) \\ &= \begin{cases} 0 & \text{for } x \in X \setminus ([\frac{1}{7},\frac{6}{7}] \times Y) \\ \sum_{j=1}^2 \left\{ \frac{\partial^2 \varphi_j}{\partial u^2}(x)\mathcal{E}_j(t;x,x')\psi_j(x') + \frac{\partial \varphi_j}{\partial u}(x)\frac{\partial \mathcal{E}_j}{\partial u}(t;x,x')\psi_j(x') \right\} & \\ & \text{for } x = (u,y) \in N \\ 0 & \text{for } d(x,x') < \frac{1}{7} \text{ (because of (22.47))}\,. \end{cases} \end{aligned}$$

To prove the estimate (22.49) we recall the corresponding bounds for $|\mathcal{E}_{\mathrm{d}}(t;x,x')|$ and $|\mathcal{E}_{\mathrm{c}}(t;u,y;v,z)|$, see (22.31) and Proposition 22.8. Then by (22.52)

$$\begin{aligned} |C(t;x,x')| &\le c_1 t^{-\frac{n+1}{2}} e^{-Cd^2(x,x')/t} \\ &\le c_1 t^{-\frac{n+1}{2}} e^{-C/7^2 t} \le c_1 e^{-c_2/t} . \end{aligned} \qquad \square$$

The preceding lemma shows that $(\partial/\partial t + \mathcal{A}^*\mathcal{A})Q$ is exponentially small as $t \to 0$. Now let us find a bound for the true heat kernel $\mathcal{E}(t;x,x')$. Using (22.51) we represent the heat kernel as a series which we can estimate. Let $\alpha * \beta(t;x,x')$ denote the *convolution*

$$\alpha * \beta(t;x,x') := \int_0^t \int_X \alpha(s;x,z)\beta(t-s;z,x')\, dzds .$$

We have the following sequence of equalities:

$$\begin{aligned} \mathcal{E}(t;x,x') &= Q(t;x,x') + (\mathcal{E} * C)(t;x,x') \\ &= Q(t;x,x') + \sum_{k=1}^{\infty} Q * C_k(t;x,x') \\ &= Q(t;x,x') + \left(Q * (\sum_{k=1}^{\infty} C_k) \right)(t;x,x') \\ &= Q(t;x,x') + (Q * \mathcal{C})(t;x,x') , \end{aligned} \tag{22.53}$$

where $C_1 := C$, $C_k := C * C_{k-1}$, and

$$\mathcal{C}(t;x,x') := \sum_{k=1}^{\infty} C_k(t;x,x') .$$

To obtain the desired estimates of the heat kernels involved, we need the following generalized triangle inequality.

Lemma 22.12. *For any $0 < s < t$ and any $x,y,z \in X$, the following inequality holds*

$$\frac{d^2(x,y)}{t} \le \frac{d^2(x,z)}{s} + \frac{d^2(z,y)}{t-s} . \tag{22.54}$$

Proof. The inequality is obvious in the case

$$d(x,z) = d(z,y) \left(\ge \frac{1}{2} d(x,y) \right) ,$$

hence assume that

$$a := d^2(x,z) < b := d^2(z,y)$$

and consider the function

$$f(s) := \frac{a}{s} + \frac{b}{t-s}\,.$$

Then f is a smooth function on the interval $(0,t)$ and

$$f'(s) = \frac{b}{(t-s)^2} - \frac{a}{s^2} = \frac{s^2 b - a(t-s)^2}{s^2(t-s)^2}\,.$$

f' is negative for $s \to 0$ and positive for $s \to t$, hence f has a minimum at a point s_0 with $f'(s_0) = 0$. There is only one such point, namely

$$s_0 = t\frac{\sqrt{ab} - a}{b-a}\,.$$

Hence the minimum of f is equal to

$$f(t\frac{\sqrt{ab}-a}{b-a}) = \frac{b-a}{t}\{\frac{a}{\sqrt{ab}-a} + \frac{b}{b-\sqrt{ab}}\} = \frac{(\sqrt{a}+\sqrt{b})^2}{t}$$

and we have

$$\frac{d^2(x,z)}{s} + \frac{d^2(z,y)}{t-s} \geq \frac{(d(x,z)+d(z,y))^2}{t} \geq \frac{d^2(x,y)}{t}\,. \qquad \square$$

Lemma 22.13. *The following estimate holds for any $x, x' \in X$ and $0 < t < T_0$:*

$$|C(t;x,x')| \leq c_3 e^{c_2 d^2(x,y)/t}\,. \tag{22.55}$$

Proof. Begin with an estimate of $C_2(t;x,x')$.

(22.56)

$$\begin{aligned}
|C_2(t;x,x')| &= |C * C(t;x,x')| \\
&= \left| \int_0^t \int_{\mathrm{supp}_z\, C(t-s;z,x')} C(s;x,z)\, C(t-s;z,x')\, dzds \right| \\
&\overset{\text{Lemma 22.11}}{\leq} c_1^2 \int_0^t \int_N e^{-c_2 d^2(x,z)/s} e^{-c_2 d^2(z,x')/(t-s)}\, dzds \\
&\overset{\text{Lemma 22.12}}{\leq} c_1 (c_1 \operatorname{vol}(Y)\, t)\, e^{-c_2 d^2(x,y)/t}\,.
\end{aligned}$$

In the same way obtain

$$|C_k(t;x,x')| \le c_1 \frac{(c_1 \operatorname{vol} Y)^{k-1}\, t^{k-1}}{(k-1)!} e^{-c_2 d^2(x,x')/t}$$

and finally

$$\begin{aligned} \mathcal{C}(t;x,x')| &\le c_1\, e^{(c_1 \operatorname{vol} Y)\, t}\, e^{-c_2 d^2(x,x')/t} \\ &\le c_1\, e^{(c_1 \operatorname{vol} Y) T_0}\, e^{-c_2 d^2(x,x')/t} \\ &\le c_3 e^{-c_2 d^2(x,x')/t} . \end{aligned}$$

□

Now we can formulate the main result of this section:

Theorem 22.14. *Let $\mathcal{A}$ denote the Atiyah-Patodi-Singer realization of a Dirac operator A^+ over a compact smooth Riemannian n-dimensional manifold X with boundary Y. Assume that the Riemannian metric is a product close to Y.*
(a) Then the fundamental solution $\mathcal{E}(t) = e^{-t\mathcal{A}^*\mathcal{A}}$ is given by a convergent series of the form

$$\mathcal{E}(t) = Q(t) + \sum_{k=1}^{\infty} Q(t) * C_k(t) , \tag{22.57}$$

where $*$ denotes convolution in t and composition of operators. Here $Q(t)$ denotes an approximate fundamental solution constructed by patching together fundamental solutions of the corresponding heat operators over the infinite cylinder $\mathbf{R}_+ \times Y$ and over the closed double $\widetilde{X}$, $C_1 := (\partial/\partial t + \mathcal{A}^*\mathcal{A})Q$, and $C_k := C_1 * C_{k-1}$.
(b) In particular, for $t > 0$, $\mathcal{E}$ has a C^∞-kernel which satisfies the bound

$$|\mathcal{E}(t;x,x')| < C\, t^{-n/2}\, e^{-C' d^2(x,x')/t} \tag{22.58}$$

as $t \to 0$ and differs from the kernel $Q(t;x,x')$ of the approximate fundamental solution on the diagonal by the term

$$|\mathcal{E}(t;x,x) - Q(t;x,x)| \le c e^{-c'/t} . \tag{22.59}$$

Proof. (a) follows from (22.53) and (22.56) which state that the series defining $\mathcal{E}$ is absolutely convergent (for bounded t). We prove (b).

We begin by proving (22.58). According to (22.53) we may estimate $|\mathcal{E}(t;x,x')|$ by the sum of $|Q(t;x,x')|$ and $|(Q*\mathcal{C})(t;x,x')|$. By definition, the first summand fits nicely with our claimed bound (22.58). Let us estimate the second summand:

(22.60)

$$\begin{aligned}
|(Q*\mathcal{C})(t;x,x')| &\le c_1c_3\int_0^t\int_X s^{-n/2}e^{-c_2d^2(x,z)/s}e^{-c_2d^2(z,y)/(t-s)}\,dzds\\
&\le c_1c_3\int_0^t\int_X s^{-n/2}e^{-c_2d^2(x,z)/(2s)}\\
&\qquad\left(e^{-c_2d^2(x,z)/(2s)}e^{-c_2d^2(z,y)/2(t-s)}\right)\,dzds\\
&\le c_1c_3e^{-c_2d^2(x,y)/(2t)}\cdot \underbrace{t\int_X s^{-n/2}e^{-c_2d^2(x,z)/(2s)}\,dz}_{<c_6}\\
&\le c_4e^{-c_5d^2(x,x')/t}\,.
\end{aligned}$$

To verify that

$$\int_X s^{-n/2}e^{-c_2d^2(x,z)/(2s)}\,dz<c_6 \tag{22.61}$$

we use a partition $\{\chi_\alpha\}$ of unity suitable to a covering of X with local charts such that

$$\int_X\ldots dz=\sum_\alpha\int_X\chi_\alpha(z)\ldots dz\,.$$

Focusing on one chart one gets

(22.62)

$$\begin{aligned}
\int_X\chi_\alpha(z)s^{-n/2}e^{-c_2d^2(x,z)/(2s)}\,dz &\le \int_{\mathbf{R}^n}\chi_\alpha(z)s^{-n/2}e^{-c_7|x-z|^2/(2s)}\,dz\\
&\le\int_{\mathbf{R}^n}s^{-n/2}e^{-c_8|z|^2/s}\,dz_1\ldots dz_n\\
&=c_8^{-n/2}\int_{\mathbf{R}^n}e^{-v^2}dv \text{ by } v:=\sqrt{\frac{c_8}{s}}z\\
&=c_8^{-n/2}(2\pi)^{n/2}\,.
\end{aligned}$$

Thus (22.61) is deduced, hence (22.60) and (22.58). It remains to show (22.59). It follows from (22.51) that

$$\mathcal{E}(t;x,x)-Q(t;x,x)=\int_0^t\int_{\operatorname{supp}_z C(t-s;z,x)}\mathcal{E}(s;x,z)\,C(t-s;z,x)\,dzds\,.$$

The integrand vanishes whenever $d(x,z) < \frac{1}{7}$. Thus we may assume that both $\mathcal{E}(t;x,z)$ and $C(t;x,z)$ are bounded by $c_9 e^{-c_{10}d^2(x,z)/t}$ for $x, z \in X$ with $d(x,z) \geq \frac{1}{7}$ and $0 < t < T_0$. We have

(22.63)

$$
\begin{aligned}
&|\mathcal{E}(t;x,x) - Q(t;x,x)| \\
&\quad \leq {c_9}^2 \int_0^t \int_{\operatorname{supp}_z C(t-s;z,x)} e^{-c_{10}d^2(x,z)/s} e^{-c_{10}d^2(z,x)/(t-s)}\, dz ds \\
&\quad \leq {c_9}^2 \int_0^t \int_{\operatorname{supp}_z C(t-s;z,x)} e^{-c_{10}1/(49s)} e^{-c_{10}1/49(t-s)}\, dz ds \\
&\quad \leq {c_9}^2 \operatorname{vol}(Y) \cdot \int_0^t e^{-c_{11}t/s(t-s)}\, ds \leq c_{12} e^{-c_{13}/t}\,.
\end{aligned}
$$

□

Remarks 22.15. (a) The result remains true for the heat kernel of the operator $e^{-t\mathcal{A}\mathcal{A}^*}$ and the corresponding parametrix $Q_*(t;x,x')$ as well.

(b) The computations here are simplified by the fact that we restricted our considerations to the bounded time interval $[0, T_0]$. This does not work in more complicated cases when we must take larger time asymptotics into account (see Douglas & Wojciechowski [1991] and Klimek & Wojciechowski [1992], [1993]).

22D. Proof of the Index Formula

Now we can prove the index formula for the operator $\mathcal{A}$. The results of the previous two sections imply the following equalities:

(22.64)

$$
\begin{aligned}
\operatorname{index}\mathcal{A} &\overset{\text{Lemma 22.2}}{=} \lim_{t\to 0_+} \{\operatorname{tr} e^{-t\mathcal{A}^*\mathcal{A}} - \operatorname{tr} e^{-t\mathcal{A}\mathcal{A}^*}\} \\
&\overset{\text{Thm. 22.14}}{=} \lim_{t\to 0_+} \{\operatorname{tr} Q(t) - \operatorname{tr} Q_*(t)\} \\
&\overset{(22.48)}{=} \lim_{t\to 0_+} \left\{ \int_0^1 \int_Y \mathcal{K}(t;(u,y))\psi_1(u)\, dy du + \int_X F(t;x)\psi_2(x) dx \right\} \\
&\overset{\text{Remark 22.10b}}{=} \lim_{t\to 0_+} \left\{ \int_0^\infty \int_Y \mathcal{K}(t;(u,y))\, dy du + \int_X F(t;x)\psi_2(x) dx \right\} \\
&\overset{\text{Def. 22.6c}}{=} \lim_{t\to 0_+} \left\{ \mathcal{K}(t) + \int_X F(t;x)\psi_2(x) dx \right\},
\end{aligned}
$$

where $\mathcal{K}(t;u,y)$ and $\mathcal{K}(t)$ are defined as in Definition 22.6c and

$$
F(t;x) := \operatorname{tr} e^{-t\widetilde{A}^-\widetilde{A}^+} - \operatorname{tr} e^{-t\widetilde{A}^+\widetilde{A}^-}\,.
$$

Recall (see Gilkey [1984], Roe [1988], or Berline, Getzler & Vergne [1992]) that for small t, we have an asymptotic expansion

$$F(t;x) \sim \sum_{k\geq -n} \alpha_k(x)\, t^{\frac{1}{2}k},$$

where the coefficients $\alpha_k(x)$ are explicit local functions of the operators $\widetilde{A}^-\widetilde{A}^+$ and $\widetilde{A}^+\widetilde{A}^-$ over $\widetilde{X}$. Since the operators $\widetilde{A}^-\widetilde{A}^+$ and $\widetilde{A}^+\widetilde{A}^-$ are isomorphic in the collar N, via G, one has $F(t;x) \sim 0$ for x in the collar. Hence $F(t;x)\psi_2(x) \sim F(t;x)$ and so (22.64) gives an asymptotic expansion for $\mathcal{K}(t)$:

$$\begin{aligned} \mathcal{K}(t) &\sim \operatorname{index}\mathcal{A} - \sum_{k\geq -n} t^{\frac{1}{2}k} \int_X \alpha_k(x)\, dx \\ &\sim \operatorname{index}\mathcal{A} - \sum_{k\geq -n} t^{\frac{1}{2}k}\, a_k\,. \end{aligned} \tag{22.65}$$

Applying Corollary 22.9, equation (22.41), we deduce that
(22.66)

$$\eta_B(2s) = -\frac{2s\sqrt{\pi}}{\Gamma(s+\frac{1}{2})}\left\{\frac{\frac{1}{2}h + \operatorname{index}\mathcal{A}}{s} - \sum_{k=-n+1}^{N} \frac{a_k}{\frac{1}{2}k+s} + \theta_N(s)\right\},$$

where $h := \dim\ker B$ and $\theta_N(s)$ is holomorphic for $\Re(s) > -\frac{N+1}{2}$. This proves the following result:

Proposition 22.16. *Let Y be a closed Riemannian manifold and let*

$$B : C^\infty(Y;\mathcal{S}) \to C^\infty(Y;\mathcal{S})$$

be a Dirac operator which extends in the following sense: There exists a smooth compact oriented Riemannian manifold X with boundary Y and a bundle S_X over X of Clifford modules such that $S_X^+|_Y = \mathcal{S}$ and the operator B is the tangential part on Y of the corresponding Dirac operator $A^+ : C^\infty(X;S_X^+) \to C^\infty(X;S_X^-)$. Then the η-function $\eta_B(s)$ of B extends meromorphically to the whole complex plane with simple poles at $s_k = -\frac{1}{2}k$ for $k \in \mathbf{N}$ and

$$Res_{s=s_k}\, \eta_B(s) = a_k\,.$$

In particular, $\eta_B(s)$ is regular at $s=0$, and we have

$$\eta_B(0) = 2\int_X \alpha_0(x)\, dx - (h + 2\operatorname{index}\mathcal{A}), \tag{22.67}$$

where $\mathcal{A}$ denotes the L^2-realization of A^+ defined by the Atiyah-Patodi-Singer boundary condition and $\alpha_0(x)$ denotes the index density for the operator $\widetilde{A}^+$ on $\widetilde{X}$.

Remark 22.17. Since that result was proved by Atiyah, Patodi & Singer, the structure of η is better known. We do not discuss that topic here, but simply refer the reader to Gilkey [1984], Bismut & Freed [1986], Bismut & Cheeger [1987], [1990a], [1990b], Branson & Gilkey [1992a], [1992b], Müller [1993], and the references given there.

Formula (22.67) is known as the *Atiyah-Patodi-Singer index formula.* We summarize the results of the whole chapter:

Theorem 22.18. (Atiyah, Patodi & Singer, 1975). *Let X be a compact oriented Riemannian manifold with boundary Y and let*

$$A^+ : C^\infty(X; S^+) \to C^\infty(X; S^-)$$

be a Dirac operator. Assume that all the structures are products in a collar N of Y. Then A^+ decomposes over N into the special form

$$A^+ = G(\frac{\partial}{\partial u} + B),$$

where u is the inward normal coordinate, G is a unitary bundle isomorphism $S^+|_Y \to S^-|_Y$, and B is the corresponding (self-adjoint) Dirac operator on Y. Moreover, A^+ (and A^-) extend canonically to operators $\widetilde{A}^+$ (and $\widetilde{A}^-$) over the closed double $\widetilde{X}$ of X. Let $C^\infty(X; S^+; \mathrm{P}_\geq)$ denote the space of spinors s of positive chirality satisfying the boundary condition

$$\mathrm{P}_\geq\, s(\cdot, 0) = 0,$$

where $\mathrm{P}_\geq$ denotes the spectral projection of B corresponding to the eigenvalues ≥ 0. Then

$$A^+_{\mathrm{P}_\geq} : C^\infty(X; S^+; \mathrm{P}_\geq) \to C^\infty(X; S^-)$$

has a finite index given by

$$\operatorname{index} A^+_{\mathrm{P}_\geq} = \int_X \alpha_0(x)\, dx - \frac{h + \eta_B(0)}{2}, \tag{22.68}$$

where α_0, h, η_B are defined as follows:

(i) *$\alpha_0(x)$ is the constant term in the asymptotic expansion (as $t \to 0$) of the trace differences*

$$\sum_{\mu \in \operatorname{spec} \widetilde{A}^+ \widetilde{A}^-} e^{-t\mu} |\varphi_\mu(x)|^2 \quad - \quad \sum_{\mu' \in \operatorname{spec} \widetilde{A}^- \widetilde{A}^+} e^{-t\mu'} |\varphi_{*\mu'}(x)|^2,$$

where μ, φ_μ *denote the eigenvalues and eigenfunctions of* $\widetilde{A}^+\widetilde{A}^-$ *on the double of* X, *and* μ', $\varphi_{*\mu'}$ *are the corresponding objects for* $\widetilde{A}^-\widetilde{A}^+$.

(ii) $h := \dim\ker B =$ *multiplicity of 0-eigenvalue of* B.

(iii) $\eta_B(s) := \sum_{\lambda\neq 0} \operatorname{sign}\lambda|\lambda|^{-s}$, *where* λ *runs over the eigenvalues of* B.

In (iii) the series converges absolutely for $\Re(s)$ *large, and then* $\eta_B(s)$ *extends to a meromorphic function on the whole s-plane with a finite value at* $s = 0$. *Moreover, if the asymptotic expansion in (i) has no negative power of* t, *then* $\eta_B(s)$ *is holomorphic for* $\Re(s) > -\frac{1}{2}$.

22E. L^2-Reformulation

To understand the Atiyah-Patodi-Singer index formula for a Dirac operator A^+ over a compact manifold X with boundary Y (and with $A = G(\partial_u + B)$ close to Y) in its correct analytical context, one has to reinterpret the action of the spectral projection $\mathrm{P}_{\geq}$ defined by the tangential part B of A^+ which constitutes the Atiyah-Patodi-Singer boundary condition. Recall the analytical context of the Calderón projector $\mathcal{P}(A^+)$, namely its relation to the Poisson type integral K_+, which provides for all sections f over Y a global solution s of $A^+s = 0$ with $\mathcal{P}(A^+)f = s|_Y$. Roughly speaking, the action of the Calderón projector is related to the *inner extension problem* of sections over the boundary to global solutions. The null space $\ker A^+_{\mathcal{P}(A^+)}$ of the corresponding boundary problem, however, is uninteresting since by definition it consists only of the zero section.

Now the remarkable fact is that the action of the spectral projection $\mathrm{P}_{\geq}$ is opposite directed, namely toward the *outer extension problem* in such a way that

- the null space $\ker \mathcal{A}$ (with $\mathcal{A} := A^+{}_{\mathrm{P}_{\geq}}$) can be identified with the square-integrable solutions of $A^+_\infty s = 0$ on the complete non-compact elongation X_∞ obtained from X by attaching the semi-infinite cylinder $(-\infty, 0] \times Y$ to the boundary (see Figure 22.5);
- the solution space $\ker \mathcal{A}^*$ (with $\mathcal{A}^* := A^-{}_{G\mathrm{P}_{<}G^*}$) can be identified with the direct sum of all square-integrable solutions of $A^-_\infty s = 0$ with a certain subspace of $\ker B$ which consists of the limiting values of all locally square-integrable solutions;
- and that the dimensions of the two spaces of limiting values coming from $A^+_\infty s = 0$ and $A^-_\infty s = 0$ add precisely to the

dimension of the null space of the tangential part B. (For the general case Müller [1988] and Melrose [1992, Chapter 7]).

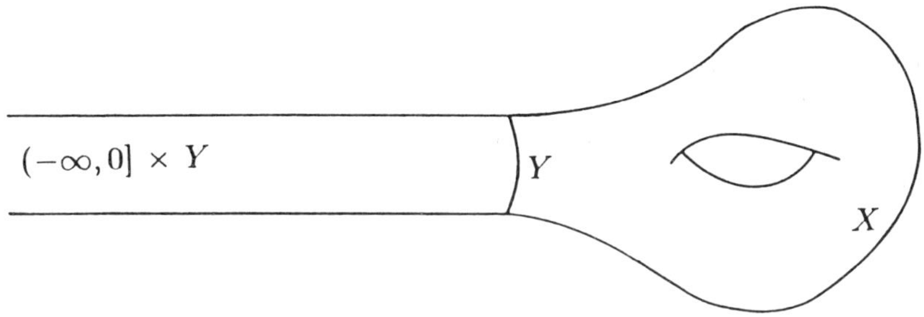

Fig. 22.5 The non-compact *elongation* X_∞

We define:

Definition 22.19. Let X be a compact oriented Riemannian manifold with boundary Y, and let

$$A^+ : C^\infty(X; S^+) \to C^\infty(X; S^-)$$

be a Dirac operator. Assume that the Riemannian structure on X and the Hermitian structure on S are products in a collar neighbourhood of Y. In particular, there we have $A = G(\partial_u + B)$ with interior normal coordinate u, unitary G, and self-adjoint B.
(a) Since X is isometric to the cylinder $Y \times [0,1]$ near the boundary, and since the Hermitian structure on S is product there, the non-compact Riemannian *elongation*

$$X_\infty := X \cup (-\infty, 0] \times Y$$

and related extensions of the bundles of Clifford modules $S^\pm$ to $S^\pm_\infty$ and of the Dirac operators $A^\pm$ to $A^\pm_\infty$ are well-defined and the unitary isomorphism G extends in the cylinder as well.
(b) We define the space of L^2-solutions of $A^+_\infty s = 0$ on X_∞:

$$L^2\text{-ker}(A^+_\infty) := \{s \in L^2(X_\infty; S^+_\infty) \mid A^+_\infty s = 0\}$$

and similarly L^2-ker(A^-_∞).
(c) By an *extended* L^2-solution of the equation $A^+_\infty s = 0$, we mean a solution which belongs to $L^2_{\text{loc}}(X_\infty; S^+_\infty)$ such that, for large negative u, we have

$$s(u, y) = s_{L^2}(u, y) + s_{\lim}(y)\,,$$

where $s_{L^2} \in L^2(X_\infty; S_\infty^+)$ and $s_{\lim} \in \ker B$. Thus s has $s_{\lim}$ as an asymptotic or limiting value as $u \to -\infty$. We shall write

$$s \in L^2_{\text{ext}}\text{-ker}A^+_\infty .$$

A similar definition holds for extended L^2-solutions $s \sim s_{L^2} + s_{\lim}$ of $A^-_\infty s = 0$, but demanding $G^* s_{\lim} \in \ker B$.

(d) Let $\mathcal{H}_+ \subset \ker B$ denote the space of all limiting values of extended L^2-solutions of $A^+_\infty s = 0$, and define in a similar way $\mathcal{H}_- \subset G \ker B$.

Proposition 22.20. *As before, let $\mathcal{A}$ denote the closed operator from $L^2(X; S^+)$ to $L^2(X; S^-)$ which acts like A^+ and is determined by* $\operatorname{dom} \mathcal{A} = \{s \in H^1(X : S^+) \mid \mathrm{P}_{\geq}(s|_y) = 0\}$. *Then we have natural isomorphisms*

$$\text{(i)} \qquad \ker \mathcal{A} \cong L^2\text{-ker}A^+_\infty \qquad \textit{and}$$

$$\text{(ii)} \qquad \ker \mathcal{A}^* \cong L^2_{\text{ext}}\text{-ker}A^-_\infty \cong L^2\text{-ker}A^-_\infty \oplus \mathcal{H}_- .$$

Note. Due to the regularity of solutions of elliptic boundary problems (Theorem 19.1) we have

$$\ker \mathcal{A} = \ker\{A^+ : C^\infty(X; S^+; \mathrm{P}_\geq) \to C^\infty(X; S^-)\}$$

and similarly

$$\ker \mathcal{A}^* = \ker\{A^- : C^\infty(X; S^-; \mathrm{P}_< G^*) \to C^\infty(X; S^+)\} .$$

Proof. *(i)* Let $s \in \ker \mathcal{A}$ and expand it near the boundary in the form

$$s(u, y) = \sum_\lambda f_\lambda(u) \varphi_\lambda(y) ,$$

where $\{\lambda; \varphi_\lambda\}_{\lambda \in \operatorname{spec}(B)}$ is a spectral resolution of $L^2(Y; S^+|_Y)$ generated by B. Since $A^+ s = 0$, we have $\partial s / \partial u + Bs = 0$, so that $f_\lambda(u) = e^{-\lambda u} f_\lambda(0)$. Since $\mathrm{P}_\geq(s|_Y) = 0$, we have $f_\lambda(0) = 0$ for $\lambda \geq 0$ and so

$$s(u, y) = \sum_{\lambda < 0} e^{-\lambda u} f_\lambda(0) \varphi_\lambda(y) . \tag{22.69}$$

This shows that s extends to a section s_∞ on X_∞ which satisfies $A^+_\infty s_\infty = 0$ and is exponentially decaying as $u \to -\infty$, hence certainly in L^2. Conversely, a solution of $A^+_\infty s_\infty = 0$ on X_∞ must obey

$\partial s_\infty/\partial u + Bs_\infty = 0$ on the semi-infinite cylinder, so that it must be of the form (22.69) if it is in L^2, because terms involving $e^{-\lambda u}$ with $\lambda \geq 0$ are not in L^2. Thus $s \mapsto s_\infty$ gives the required isomorphism.
(ii) Let $s \in \ker \mathcal{A}^*$, then instead of (22.69) we get

$$s(u,y) = \sum_{\lambda \geq 0} e^{\lambda u} f_\lambda(0)\, \varphi_\lambda(y) \tag{22.70}$$

near the boundary. Here, for simplicity, we have suppressed the action of G or, put otherwise, set $G = \mathrm{Id}$. The natural extension of s to a section s_∞ on X_∞ is now an extended L^2-solution (as defined above in Definition 22.19c) with limiting value

$$s_{\lim}(y) = \sum_{\lambda = 0} f_\lambda(0)\, \varphi_\lambda(y)\,.$$

(Note that the summation is over the finite multiplicity of the 0-eigenvalue). Conversely, any $s_\infty \in L^2_{\mathrm{ext}}\text{-}\ker A^-_\infty$ is necessarily of the form (22.70) so that $s \mapsto s_\infty$ also gives an isomorphism. □

Since the cokernel of an elliptic boundary problem can be identified with the null space of the adjoint boundary problem (Theorem 20.8), we deduce

Corollary 22.21. *The index of the Atiyah-Patodi-Singer boundary problem $\mathcal{A} = A^+{}_{P_\geq}$ of a Dirac operator A^+ over a compact manifold X with boundary Y can be expressed by*

$$\operatorname{index} \mathcal{A} = L^2\text{-index}(A^+_\infty) - h_-\,, \tag{22.71}$$

where $h_- := \dim \mathcal{H}_-$ and $\mathcal{H}_-$ denotes the subspace of $G(\ker B)$ consisting of the limiting values of extended L^2-solutions of $A^-_\infty s = 0$ (see Definition 22.19d).

Let now $h_+ := \dim \mathcal{H}_+$, where $\mathcal{H}_+$ denotes the subspace of $\ker B$ consisting of the limiting values of extended L^2-solutions of $A^+_\infty s = 0$, and recall that $h := \dim \ker B$. Then the non-negative integers $h_\pm$ are less than h and in fact add to h.

Proposition 22.22. *For the three dimensions defined here, one has the formula*

$$h = h_+ + h_- \,. \tag{22.72}$$

Proof. First merge (22.68) and (22.71) yielding

$$\int_X \alpha_0(x)\,dx - \frac{h + \eta_B(0)}{2} = L^2\text{-index}(A_\infty^+) - h_- \,. \tag{22.73}$$

Then deduce a formula similar to (22.73) with some but not all of the signs reversed. Consider the operator $\widehat{\mathcal{A}} := A^-_{\mathrm{P}_{\leq 0}\,G^*}$. It has a slightly larger domain than $\mathcal{A}^* = A^-_{\mathrm{P}_<\,G^*}$ and hence a different index. Applying (22.71), one gets

$$\operatorname{index}\widehat{\mathcal{A}} = L^2\text{-index}(A_\infty^-) - h_+ \,. \tag{22.74}$$

Now $\widehat{\mathcal{A}}$ acts like A^- which near the boundary takes the form

$$A^- = -G^*\left(\frac{\partial}{\partial u} - B\right).$$

From the construction of the index density it follows that $\alpha_0(A^-) = -\alpha_0(A^+)$; hence applying (22.68) gives

$$\operatorname{index}\widehat{\mathcal{A}} = -\int_X \alpha_0(x)\,dx - \frac{h + \eta_{-B}(0)}{2} \,. \tag{22.75}$$

Note that $h = \dim\ker B = \dim\ker(-B)$ is unaltered and that $\eta_{-B}(0) = -\eta_B(0)$. Thus (22.74) and (22.75) yield

$$-\int_X \alpha_0(x)\,dx - \frac{h - \eta_B(0)}{2} = L^2\text{-index}(A_\infty^-) - h_+ \,. \tag{22.76}$$

Since by definition $L^2\text{-index}(A_\infty^-) = -L^2\text{-index}(A_\infty^+)$, we deduce (22.72) from adding (22.76) and (22.73). □

There is one more interesting observation regarding the L^2-index and the $h_\pm$-dimensions:

Proposition 22.23. *For any Dirac operator A^+ on a compact manifold X with boundary Y, we have on the non-compact elongation X_∞*

(i) $$L^2\text{-ker}A^+_\infty = L^2\text{-ker}A^-_\infty A^+_\infty \qquad \textit{and}$$

(ii) $$L^2_{\text{ext}}\text{-ker}A^+_\infty = L^2_{\text{ext}}\text{-ker}A^-_\infty A^+_\infty\,.$$

The same also holds with A^+ and A^- interchanged.

Note. On the compact manifold X we have already seen in Lemma 22.1d that $\ker \mathcal{A}$ coincides with $\ker \mathcal{A}^*\mathcal{A}$ taking regard of the respective boundary conditions. The preceding proposition claims that similar results hold for L^2-sections on X_∞.

Proof. We argue as in the proof of Proposition 22.20. Any solution of $A^-_\infty A^+_\infty s = 0$ is, expanded on the cylinder $(-\infty, 0] \times Y$, necessarily of the form

$$s(u,y) = \sum_\lambda (a_\lambda e^{\lambda u} + b_\lambda e^{-\lambda u})\varphi_\lambda(y)\,. \tag{22.77}$$

If s is in L^2, it must be rapidly decreasing as $u \to -\infty$. More precisely, let X_u denote the compact manifold $X \cup ([u,0] \times Y) \subset X_\infty$ and let Y_u denote its boundary $\{u\} \times Y$ which is isometric to Y. Then we can estimate

$$\|s|_{Y_u}\| < C\,e^{\alpha u} \qquad \text{with } \alpha > 0\,, \tag{22.78}$$

where $\|\cdot\|$ denotes the L^2-norm on Y and C is independent of u, and also

$$\|A^+_\infty s|_{Y_u}\| < C\,e^{\alpha u} \qquad \text{for all } u \le 0 \tag{22.79}$$

for a suitable choice of α. Since A^- (and A^-_∞) are the formal adjoints of A^+ (and A^+_∞ - we suppress the subscript ∞ in the following), and since $A^+ = G(\partial_u + B)$ and $A^- = -G^*(\partial_u - B)$ also near Y_u, we get from $A^-A^+s = 0$ and Green's formula (cf. Proposition 3.4b)

$$\begin{aligned}
-\langle A^+s; A^+s\rangle_{X_u} &= \langle A^-A^+s; s\rangle_{X_u} - \langle A^+s; A^+s\rangle_{X_u} \\
&= \int_{Y_u} \langle G^*(y)A^+s(y); s(y)\rangle dvol(y) \\
&= \int_{Y_u} \langle A^+s(y); G(y)s(y)\rangle dvol(y) \\
&\overset{(22.78),(22.79)}{\sim} 0 \qquad \text{as } u \to -\infty\,.
\end{aligned} \tag{22.80}$$

This proves (i). Now consider a section $s \in L^2_{\text{ext}}\text{-ker}A^+$. Then its expansion (22.77) can have non-zero terms a_0 and b_0; so instead of (22.78) we can only assert that $\|s|_{Y_u}\|$ is bounded as $u \to -\infty$. However, applying A^+ removes these terms and so (22.79) holds. This is enough to show that the integral in (22.80) tends to zero as $u \to -\infty$. As before we conclude that $A^+s = 0$, proving (ii). The corresponding statements with A^+ and A^- interchanged are proved in exactly the same way. □

22F. The Odd-Dimensional Case. A Three-Dimensional Example

The proof given for the Atiyah-Patodi-Singer index theorem, Theorem 22.18, is valid in the odd-dimensional case as well, although in this case the index formula for $\mathcal{A}$ is simpler.

Let X be an odd-dimensional compact smooth Riemannian manifold with boundary Y. Assume that, near Y, X is isometric to the product $[0,1] \times Y$; here $\{0\} \times Y = \partial X$. Let S be a bundle of Clifford modules over X. Let $A : C^\infty(X;S) \to C^\infty(X;S)$ denote the corresponding Dirac operator. Recall from Theorem 21.5a that then A has a decomposition near Y of the form

$$A = \Gamma(\partial_u + B)\,,$$

where Γ is a unitary bundle automorphism with $\Gamma^2 = -\operatorname{Id}_{S|_Y}$ and hence Γ defines a decomposition of $S|_Y$ into the direct sum $S^+ \oplus S^-$ of the subbundles of the $\pm i$-eigenvalues of $\{\Gamma_y\}_{y\in Y}$. With respect to this decomposition the operator A takes the following form near Y:

$$A = \begin{pmatrix} i & 0 \\ 0 & -i \end{pmatrix} \left(\partial_u + \begin{pmatrix} 0 & B^- \\ B^+ & 0 \end{pmatrix}\right).$$

Observe that the formal self-adjointness of $A = \Gamma(\partial_u + B)$ implies $\Gamma B = -B\Gamma$. That leads to the following simplification of Theorem 22.18:

Theorem 22.24. *Let $\mathcal{A} := A_{P_\geq(B)}$ denote the standard Atiyah-Patodi-Singer realization of a Dirac operator over an odd-dimensional compact manifold X with boundary Y. Then*

$$\operatorname{index} \mathcal{A} = -\dim\ker B^+\,, \tag{22.81}$$

where B^+ comes from the splitting of A near Y into

$$A = \Gamma \left(\partial_u + \begin{pmatrix} 0 & B^- \\ B^+ & 0 \end{pmatrix} \right) .$$

Proof. If φ_λ is an eigenfunction of B, $B\varphi_\lambda = \lambda\varphi_\lambda$ with $\lambda \neq 0$, then $\Gamma\varphi_\lambda$ is an eigenfunction of B with eigenvalue $-\lambda$. That means that the spectrum of B is symmetric with respect to 0, hence

$$\eta_B(s) = \sum_{\lambda \in \operatorname{spec} B,\ \lambda \neq 0} \operatorname{sign} \lambda \, |\lambda|^{-s} = 0 .$$

Moreover, we have proved above that $\operatorname{index} B^+ = 0$ (Theorem 21.5b). That means that $\dim\ker B = \dim\ker B^+ + \dim\ker B^- = 2\dim\ker B^+$. Then we get for the standard Atiyah-Patodi-Singer realization of B, from the results of Atiyah, Patodi, and Singer mentioned above, that

$$\operatorname{index} A_{\mathrm{P}_{\geq}(B)} = \int \alpha_0(x)\, dx - \frac{1}{2}(\eta_B(0) + \dim\ker B) = -\dim\ker B^+ ,$$

since the index density α_0 is equal 0, when $\dim X$ is odd (see e.g. Gilkey [1984; Theorem 1.7.6.a] and Berline, Getzler & Vergne [1992; Section 4.1]). □

Remark 22.25. From our previous discussion (Theorem 20.8), we have

$$\operatorname{index} A_{\mathrm{P}_{\geq}(B)} = \mathbf{i}(\mathrm{P}_{\geq}(B), \mathcal{P}_+(A))$$

and the index vanishes, if and only if $\mathrm{P}_{\geq}(B)$ and $\mathcal{P}_+(A)$ belong to the same connected component of Gr_{p_+}. Hitchin [1974] showed that $\dim\ker B$ varies when we deform the metric (the most simple non-trivial examples are all Riemann surfaces Z with genus $g > 3$). This means that $\mathrm{P}_{\geq}(B)$ and $\mathcal{P}_+(A)$ change their connected components independently of each other and that under a smooth change of the metric the operator $A_{\mathrm{P}_{\geq}(B)}$ varies in a non-continuous way.

So the homotopy properties of the Atiyah-Patodi-Singer boundary value problem differ completely from the familiar homotopy invariance of the index of local elliptic boundary value problems. For them the homotopy invariance follows from Theorem 20.13.

We illustrate the application of the Atiyah-Patodi-Singer index formula with an example which has gained some prominence in recent geometry and gauge theory, see e.g. Taubes [1990; Section 2, Proposition 4.9, and Lemma A.4] and Yoshida [1991; Sections 1-4]. Let X be a compact oriented Riemannian manifold of dimension $\dim X = 3$ with connected boundary Y. Assume that, near Y, X is isometric to the product $[0,1] \times Y$; here $\{0\} \times Y = \partial X$. Assume also that the genus g of Y is ≥ 2. We define a *twisted signature operator* on the space

$$W := \left(\Omega^0(X) \otimes \mathrm{su}(2)\right) \oplus \left(\Omega^1(X) \otimes \mathrm{su}(2)\right) \tag{22.82}$$

of differential 0- and 1-forms with coefficients in the bundle $X \times \mathrm{su}(2)$. Here su(2) denotes the Lie algebra of SU(2) and $\Omega^p(X) = C^\infty(X; \Lambda^p T^* X)$, $p = 1,2,3,4$ denotes the space of smooth differential p-forms on X. Recall that every principal SU(2)-bundle over X is isomorphic to the trivial bundle $X \times \mathrm{SU}(2)$, and that the space of smooth connections on $X \times \mathrm{SU}(2)$ can be identified with $\Omega^1(X) \otimes \mathrm{su}(2)$. The Riemannian metric on X and the associated Hodge star operator $* : \Lambda^p(T^*X) \to \Lambda^{3-p}(T^*X)$ define an L^2 inner product

$$\langle a; b\rangle := -\int_X \mathrm{tr}(a \wedge *b) \qquad \text{for } a,\ b \in \Omega^p(X) \otimes \mathrm{su}(2)\,. \tag{22.83}$$

Example 22.26. Let A be a smooth flat connection on $X \times \mathrm{SU}(2)$ which restricts to a product connection $B \times \mathrm{Id}$ in the collar neighbourhood $[0,1] \times Y \times \mathrm{SU}(2)$ of $Y \times \mathrm{SU}(2)$ for an irreducible flat connection B on $Y \times \mathrm{SU}(2)$. We define the corresponding *twisted signature operator* $\not{D}_A : W \to W$ with coefficients in $X \times \mathrm{su}(2)$ by

$$\not{D}_A(a,b) := (d_A^* b\,,\ *d_A a + d_A b)\,, \tag{22.84}$$

where a is a 0-form, b is a 1-form, and d_A denotes the covariant derivative.

Here we follow the terminology of gauge theory and denote connections by the letters A and B which otherwise were reserved for elliptic operators in this book. The twisted signature operator is an operator of Dirac type (i.e. with principal symbol defining a $C\ell(X)$-module structure on the bundle $\Lambda^* X \times \mathrm{su}(2)$) and hence elliptic: Its principal symbol is not inflicted by the choice of a flat connection and therefore coincides with the principal symbol of the (untwisted) signature operator. It is, however, not a true Dirac operator in our sense, since it is not defined by a compatible connection. Actually, we have not used any compatibility of the connections with Clifford multiplication for the proof of the Atiyah-Patodi-Singer index theorem and may therefore now conclude:

Proposition 22.27. *The index of the twisted signature operator* $\not{D}_A$ *with Atiyah-Patodi-Singer boundary condition is well-defined, does not depend of the connection* A*, and can be expressed by*

$$\operatorname{index}(\not{D}_A)_{P_\geq} = 3 - 3g\,, \tag{22.85}$$

where g *denotes the genus of* Y.

Proof. In the collar $[0,1]\times Y$, any element $(a(u), b(u))$ decomposes as a triple

$$(a,b) = (a(u)\,,\, b(u) = p(u) + q(u)du) =: (p,q,a)\,. \tag{22.86}$$

Then $\not{D}_A$ can be written in matrix form

$$\not{D}_A \begin{pmatrix} p \\ q \\ a \end{pmatrix} = \Gamma\,(\partial_u - \not{D}_B) \begin{pmatrix} p \\ q \\ a \end{pmatrix}, \tag{22.87}$$

where
(22.88)

$$\Gamma := \begin{pmatrix} * & 0 & 0 \\ 0 & 0 & \mathrm{Id} \\ 0 & -\mathrm{Id} & 0 \end{pmatrix} \quad \text{and} \quad \not{D}_B := \begin{pmatrix} 0 & d_B & *d_B \\ (d_B)^* & 0 & 0 \\ -*(d_B)^* & 0 & 0 \end{pmatrix}.$$

Here $*$ denotes the Hodge star operator on $\Lambda^j(T^*Y)$ for $j = 0,1$ and its section space $\Omega^j(Y)$, and d_B denotes the covariant derivative. Clearly we have

$$\Gamma\not{D}_B + \not{D}_B\Gamma = 0 \quad \text{and} \quad \Gamma^2 = -\mathrm{Id}\,. \tag{22.89}$$

Hence we can apply Theorem 22.24 and get for the corresponding Atiyah-Patodi-Singer boundary problem

$$\operatorname{index}(\not{D}_A)_{P_\geq} = -\frac{1}{2}\dim\ker\not{D}_B\,. \tag{22.90}$$

It is well-known that

$$\ker\not{D}_B = \mathcal{H}^0_B \oplus \mathcal{H}^1_B\,,$$

where

$$\mathcal{H}^j_B := \{\omega \in \Omega^j(Y) \otimes \mathrm{su}(2) \mid d_B\omega = 0 = d^*_B\omega\}$$

(see Taubes [1990], Yoshida [1991] and the references given there). Moreover, since B is irreducible, the covariant derivative d_B has no non-trivial kernel in Ω^0, and $\mathcal{H}^1_B$ is a $(6g-6)$-dimensional real vector space isomorphic to the de Rham cohomology groups $H^1_B(Y;\mathrm{su}(2))$ with su(2)-valued local coefficient system defined by the holonomy representation of B. This proves the formula (22.85). □

23. Some Remarks on the Index of Generalized Atiyah-Patodi-Singer Problems

By combining the Atiyah-Patodi-Singer index formula with our theory of boundary integrals we obtain explicit formulas for all boundary problems in the infinite-dimensional Grassmannian and a non-additivity theorem for the index of Atiyah-Patodi-Singer boundary problems under pasting. We introduce an alternative *adiabatic* Atiyah-Patodi-Singer boundary problem which returns the signature as its true index.

We have the following immediate corollary of the Agranovič-Dynin formula (Proposition 21.4) and the index formula for Atiyah-Patodi-Singer boundary problems (Theorem 22.18).

Theorem 23.1. *Let $P \in \mathrm{Gr}_{p_+}$, where p_+ denotes the principal symbol of the Calderón projector (and the spectral projection $\mathrm{P}_\geq$ as well). Then the following formula holds:*

$$\text{(23.1)} \qquad \operatorname{index} A_P = \int_X \alpha(x) - \frac{1}{2}\left(\eta_B(0) + \dim\ker B\right) + \mathbf{i}(P, \mathrm{P}_\geq)\,.$$

Proof. Recall the definition of the virtual codimension (Definition 15.8)

$$\mathbf{i}(P, \mathrm{P}_\geq) := \operatorname{index}\left\{P\,\mathrm{P}_\geq : \operatorname{range}\mathrm{P}_\geq \to \operatorname{range} P\right\}\,.$$

Then (23.1) follows from the Agranovič-Dynin formula (Proposition 21.4) which gives

$$\operatorname{index} A_P \quad = \quad \operatorname{index} A_{\mathrm{P}_\geq} + \mathbf{i}(P, \mathrm{P}_\geq)\,,$$

and the Atiyah-Patodi-Singer formula for $\operatorname{index} A_{\mathrm{P}_\geq}$. □

Note. In Kori [1993; Theorem 6.4] the situation on S^3 and S^4 is discussed and the following formula is derived for $X = D^4$:

$$\text{(23.2)} \qquad \operatorname{index} A_P = \mathbf{i}(P, \mathrm{P}_\geq)\,.$$

This is in accordance with our result: The index formula depends on the metric on S^4, and in Kori's metric, the boundary component operator B is invertible. Moreover, he proved directly that, in the situation he considered, the spectral projection $\mathrm{P}_{\geq}$ and the Calderón projector $\mathcal{P}_+$ coincide. We refer to Kori [1993] for further details.

We use Theorem 23.1 to discuss the non-additivity of the index. From a topological point of view, one special feature of the operator $A_{\mathrm{P}_{\geq}}$ is that its index is non-additive, i.e. two *consistent* Atiyah-Patodi-Singer boundary problems do not in general sum up to the index of the corresponding Dirac operator over a closed manifold:

Let X_1, X_2 be two compact Riemannian manifolds with boundary Y, and let A_j, $j = 1,2$ be Dirac operators on X_j such that, in a certain collar of Y, they are *consistent* in the following form:

$$A_1 = G(\partial_u + B)\,, \qquad A_2 = G^{-1}(\partial_v - GBG^{-1})\,, \tag{23.3}$$

where v is the normal inward coordinate on X_2 which we can identify with $-u$. We can apply the glueing procedure from Chapter 9. Notice that A_2 is formally equal to A_1^* on the collar. As a result of the pasting, we obtain an operator $\widetilde{A} = A_1 \cup A_2$ on the manifold $X_1 \cup X_2$. One would like to have an equality

$$\operatorname{index}(A_1 \cup A_2) \stackrel{?}{=} \operatorname{index}(A_1)_{\mathrm{P}^1_{\geq}} + \operatorname{index}(A_2)_{\mathrm{P}^2_{\geq}}\,. \tag{23.4}$$

Here $\mathrm{P}^j_{\geq} := \mathrm{P}(A_j)_{\geq 0}$, $j = 1,2$ denotes the Atiyah-Patodi-Singer boundary operator corresponding to the operator A_j. Recall that for any real a, $\mathrm{P}(A_j)_{\geq a}$ denotes the orthogonal projection onto the subspace of $L^2(Y; S^+|_Y)$, resp. onto the subspace of $L^2(Y; S^-|_Y)$ spanned by the eigenfunctions of B, resp. the eigenfunctions of $-GBG^{-1}$ corresponding to eigenvalues greater or equal a. Let $\{\lambda; \varphi_\lambda\}$ be a spectral resolution of $L^2(Y; S^+|_Y)$ generated by B and $\{-\lambda; \psi_{-\lambda} := G\varphi_\lambda\}$ the corresponding resolution of $L^2(Y; S^-|_Y)$ generated by $-GBG^{-1}$. $\mathrm{P}_a := \mathrm{P}^1_a$ in particular then denotes the orthogonal projection onto the a-eigenspace $V_a := \mathcal{L}_{\mathbf{C}}\{\varphi_\lambda \mid \lambda = a\}$ of B. Then

$$G\,\mathrm{P}_0\,G^{-1}(\psi_{-\lambda}) = \begin{cases} \psi_{-\lambda} & \text{if } \lambda = 0 \\ 0 & \text{otherwise} \end{cases}$$

and

$$G(\mathrm{Id} - \mathrm{P}^1_{\geq})G^{-1}(\psi_{-\lambda}) = \begin{cases} 0 & \text{if } \lambda \geq 0 \\ \psi_{-\lambda} & \text{otherwise}\,, \end{cases}$$

hence

$$\mathrm{P}^2_{\geq} = G(\mathrm{Id} - \mathrm{P}^1_{\geq} + \mathrm{P}_0)G^{-1}\,. \tag{23.5}$$

The conjecture of equation (23.4) does not hold in general, or, more precisely, it holds if and only if $\ker B = \{0\}$.

Proposition 23.2. *For Atiyah-Patodi-Singer boundary value problems, the following non-additivity holds under glueing:*

$$\mathrm{index}(A_1)_{\mathrm{P}^1_{\geq}} + \mathrm{index}(A_2)_{\mathrm{P}^2_{\geq}} = \mathrm{index}(A_1 \cup A_2) - \dim\ker B\,. \tag{23.6}$$

Proof. This follows directly from Theorem 22.18 which yields

$$\mathrm{index}(A_1)_{\mathrm{P}^1_{\geq}} = \int_{X_1} \alpha_1(x) - \frac{1}{2}\cdot(\dim\ker B + \eta_B(0))$$

and

$$\begin{aligned}\mathrm{index}(A_2)_{\mathrm{P}^2_{\geq}} &= \int_{X_2} \alpha_2(x) - \frac{1}{2}\cdot\big(\dim\ker GBG^{-1} + \eta_{-GBG^{-1}}(0)\big)\\ &= \int_{X_2} \alpha_2(x) - \frac{1}{2}\cdot(\dim\ker B - \eta_B(0))\,.\end{aligned}$$

□

We give the most general result of this type. Much deeper non-additivity results can be obtained for spectral invariants like the analytic torsion and the η-invariant (see Klimek & Wojciechowski [1992], Matteo & Melrose [1992], Booß-Bavnbek & Wojciechowski [1993], Bunke [1993], Lück [1993], and Müller [1993]). The following result also generalizes the calculations about exchanges on the boundary for the Cauchy-Riemann operator given in Example 21.1.

Theorem 23.3. *Let X_1, X_2 be two compact Riemannian manifolds with common boundary Y and let A_j, $j = 1,2$ be Dirac operators on X_j satisfying the consistency condition (22.1). Let $a \in \mathbf{R}$ and let $(A_j)_a$ denote the operator $(A_j)_{\mathrm{P}^j_{\geq a}}$. We have*

$$\mathrm{index}(A_1)_a + \mathrm{index}(A_2)_{-a} = \mathrm{index}(A_1 \cup A_2) - \dim V_a \tag{23.7}$$

and

(23.8)

$$\mathrm{index}(A_1)_a + \mathrm{index}(A_2)_a = \mathrm{index}(A_1 \cup A_2) + \mathbf{i}(\mathrm{P}^1_{\geq a}, \mathrm{P}^1_{\geq -a}) - \dim V_{-a}\,.$$

Remark 23.4. Recall that $\mathrm{P}^1_{\geq a}$ denotes the weighted spectral projection of the operator B, namely the orthogonal projection onto the subspace $V_{\geq a}$ spanned by the eigensections of B corresponding to eigenvalues in $[a, \infty)$, and that $\mathrm{P}^2_{\geq a}$ denotes the corresponding weighted spectral projection of the operator $-GBG^{-1}$. Moreover, V_a denotes the a-eigenspace of B (if $a \in \mathrm{spec}(B)$; otherwise $V_a := \{0\}$). By definition we have $\mathrm{P}^j_{\geq} = \mathrm{P}^j_{\geq 0}$. Moreover, for $a \geq b \geq c$ we have $\mathrm{P}^j_{\geq a}\,\mathrm{P}^j_{\geq b} = \mathrm{P}^j_{\geq a}$ and by Proposition 15.15

$$\mathbf{i}(\mathrm{P}^j_{\geq a}, \mathrm{P}^j_{\geq b}) + \mathbf{i}(\mathrm{P}^j_{\geq b}, \mathrm{P}^j_{\geq c}) = \mathbf{i}(\mathrm{P}^j_{\geq a}, \mathrm{P}^j_{\geq c}). \tag{23.9}$$

Finally, for $a > b$, we have two short exact sequences

$$\mathrm{range}(\mathrm{P}^1_{\geq b}) \xrightarrow{\mathrm{P}^1_{\geq a}\,\mathrm{P}^1_{\geq b}} \mathrm{range}(\mathrm{P}^1_{\geq a}) \longrightarrow \{0\}$$

and

$$\{0\} \longrightarrow \mathrm{range}\,\mathrm{P}^1_{\geq a} \xrightarrow{\mathrm{P}^1_{\geq b}\,\mathrm{P}^1_{\geq a}} \mathrm{range}\,\mathrm{P}^1_{\geq b} \longrightarrow \bigoplus_{b \leq \lambda < a} V_\lambda \longrightarrow \{0\},$$

which yield

$$\mathbf{i}(\mathrm{P}^1_{\geq a}, \mathrm{P}^1_{\geq b}) = \sum_{b \leq \lambda < a} \dim V_\lambda \quad \text{and} \quad \mathbf{i}(\mathrm{P}^1_{\geq b}, \mathrm{P}^1_{\geq a}) = -\sum_{b \leq \lambda < a} \dim V_\lambda .$$

In particular we get that $\mathbf{i}(\mathrm{P}^1_{\geq a}, \mathrm{P}^1_{\geq -a})$ is equal to the dimension of the direct sum of eigenspaces corresponding to the eigenvalues of B in the interval $[-a, a)$.

Proof of Theorem 23.3. First we express the spectral projections of $-GBG^{-1}$ in terms of the spectral projections of B, as in the calculations for (23.5). From $-GBG^{-1}(G\varphi_\lambda) = -\lambda(G\varphi_\lambda)$, we get

$$\mathrm{P}^2_{\geq a} = G(\mathrm{Id} - \mathrm{P}^1_{\geq -a} + \mathrm{P}_a)G^{-1}. \tag{23.10}$$

We have already proved (Proposition 23.2) that

$$\mathrm{index}(A_1)_0 + \mathrm{index}(A_2)_0 = \mathrm{index}(A_1 \cup A_2) - \dim\ker B .$$

Let us choose $\varepsilon > 0$ such that

$$\mathrm{spec}\, B \cap (-2\varepsilon, 2e) \subset \{0\} \quad \text{and} \quad \mathrm{spec}\, B \cap (-a-2\varepsilon, -a+2e) \subset \{-a\}.$$

Then clearly

$$\mathbf{i}(\mathrm{P}^1_{\geq\varepsilon}, \mathrm{P}^1_{\geq-\varepsilon}) = \dim\ker B \qquad \text{and} \qquad \mathbf{i}(\mathrm{P}^1_{\geq a}, \mathrm{P}^1_{\geq 0}) = \mathbf{i}(\mathrm{P}^1_{\geq a}, \mathrm{P}^1_{\geq-\varepsilon}).$$

We also notice

$$\operatorname{index}(A_1)_0 = \operatorname{index}(A_1)_\delta + \mathbf{i}(\mathrm{P}^1_{\geq 0}, \mathrm{P}^1_{\geq\delta}) = \operatorname{index}(A_1)_\delta$$

for $-2\varepsilon < \delta \leq 0$.

With (23.10) we obtain

(23.11)

$$\begin{aligned}
\mathbf{i}(\mathrm{P}^2_{\geq a}, \mathrm{P}^2_{\geq 0}) &= \mathbf{i}(G(\mathrm{Id} - \mathrm{P}^1_{\geq -a} + \mathrm{P}_{-a})G^{-1}, G(\mathrm{Id} - \mathrm{P}^1_{\geq 0} + \mathrm{P}_0)G^{-1}) \\
&= \mathbf{i}(\mathrm{Id} - (\mathrm{P}^1_{\geq -a} - \mathrm{P}_{-a}), \mathrm{Id} - (\mathrm{P}^1_{\geq 0} - \mathrm{P}_0)) \\
&= \mathbf{i}(\mathrm{P}^1_{\geq 0} - \mathrm{P}_0, \mathrm{P}^1_{\geq -a} - \mathrm{P}_{-a}) = \mathbf{i}(\mathrm{P}^1_{\geq 0} - \mathrm{P}_0, \mathrm{P}^1_{\geq -a+\varepsilon}) \\
&= \mathbf{i}(\mathrm{P}^1_{\geq 0} - \mathrm{P}_0, \mathrm{P}^1_{\geq -a}) + \mathbf{i}(\mathrm{P}^1_{\geq -a}, \mathrm{P}^1_{\geq -a+\varepsilon}) \\
&= \mathbf{i}(\mathrm{P}^1_{\geq 0} - \mathrm{P}_0, \mathrm{P}^1_{\geq -a}) - \dim V_{-a} \\
&= \mathbf{i}(\mathrm{P}^1_{\geq \varepsilon}, \mathrm{P}^1_{\geq -a}) - \dim V_{-a}\,.
\end{aligned}$$

We use the generalized Agranovič-Dynin formula (Proposition 21.4) and obtain

$$\begin{aligned}
\operatorname{index}(A_1)_a &= \operatorname{index}(A_1)_0 + \mathbf{i}(\mathrm{P}^1_{\geq a}, \mathrm{P}^1_{\geq 0}) \qquad \text{and} \\
\operatorname{index}(A_2)_{\geq a} &= \operatorname{index}(A_2)_{\geq 0} + \mathbf{i}(\mathrm{P}^2_{\geq a}, \mathrm{P}^2_{\geq 0})\,.
\end{aligned} \tag{23.12}$$

We have

$$\begin{aligned}
\operatorname{index}(A_1)_a + \operatorname{index}(A_2)_a &= \{\operatorname{index}(A_1 \cup A_2) - \dim\ker B\} \\
&\qquad + \mathbf{i}(\mathrm{P}^1_{\geq a}, \mathrm{P}^1_{\geq 0}) + \mathbf{i}(\mathrm{P}^2_{\geq a}, \mathrm{P}^2_{\geq 0}) \\
&\overset{(23.11)}{=} \{\dots\} + \mathbf{i}(\mathrm{P}^1_{\geq a}, \mathrm{P}^1_{\geq 0}) + \mathbf{i}(\mathrm{P}^1_{\geq\varepsilon}, \mathrm{P}^1_{\geq -a}) - \dim V_{-a} \\
&= \{\dots\} + \mathbf{i}(\mathrm{P}^1_{\geq a}, \mathrm{P}^1_{\geq -\varepsilon}) \\
&\qquad + \mathbf{i}(\mathrm{P}^1_{\geq\varepsilon}, \mathrm{P}^1_{\geq -\varepsilon}) + \mathbf{i}(\mathrm{P}^1_{\geq -\varepsilon}, \mathrm{P}^1_{\geq -a}) - \dim V_{-a} \\
&= \operatorname{index}(A_1 \cup A_2) + \mathbf{i}(\mathrm{P}^1_{\geq a}, \mathrm{P}^1_{\geq -\varepsilon}) \\
&\qquad + \mathbf{i}(\mathrm{P}^1_{\geq -\varepsilon}, \mathrm{P}^1_{\geq -a}) - \dim V_{-a} \\
&= \operatorname{index}(A_1 \cup A_2) + \mathbf{i}(\mathrm{P}^1_{\geq a}, \mathrm{P}^1_{\geq -a}) - \dim V_{-a}\,,
\end{aligned}$$

which proves (23.8). In the same way we prove (23.7):

$$
\begin{aligned}
\operatorname{index}(A_1)_a + \operatorname{index}(A_2)_{-a} \\
&= \{\operatorname{index}(A_1 \cup A_2) - \dim\ker B\} \\
&\qquad + \mathbf{i}(\mathrm{P}^1_{\geq a}, \mathrm{P}^1_{\geq -\varepsilon}) + \mathbf{i}(\mathrm{P}^2_{\geq -a}, \mathrm{P}^2_{\geq -\varepsilon}) \\
&= \{\dots\} + \mathbf{i}(\mathrm{P}^1_{\geq a}, \mathrm{P}^1_{\geq -\varepsilon}) + \mathbf{i}(\mathrm{P}^1_{\geq \varepsilon}, \mathrm{P}^1_{\geq a}) - \dim V_a \\
&= \{\dots\} + \mathbf{i}(\mathrm{P}^1_{\geq \varepsilon}, \mathrm{P}^1_{\geq -\varepsilon}) - \dim V_a \\
&= \operatorname{index}(A_1 \cup A_2) - \dim V_a\,. \qquad \square
\end{aligned}
$$

Theorem 23.3 shows that we can not expect additivity of the index for Atiyah-Patodi-Singer problems. In the following we discuss the important case of the signature operator. For this operator we give a natural boundary condition which provides the desired additivity.

First we notice that the signature of X is not the index of a true Atiyah-Patodi Singer problem $A_{\mathrm{P}_\geq}$. From Atiyah, Patodi & Singer [1975, Section 4, Theorem 4.14 and Equation (17.9)] we extract the two following formulas:

$$
\begin{aligned}
\operatorname{sign} X &= \int_X \mathcal{L}(x) - \frac{1}{2}\eta_B(0) \quad \text{and} \\
\operatorname{index} A_{\mathrm{P}_\geq} &= \int_X \mathcal{L}(x) - h - \eta_{B^{\mathrm{ev}}}(0),
\end{aligned}
$$

and hence the *signature deficiency formula:*

$$
\operatorname{sign} X = \operatorname{index} A_{\mathrm{P}_\geq} + h\,. \tag{23.13}
$$

Here X is an oriented Riemannian manifold of dimension $4k$ with smooth boundary Y (and X is isometric to a product near Y), $A : \Omega_+ \to \Omega_-$ denotes the signature operator introduced in Chapter 7 earlier. The signature operator takes the form $A = G(\partial_u + B)$ close to Y. The tangential component B splits into $B = B^{\mathrm{ev}} + B^{\mathrm{odd}}$ with B^{ev} isomorphic to B^{odd}. Hence

$$
\eta_{B^{\mathrm{ev}}}(0) = \frac{1}{2}\eta_B(0) \quad \text{and} \quad h := \dim\ker B^{\mathrm{ev}} = \frac{1}{2}\dim\ker B\,. \tag{23.14}
$$

Moreover, $\mathcal{L}(x) := L_k(p_1(x), \dots, p_k(x))$, where L_k denotes the k-th Hirzebruch L-polynomial and the p_j denote the Pontrjagin forms of the Riemannian metric. We introduce the notation

$$
V_\geq := \mathcal{L}_{\mathbf{C}}\{\varphi_\lambda \mid B\varphi_\lambda = \lambda\varphi_\lambda \text{ and } \lambda \geq 0\}\,,
$$

so

$$\mathrm{P}_{\geq} : \Omega_+|_Y \xrightarrow{\text{orth.}} V_{\geq}\,.$$

There exists a *natural projection* $\mathrm{P}_\Sigma : \Omega_+|_Y \to \operatorname{range} \mathrm{P}_\Sigma$ belonging to the Grassmannian Gr_{p_+} such that

$$\operatorname{sign} X = \operatorname{index} A_{\mathrm{P}_\Sigma}, \tag{23.15}$$

or equivalently, by (23.1), such that
(23.16)
$$\operatorname{index}\{\mathrm{P}_\Sigma \mathrm{P}_{\geq} : \operatorname{range} \mathrm{P}_{\geq} \to \operatorname{range} \mathrm{P}_\Sigma\} = \operatorname{index} A_{\mathrm{P}_\Sigma} - \operatorname{index} A_{\mathrm{P}_{\geq}} = h\,.$$

Clearly the projection P_Σ onto any subspace Σ of $\operatorname{range} \mathrm{P}_{\geq}$ of codimension h will provide (23.15), since then

$$\mathrm{P}_\Sigma \mathrm{P}_{\geq} : \operatorname{range} \mathrm{P}_{\geq} \to \Sigma$$

is surjective and $\ker P\mathrm{P}_{\geq} \cong \operatorname{range} \mathrm{P}_{\geq}/\Sigma$.

We use results of Section 22E to describe a choice of Σ which appears in this context in a natural way. There is actually a very natural subspace of

$$\operatorname{range} \mathrm{P}_{\geq} = V_{\geq} = V_{>} + \ker B$$

of codimension h, namely the space $V_{>} + \mathcal{H}_+$, where

$$V_{>} := \mathcal{L}_{\mathbf{C}}\{\varphi_\lambda \mid \lambda > 0\} \tag{23.17}$$

and $\mathcal{H}_+$ denotes the space of limiting values of the *extended* solutions, as introduced in Definition 22.19d. Then the projection onto $V_{>} + \mathcal{H}_+$ will do the job of our desired alternative *signature*-amended Atiyah-Patodi-Singer boundary condition P_Σ.

We recall from Section 22E the concept of a non-compact Riemannian *elongation* X_∞; of related extensions, especially of the signature operator A to A_∞; and the concept of an *extended solution (extended square-integrable harmonic form of even parity)*. As seen above, these concepts are meaningful in much more general situations. The following lemma, however, depends heavily on the internal symmetry of the tangential part of the signature operator. The lemma is proved in Atiyah, Patodi & Singer [1975; pp 64–66 (before Theorem 4.14)] but not explicitly stated there:

Lemma 23.5. *Let $\mathcal{H}_+$ denote the subspace of* $\ker(B)$ *which consists of the limiting values of extended L^2-harmonic forms of even parity. It has dimension* $h = \dim\ker B^{\mathrm{ev}} = \frac{1}{2}\dim\ker B$.

We fix the orthogonal projection $\Theta : \ker B \to \ker B$ of $\ker B$ onto $\mathcal{H}_+$ and introduce the projection $\mathrm{P}_>$ of the L^2 differential forms of even parity over Y onto the subspace $V_>$ defined in (23.17). Now we define

$$\mathrm{P}_\Sigma := \mathrm{P}_> + \Theta\,. \tag{23.18}$$

Since P_Σ is only a perturbation of the pseudo-differential projection $\mathrm{P}_\geq$ by an operator of finite range, we obtain from equation (23.1) with Lemma 23.5 and equation (23.16) the following theorem:

Theorem 23.6. *The projection* P_Σ *defined in equation (23.18) belongs to the Grassmannian* Gr_{p_+} *and defines a generalized Atiyah-Patodi-Singer boundary value problem with*

$$\operatorname{index} A_{\mathrm{P}_\Sigma} = \operatorname{sign} X\,.$$

Corollary 23.7. *We have*

$$L^2\text{-index}(A) = \int_{X_\infty} \alpha(x) - \frac{1}{2}\,\eta_{B^{\mathrm{ev}}}(0)\,, \tag{23.19}$$

where X_∞ denotes the non-compact elongation of X, the L^2-index is defined as the difference of the dimensions of the spaces of 'extended solutions' of even and odd parity over the elongation (see Definition 22.19d above), A denotes the signature operator of X, and $\alpha(x)$ denotes the index density of A which is naturally extended over the whole of the elongation.

Remarks 23.8. (a) The preceding theorem permits a reformulation of the cutting and pasting invariance of the signature which we shall discuss in Chapter 25. Moreover, it gives a nice frame for the analytical explanation of the Novikov additivity of the signature (see e.g. Jänich [1968], Karras et al. [1973], and Booß & Rempel [1982]) and, when treating Riemannian manifolds with corners, an analytical explanation of the Wall non-additivity, see Wall [1969], Rees [1983], and Melrose [1992].

(b) Theorem 23.6 and Corollary 23.7 do not hold for arbitrary Dirac operators. The reason is that in general $h = h(A) \neq h(A^*)$. However, one has the following formula for the L^2-index (Corollary 22.21):

$$\text{(23.20)} \qquad \operatorname{index} A_{P_\geq} = L^2\text{-index}(A) - h(A^*)\,,$$

and more generally

$$\text{(23.21)} \qquad L^2\text{-index}(A) = \int_X \alpha(x) - \frac{1}{2}\,\eta(0) - \frac{1}{2}\,(h(A) - h(A^*))\,.$$

See also Melrose [1981] and Roe [1988].

Anyway, we have no problems calculating the L^2-index in the case of vanishing $\ker B$.

Corollary 23.9. *Let A be a Dirac operator with non-singular tangential part B. Then*

$$\operatorname{index} A_{P_\geq} = L^2\text{-index}(A) = \int_X \alpha(x) - \frac{1}{2}\,\eta(0)\,.$$

Notice that we used the preceding corollary in Section 17D above.

One more reason for the choice of our alternative boundary condition is the odd-dimensional case. It is well-known that for $\dim X$ odd, $A_{P_\geq}$ is not a self-adjoint operator whenever $\ker B \neq \{0\}$. See Booß & Wojciechowski [1989; Proposition 5]. However, one obtains that A_{P_Σ} is an elliptic self-adjoint problem in this case.

24. Bojarski's Theorem. General Linear Conjugation Problems

We recover the index of a Dirac operator A over a closed partitioned manifold $M = X_+ \cup X_-$ with $\partial X_+ = \partial X_- = X_+ \cap X_- = Y$ from the *Fredholm pair* of Cauchy data spaces along Y. Similarly, the index of the linear conjugation (or *transmission*) problem $As_\pm = 0$ in $X_\pm \setminus Y$ and $s_-|_Y = \Phi\,(s_+|_Y)$ is given by twisting the Cauchy data spaces with Φ. Related local elliptic boundary conditions for systems of Dirac operators are considered.

In this chapter we present some applications of the theory developed in the previous chapters and explain the geometric meaning of the spaces of Cauchy data and of the Calderón projector, i.e. objects which *live* on the submanifold Y but originate from global data.

We deal with a closed Riemannian manifold M which is partitioned into two manifolds with joint boundary:

$$M = X_- \cup X_+,\ X_- \cap X_+ = Y = \partial X_- = \partial X_+. \tag{24.1}$$

Let $A : C^\infty(M;E) \to C^\infty(M;F)$ be a Dirac operator on M. We assume the splitting form (9.3) in a bicollar neighbourhood of Y (see Figure 24.1). Then on Y and close to Y we have the objects and relations introduced earlier in a slightly more restrictive situation, namely for the closed double $M := \widetilde{X}_+$ and $A := \widetilde{A}_+ \approx A_+ \cup (A_+)^*$.

Theorem 24.1. *Let $A_\pm$ denote the restrictions of the operator A to the manifolds $X_\pm$ with M, $X_\pm$, A as above. Then the following equality holds:*

$$\operatorname{index} A = \operatorname{index}\{(\mathrm{Id} - \mathcal{P}(A_-))\mathcal{P}(A_+) : H(A_+) \to H(A_-)^\perp\}. \tag{24.2}$$

The theorem was conjectured by Bojarski in the middle of the seventies. He envisaged it as the first step toward a *surgery type* proof of the Atiyah-Singer index theorem via a double induction argument: with respect to the number of handles in a Morse decomposition of

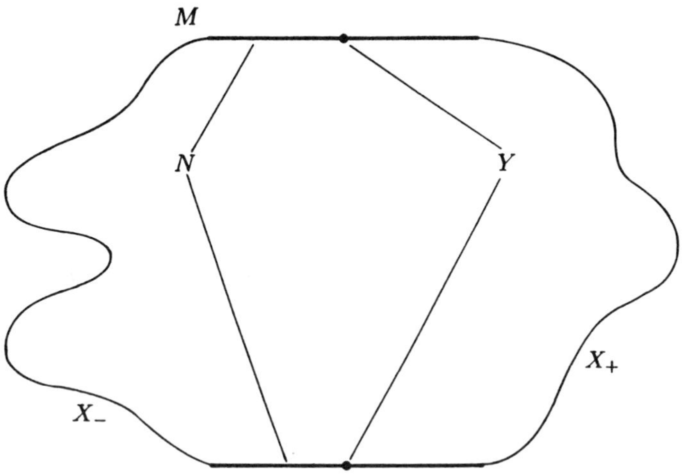

Fig. 24.1 The partitioned manifold M with bicollar neighbourhood M of Y

the manifold, and with respect to the dimension of the manifold. This will be discussed (to some extent) in the next chapter.

Bojarski formulated his conjecture as the equality of the index of A and the index of the *Fredholm pair* of Cauchy data of the operator A along the submanifold Y. Let us explain his formulation. We begin with the concept of a Fredholm pair of subspaces.

Definition 24.2. Let H_1, H_2 be closed infinite-dimensional subspaces of a separable Hilbert space H. We call (H_1, H_2) a *Fredholm pair of subspaces*, if the following conditions are fulfilled:

$$\alpha = \dim(H_1 \cap H_2) < \infty \text{ and} \tag{24.3}$$

$$\beta = \dim(H/H_1 + H_2) < \infty\,. \tag{24.4}$$

The integer $\alpha - \beta$ is called the *index* of the pair (H_1, H_2).

The notion was introduced by Kato [1976; IV.4.1] and used by Bojarski in his paper [1979] about the abstract linear conjugation problem for elliptic operators.

Clearly we have a natural embedding of the space of Fredholm operators on H into the space of Fredholm pairs of subspaces of $H \times H$ given by $F \mapsto (H \times 0, \operatorname{graph} F)$ which preserves the index. More generally we have:

Lemma 24.3. *A pair (H_1, H_2) of closed infinite-dimensional subspaces of a (separable) Hilbert space H is a Fredholm pair, if and only* *if the operator*

$$(\mathrm{Id} - P_2)P_1 : H_1 \to {H_2}^{\perp}$$

is a Fredholm operator. Here P_i denotes the orthogonal projection onto H_i, $i = 1, 2$. In this case we have

$$\text{(24.5)} \qquad \operatorname{index}(H_1, H_2) = \operatorname{index}(\mathrm{Id} - P_2)P_1 = \mathbf{i}((\mathrm{Id} - P_2)P_1)\,.$$

Proof. We have

$$\ker(\mathrm{Id} - P_2)P_1 = \{f \in H_1 \mid (\mathrm{Id} - P_2)f = 0\} = H_1 \cap H_2$$

and

$$\operatorname{coker}(\mathrm{Id} - P_2)P_1 = \{f \in H_2^{\perp} \mid P_1 f = 0\} = H_1^{\perp} \cap H_2^{\perp} = (H_1 + H_2)^{\perp},$$

which proves the lemma. □

Proposition 24.4. *Let (H_+, H_-) be a pair of orthogonally complementary subspaces of a fixed Hilbert space H with orthogonal projections $P_\pm$ of H onto $H_\pm$. Then for any bounded invertible operator Φ on H, which commutes with $\mathcal{P}_+$ modulo compact operators, the pair $(\Phi H_+, H_-)$ is a Fredholm pair of subspaces and we have*

$$\text{(24.6)} \qquad \operatorname{index}(\Phi H_+, H_-) = \operatorname{index}(\mathcal{P}_- - \Phi\mathcal{P}_+)\,.$$

Note. If (H_+, H_-) is just a Fredholm pair of (not necessarily orthogonally complementary) subspaces of H, we also get that $(\Phi H_+, H_-)$ is a Fredholm pair by compact perturbation. However, formula (24.6) has to be replaced by

$$\begin{aligned} \operatorname{index}(\Phi H_+, H_-) &= \mathbf{i}(\Phi^{-1}(\mathrm{Id} - \mathcal{P}_-)\Phi, \mathcal{P}_+) \\ &= \operatorname{index}\{(\mathrm{Id} - \mathcal{P}_-)\Phi\mathcal{P}_+ : H_+ \to H_-^{\perp}\} \qquad \text{(24.7)} \\ &= \operatorname{index}(H_+, H_-) + \operatorname{index}(\mathcal{P}_- - \Phi\mathcal{P}_+)\,. \end{aligned}$$

See also Theorem 24.9 below.

Proof. From Lemma 24.3 we obtain

$$\begin{aligned} \operatorname{index}(\Phi H_+, H_-) &= \operatorname{index}\{(\mathrm{Id} - \mathcal{P}_-)\Phi\mathcal{P}_+\Phi^{-1} : \Phi H_+ \to H_-^{\perp}\} \\ &= \operatorname{index}\{\mathcal{P}_+\Phi\mathcal{P}_+\Phi^{-1} : \Phi H_+ \to H_+\} \\ &= \operatorname{index}\{\mathcal{P}_+\Phi\mathcal{P}_+ : H_+ \to H_+\} \\ &= \operatorname{index}\{(\mathrm{Id} - \mathcal{P}_+) - \mathcal{P}_+\Phi\mathcal{P}_+ : H \to H\}\,. \end{aligned}$$

But $(\mathrm{Id} - \mathcal{P}_+) - \mathcal{P}_+\Phi\mathcal{P}_+$ differs from $(\mathrm{Id} - \mathcal{P}_+) - \Phi\mathcal{P}_+ = \mathcal{P}_- - \Phi\mathcal{P}_+$ only by a compact operator which proves our assertion. □

Remark 24.5. Lemma 24.3 has a series of nice consequences for the homotopy type of the space of Fredholm pairs of subspaces when this space is endowed with a metric inherited from the operator norm of the corresponding orthogonal projections by the formula

$$\rho((H_1, H_2), (H_1', H_2')) := \|P_1 - P_1'\| + \|P_2 - P_2'\| . \tag{24.8}$$

All the appropriate technical results were proved in Booß & Wojciechowski [1986, Section 2].

For abbreviation, we shall write $\mathrm{Fred}^2(H)$ for the space of Fredholm pairs of subspaces of a Hilbert space H, and $\mathrm{Fred}^2{}_{H_1}(H)$ for the subspace of all Fredholm pairs (H_1, H_2) with fixed H_1. Then we have:

(a) A pair (H_1, H_2) of closed infinite-dimensional subspaces is Fredholm if and only if the difference $(\mathrm{Id} - P_2) - P_1$ of the corresponding projections is a compact operator. In fact, we rewrite $\mathrm{Id} - P_2 = g^{-1} P_1 g$ with a suitable unitary operator g, and in particular, we obtain $H_2^\perp = g^{-1} H_1$ and

$$(\mathrm{Id} - P_2) P_1 = g^{-1} P_1 g P_1 : H_1 \to g H_1 .$$

Hence (H_1, H_2) is a Fredholm pair if and only if the operator $P_1 g P_1 : H_1 \to H_1$ is Fredholm. That is the case if and only if g (and hence g^{-1}) commutes with P_1 modulo compact operators. Hence

$$(\mathrm{Id} - P_2) - P_1 = g^{-1} P_1 g - P_1 = (g^{-1} P_1 - P_1 g^{-1}) g$$

is a compact operator.

We also see that then $(\mathrm{Id} - P_1) + g P_1$ is a Fredholm operator (with $(\mathrm{Id} - P_1) + P_1 g^{-1}$ as parametrix) and obtain as in the proof of Proposition 24.4

$$\mathrm{index}(H_1, (g^{-1} H_1)^\perp) = \mathrm{index}\left((\mathrm{Id} - P_1) + g P_1\right) . \tag{24.9}$$

(b) Now let $\mathcal{P}_+$ be a fixed orthogonal projection with infinite-dimensional range and kernel. We have seen that any Fredholm pair (H_1, H_2) with $H_1 = \mathcal{P}_+ H =: H_+$ is of the form $(H_+, (\mathrm{Id} - g^{-1} \mathcal{P}_+ g) H)$ for some unitary operator g which commutes with $\mathcal{P}_+$ modulo compact operators. This provides an identification of $\mathrm{Fred}^2{}_{H_+}$ with the homogeneous space

$$\mathcal{U}_{\mathcal{P}_+} / \mathcal{U}(H_+) \oplus \mathcal{U}(H_+^\perp) , \tag{24.10}$$

where $\mathcal{U}$ denotes the spaces of unitary operators and $\mathcal{U}_{\mathcal{P}_+}$ denotes the space of unitary operators on H which commute with $\mathcal{P}_+$ modulo compact operators.

The space of (24.10) has the homotopy type of the space of Fredholm operators on H. This follows from the fact that the natural projection

$$\mathcal{U}_{\mathcal{P}_+} \to \mathcal{U}_{\mathcal{P}_+}/\mathcal{U}(H_+) \oplus \mathcal{U}(H_+^\perp)$$

is a principal fibre bundle with contractible fibre and structural group $\mathcal{U}(H_+) \oplus \mathcal{U}(H_+^\perp)$. We do not reproduce all the details of the calculations here, but only mention that the total space $\mathcal{U}_{\mathcal{P}_+}$ is homotopically equivalent to the space $GL_{\mathcal{P}_+}$ of all bounded invertible operators in H which commute with $\mathcal{P}_+$ modulo compact operators. And that group is of the same homotopy type as the space of Fredholm operators.

Finally, we recognize that the whole space $\mathrm{Fred}^2(H)$ of Fredholm pairs of subspaces is homeomorphic to the product space

$$Proj^\infty(H) \times \mathcal{U}_{\mathcal{P}_+}/\mathcal{U}(H_+) \oplus \mathcal{U}(H_+^\perp)\,,$$

where $Proj^\infty(H)$ denotes the space of projections in H with infinite-dimensional kernel and range. Since $Proj^\infty(H)$ is contractible by Kuiper's theorem, we have a homotopy equivalence between $\mathrm{Fred}^2(H)$ and the space of Fredholm operators on H.

(c) Clearly the total Grassmannian $\overline{\mathrm{Gr}(\mathcal{P}_+)}$ of (not necessarily pseudo-differential) projections in H, which differ from $\mathcal{P}_+$ by a compact operator, is nothing but the space $\mathrm{Fred}^2{}_{H_+}$ of Fredholm pairs of subspaces with fixed first component $H_+ = \mathcal{P}_+ H$. The identification is given by the map $P \mapsto (H_+, (\mathrm{Id} - P)H)$.

Theorem 24.1 can be reformulated in the language of Fredholm pairs by applying Lemma 24.3:

Theorem 24.1. (Reformulated). *The pair of the L^2-closures of the Cauchy data spaces $H(A_\pm)$ of any Dirac operator A over a partitioned closed manifold $M = X_- \cup X_+$, taken at the cut submanifold $Y = X_- \cap X_+ = \partial X_- = \partial X_+$, is a Fredholm pair of subspaces of the corresponding $L^2(Y)$ section space and we have*

$$\text{(24.11)} \qquad \operatorname{index} A = \operatorname{index}(H(A_+), H(A_-))\,.$$

To keep our notation handy, in the rest of this chapter $H(A_\pm)$ denotes the closure in $L^2(Y; E|_Y)$ of

$$\text{(24.12)} \quad \{u_\pm|_Y \mid u_\pm \in C^\infty(X_\pm; E|_{X_\pm}) \text{ and } A_\pm u_\pm = 0 \text{ in } X_\pm \setminus Y\}\,.$$

Proof. It follows from the unique continuation property for Dirac operators (Theorem 8.2) that

$$\ker A \cong H(A_+) \cap H(A_-)\,, \tag{24.13}$$

see the intersection lemma 12.3. That is the trivial part of Theorem 24.1.

Now we apply our Corollary 12.6 and get

$$\begin{aligned} \operatorname{coker} A = \ker A^* &\cong H(A_+^*) \cap H(A_-^*) \\ &= G\left(H(A_+)^\perp\right) \cap G\left(H(A_-)^\perp\right) \cong H(A_+)^\perp \cap H(A_-)^\perp. \end{aligned} \tag{24.14}$$

The last space is evidently the orthogonal complement of $H(A_+) + H(A_-)$ in $L^2(Y; E|_Y)$. □

Remark 24.6. As long as we are only interested in the index of Dirac operators, we may restrict ourselves to the case of even-dimensional manifolds, since the index of differential operators vanishes over (closed) odd-dimensional manifolds. This fits nicely with formula (24.11) because Dirac operators over odd-dimensional manifolds are self-adjoint, so

$$H(A_\pm) = H(A_\pm^*) = G\left(H(A_\pm)^\perp\right) \tag{24.15}$$

and hence

$$\begin{aligned} H(A_+) \cap H(A_-) &= G(H(A_+)^\perp) \cap G(H(A_-)^\perp) \\ &\cong H(A_+)^\perp \cap H(A_-)^\perp = (H(A_+) + H(A_-))^\perp \end{aligned} \tag{24.16}$$

which means $\operatorname{index}(H(A_+), H(A_-)) = 0$.

To obtain a truly *odd* variant of Bojarski's theorem, some replacements must be made on both sides of formula (24.11): The left side must be replaced by the spectral flow of a suitable family of self-adjoint elliptic boundary problems over a fixed cylindrical collar of Y in M. The right side must be replaced by the Maslov index of the corresponding family of Cauchy data spaces. This was worked out by L. Nicolaescu [1993].

More generally, we shall investigate *the linear conjugation problem* for a Dirac operator $A = A_+ \cup A_- : C^\infty(M; E) \to C^\infty(M; F)$ over a partitioned manifold $M = X_+ \cup X_-$ along $X_+ \cap X_- = Y$.

Definition 24.7. (a) Under the preceding conditions we denote by $\text{LCP}(A, \Phi)$ the following problem: We look for couples (u_+, u_-), where $u_\pm \in H^1(X_\pm; E|_{X_\pm})$ such that

$$A_\pm u_\pm = 0 \text{ in } X_\pm \setminus Y \quad \text{and} \quad u_-|_Y = \Phi(u_+|_Y)\,. \tag{24.17}$$

Here Φ shall be an automorphism of $E|_Y$ with the consistency condition

$$\Phi p_+ = p_+ \Phi \tag{24.18}$$

of Φ with the symbol p_+. To simplify the notation we assume that Φ is unitary.
(b) We let

$$\text{index}_{\text{LCP}}(A, \Phi)$$

denote the difference between the dimensions of the kernel of the original linear conjugation problem (24.17) and of the kernel of the *adjoint linear conjugation problem*, where we look for couples (v_+, v_-) with $v_\pm \in H^1(X_\pm; F|_{X_\pm})$ satisfying

$$(A_\pm)^*(v_\pm) = 0 \quad \text{in } X_\pm \text{ and} \quad v_-|_Y = \Phi(v_+|_Y)\,. \tag{24.19}$$

In the case of a Dirac operator with coefficients in an auxiliary bundle V, we have $E := S^+ \otimes V$ and $F := S^- \otimes V$, and condition (24.18) is satisfied by $\Phi := \text{Id} \otimes g$ for any automorphism $g : V|_Y \to V|_Y$ of V on Y. This raises the following:

Problem 24.8. *Describe the subgroup* $\text{Aut}_{p_+}(E|_Y)$ *of the automorphism group* $\text{Aut}(E|_Y)$ *of the vector bundle* $E|_Y$ *(with shift permitted in the base) which preserves the symbol* p_+*, or, in other words, acts nicely on the Grassmannian* Gr_{p_+} *by*

$$\begin{array}{ccc} \text{Aut}_{p_+}(E|_Y) & \hookrightarrow & \text{Aut}(\text{Gr}_{p_+}) \\ \Phi & \mapsto & (P \mapsto \Phi P \Phi^{-1}). \end{array}$$

Now we turn to the index problem for linear conjugation problems. There we are confronted with two methodological problems: First, there is no well-defined operator in $L^2(M; E)$ which contains all information about a given linear conjugation problem. Second, the obvious question is whether we have a topological formula for the index of linear conjugation problems. The answer to both problems is given in the next theorem.

Theorem 24.9. *Let* $A : C^\infty(M;E) \to C^\infty(M;F)$ *be a Dirac operator over a partitioned manifold* $M = X_+ \cup X_-$ *and* $\Phi \in \operatorname{Aut}_{p_+}(E|_Y)$, *where* Y *denotes the cutting submanifold. Then the pair* $(\Phi H(A_+), H(A_-))$ *is a Fredholm pair of subspaces of* $L^2(Y;E|_Y)$, *and we get the following equalities for the index of the linear conjugation problem* ${}_{\mathrm{LCP}}(A,\Phi)$:

$$\begin{aligned}(24.20)\qquad & \operatorname{index}_{\mathrm{LCP}}(A,\Phi)\\ &= \operatorname{index}(\Phi H(A_+), H(A_-))\\ &= \operatorname{index}\{(\mathrm{Id} - \mathcal{P}(A_-))\Phi\mathcal{P}(A_+) : H(A_+) \to H(A_-)^\perp\}\\ &= \operatorname{index} A + \operatorname{index}(\mathcal{P}(A_-) - \Phi\mathcal{P}(A_+)).\end{aligned}$$

Proof. It is not difficult to see that the pair $(\Phi H(A_+), H(A_-))$ is a Fredholm pair. This is a consequence of the p_+-invariance of Φ which implies that the projection $\Phi\mathcal{P}(A_+)\Phi^{-1}$ of $H := L^2(Y;E|_Y)$ onto $\Phi H(A_+)$ belongs to Gr_{p_+}. Therefore, since $(\mathcal{P}(A_+)H, H(A_-))$ is a Fredholm pair by Theorem 24.1, so is the compact perturbation $(\Phi\mathcal{P}(A_+)\Phi^{-1}H, H(A_-)) = (\Phi H(A_+), H(A_-))$ (see also Remark 24.5a).

By the 1-1 correspondence between solutions and boundary data (established in Theorem 12.4a), $\Phi H(A_+) \cap H(A_-)$ can be identified with the space of solutions of (24.17) and alternatively with the kernel of the operator

$$(24.21)\qquad T_\Phi := (\mathrm{Id} - \mathcal{P}(A_-))\Phi\mathcal{P}(A_+) : H(A_+) \to H(A_-)^\perp.$$

The cokernel of this *generalized* Toeplitz operator is equal to the space $(\Phi H(A_+) + H(A_-))^\perp$, namely:

$$(24.22)\quad \operatorname{coker} T_\Phi = (\Phi H(A_+))^\perp \cap H(A_-)^\perp = (\Phi H(A_+) + H(A_-))^\perp.$$

This proves

$$(24.23)\qquad \operatorname{index}(\Phi H(A_+), H(A_-)) = \operatorname{index} T_\Phi.$$

We elaborate (24.22) further. Recall from Corollary 12.6 the *gauge* complementarities

$$H(A_-)^\perp = G^{-1}H(A_-^*) \qquad \text{and} \qquad H(A_+)^\perp = G^{-1}H(A_+^*).$$

Since Φ is unitary, we have $(\Phi H(A_+))^\perp = \Phi(H(A_+))^\perp$. Then the first equality of (24.21) yields the identity

$$\operatorname{coker} T_\Phi = G^{-1}(H(A_-^*)) \cap \Phi(H(A_+^*))$$

which is isomorphic to the space of solutions of the adjoint linear conjugation problem defined in (24.19). The isomorphy follows from the 1-1 correspondence between solutions and boundary data. Thus the first two equations of (24.20) are proved.

To prove the third equation of (24.20), we recall that the index really identifies the connected components in the Grassmannian, and that, by Proposition 15.15, for any $R, S, T \in \mathrm{Gr}_{p_+}$

$$\begin{aligned}(24.24)\quad &\operatorname{index}\{RT : \operatorname{range} T \to \operatorname{range} R\}\\ &= \operatorname{index}\{RS : \operatorname{range} S \to \operatorname{range} R\} + \operatorname{index}\{ST : \operatorname{range} T \to \operatorname{range} S\}.\end{aligned}$$

From that we obtain

$$\begin{aligned}\operatorname{index} T_\Phi &= \operatorname{index}\{(\mathrm{Id} - \mathcal{P}(A_-))\Phi\mathcal{P}(A_+) : H(A_+) \to H(A_-)^\perp\}\\ &\overset{(24.24)}{=} \operatorname{index}\{(\mathrm{Id} - \mathcal{P}(A_-))\mathcal{P}(A_+) : H(A_+) \to H(A_-)^{-1}\}\\ &\qquad + \operatorname{index}\{\mathcal{P}(A_+)\Phi\mathcal{P}(A_+) : H(A_+) \to H(A_+)\}.\end{aligned}$$

The first summand is $\operatorname{index} A$ by Theorem 24.1. The operator of the second summand can first be extended to $(\mathrm{Id} - \mathcal{P}(A_+)) + \mathcal{P}(A_+)\Phi\mathcal{P}(A_+)$ over the whole Hilbert space, and then compact perturbated to

$$\begin{aligned}(\mathrm{Id} - \mathcal{P}(A_+)) + \mathcal{P}(A_+)\Phi\mathcal{P}(A_+) + (\mathrm{Id} - \mathcal{P}(A_+))\Phi\mathcal{P}(A_+)&\\ = (\mathrm{Id} - \mathcal{P}(A_+)) - \Phi\mathcal{P}(A_+)&\end{aligned}$$

without changing the index. Here we have used the p_+-invariance of Φ which gives that Φ commutes with $\mathcal{P}(A_+)$ modulo compact operators.

□

Remarks 24.10. (a) The main point of Theorem 24.9 is that (24.20) separates the index of the linear conjugation problem into two contributions, one which comes from the *global* data (the operator A on the manifold M) and one which comes from data that live on the cutting submanifold Y.

(b) The first summand in the final formula of (24.20), the index of A, is given by the Atiyah-Singer theorem. In the next chapter we

explain why the second summand, the index of the Toeplitz operator $\mathcal{P}(A_+)\Phi\mathcal{P}(A_+)$, or equivalently the index of $\mathcal{P}(A_-)-\Phi\mathcal{P}(A_+)$, is also given by the index theorem, namely as the index of a corresponding Dirac operator on the mapping torus.
(c) If A is invertible, its index vanishes, its Cauchy data spaces are orthogonally complementary, and we recover Proposition 24.4 from Theorem 24.9.
(d) Over the closed manifold M, the index of the adjoint problem equals the index of the following *orthogonal* Fredholm pairs

$$\begin{aligned}(H(A^*|_{X_+}), H(A^*|_{X_-})) &= (G(H(A_+)^\perp), G(H(A_-)^\perp)) \\ &\cong (H(A_+)^\perp, H(A_-)^\perp)\,.\end{aligned}$$

Similar calculations can be made for the linear conjugation problems.

We give an example which belongs to the sphere of prominent classical problems of mathematics and is called, at least in the Russian mathematical literature, *the Riemann-Hilbert problem*, see Mushelišvili [1962], Bitsadze [1966], Bojarski [1979], and Meister [1983, pp 123–172].

Example 24.11. Our model situation is the two-dimensional case, where we deal with the Cauchy-Riemann operator. We are looking for a couple of functions

$$f_+ \quad \text{on } D^2, \qquad f_- \quad \text{on } \mathbf{C}\setminus D^2$$

which satisfy

$$\bar{\partial} f_+ = 0,\ \bar{\partial} f_- = 0,\ f_-(\infty) = 0, \tag{24.25a}$$

and

$$f_-(z) = g(z) f_+(z) \text{ for } z \in S^1. \tag{24.25b}$$

We assume that g is a continuous complex-valued function on S^1 such that $g(y) \neq 0$ for any $y \in S^1$. This problem seems to differ from the problem (24.17), since our manifold $\mathbf{C}$ is not compact. We explain how to transform (24.25) into (24.17) in Chapter 26.

Clearly (24.25) with $g(z) = z^k$ has no solutions for $k \geq 0$, and it has a $(-k)$-dimensional space of solutions for $k < 0$. More generally our Theorem 24.9 gives

$$\mathrm{index}_{\mathrm{LCP}}(\bar{\partial}, g) = -\deg g\,. \tag{24.26}$$

Now consider the special *non-shift* case of our situation. Assume that M is an even-dimensional manifold, that $E := S^+ \otimes V$, $F := S^- \otimes V$, and that g is a unitary automorphism of $V|_Y$. In this case $T := \mathcal{P}_- - g\mathcal{P}_+$ is a pseudo-differential operator and its index is a topological invariant given by the Atiyah-Singer formula. Let us restrict ourselves to an even more special situation, where g is a unitary automorphism of the auxiliary bundle V. Then we can collect a number of useful formulas:

Proposition 24.12. *We have the following equalities, if the linear conjugation problem* $_{\mathrm{LCP}}(A, g)$ *is defined by an automorphism* g *of the auxiliary bundle* V*:*

$$
\begin{aligned}
\operatorname{index}_{\mathrm{LCP}}(A, g) - \operatorname{index} A &= \operatorname{index}\{(\mathrm{Id} - \mathrm{P}(B)) - g\ \mathrm{P}(B)\} \\
&= \operatorname{index} \mathrm{P}(B)\, g\ \mathrm{P}(B) \\
&= \operatorname{index} \mathcal{P}(A_+)\, g\, \mathcal{P}(A_+) \\
&= \operatorname{index}(\mathcal{P}(A_-) - g\,\mathcal{P}(A_+)) \\
&= \int_{T^*Y} \operatorname{ch}([B] \cdot [g]) \cup \tau(Y)\,.
\end{aligned}
\tag{24.27}
$$

Here the integration is over the Thom space of the cotangent bundle of Y; $[B]$ denotes the class of the principal symbol of the tangential operator B in $K^{-1}(T^*Y)$; $[g]$ is the class in $K^{-1}(Y)$ defined by g; $\tau(Y) \in H^*(T^*Y)$ the (lifted) Todd class of Y; $\operatorname{ch} : K(T^*Y) \to H^{\mathrm{even}}_{\mathrm{comp}}(T^*Y; \mathbf{Q})$ the Chern character; and

$$
K^{-1}(T^*Y) \otimes K^{-1}(Y) \to K(T^*Y)
$$

the usual multiplication of K-theory. $\mathcal{P}(A_\pm)$ and $\mathrm{P} := \mathrm{P}_\geq$ denote, as before, the Calderón projectors and the spectral projection.

The formulas (24.27) are all consequences of the fact that $\mathcal{P}(A_-) - g\mathcal{P}(A_+)$ is an elliptic pseudo-differential operator of order 0. We introduce one more formula which makes a direct connection with the index theory of a certain boundary problem on $X := X_+$.

Theorem 24.13. *Let* $A : C^\infty(X; E = S^+ \otimes V) \to C^\infty(X; F = S^- \otimes V)$ *be a Dirac operator over a compact manifold* X *with boundary* Y*, where* A *takes the form* $G(\partial_u + B)$ *with unitary* G*. Let* g *be a unitary automorphism of* V*. Denote by* $\mathbf{A}$ *the operator*

$$
\mathbf{A} := \begin{pmatrix} A & 0 \\ 0 & A^* \end{pmatrix} : C^\infty(X; E) \oplus C^\infty(X; F) \to C^\infty(X; F) \oplus C^\infty(X; E)\,.
$$

Denote by $\mathbf{R}$ *the boundary condition*

$$\mathbf{R} : C^\infty(Y; E|_Y) \oplus C^\infty(Y; F|_Y) \to C^\infty(Y; F|_Y)$$

defined by

$$\mathbf{R}(s, r) := G(gs) - r\,. \tag{24.28}$$

Then $(\mathbf{A}, \mathbf{R})$ *is an elliptic boundary problem of local type which has the same index as the corresponding linear conjugation problem:*

$$\operatorname{index}(\mathbf{A}, \mathbf{R}) = \operatorname{index}_{\mathrm{LCP}}(A \cup A^*, Gg)\,. \tag{24.29}$$

Note. In combination with Proposition 24.12 and from some lines of the following proof, we get again a number of formulas to insert in (24.29):

$$\begin{aligned} \cdots = \operatorname{index} \mathbf{A}_{\mathbf{R}} &= \operatorname{index}((\mathrm{Id} - \mathrm{P}(B)) - g \cdot \mathrm{P}(B)) \\ &= \operatorname{index}((\mathrm{Id} - \mathcal{P}(A)) - g\mathcal{P}(A))\,. \end{aligned} \tag{24.30}$$

Proof. The ellipticity is obvious. $\mathbf{R}$ is a surjective condition; hence by Corollary 20.14 we have the equality $\operatorname{index}(\mathbf{A}, \mathbf{R}) = \operatorname{index} \mathbf{A}_{\mathbf{R}}$. It is also obvious that

$$\ker \mathbf{A}_{\mathbf{R}} = \ker \mathcal{P}(A) g \mathcal{P}(A)\,, \tag{24.31}$$

(see, however, the calculations of the cokernel below). Now we can rewrite the boundary condition $\mathbf{R}$ as a projection $\mathbf{P}$ such that $\mathbf{A}_{\mathbf{P}} = \mathbf{A}_{\mathbf{R}}$:

$$\mathbf{P} := \frac{1}{2}\begin{pmatrix} \mathrm{Id} & -g^{-1}G^{-1} \\ -Gg & \mathrm{Id} \end{pmatrix}. \tag{24.32}$$

This shows that the adjoint condition is (by Proposition 20.3):
(24.33)

$$\mathbf{T} := \begin{pmatrix} G & 0 \\ 0 & G^{-1} \end{pmatrix} (\mathrm{Id} - \mathbf{P}) \begin{pmatrix} G^{-1} & 0 \\ 0 & G \end{pmatrix} = \frac{1}{2}\begin{pmatrix} \mathrm{Id} & Gg \\ gG^{-1} & \mathrm{Id} \end{pmatrix}.$$

We can now calculate the cokernel of $\mathbf{A}_{\mathbf{R}}$ which is equal to the kernel of the operator $(\mathbf{A}^*)_{\mathbf{T}}$. Therefore it is the set of pairs

$$(r, s) \in C^\infty(X; F) \oplus C^\infty(X; E)$$

which satisfy the following equations:

$$A^*r = 0,\ As = 0,\ \text{and } s|_Y = -gG^{-1}(r|_Y)\,. \tag{24.34}$$

The first two equations show that $r|_Y \in H(A^*)$ and $s|_Y \in H(A)$. Now we use Corollary 12.6 and the unique continuation property for Dirac operators. That allows us to identify $\ker \mathbf{A}^*_{\mathbf{T}}$ with the kernel of the operator $\mathcal{P}(A)g^{-1}\mathcal{P}(A)$. Therefore we have:

$$\begin{aligned}
&\operatorname{index}(\mathbf{A},\mathbf{R}) = \operatorname{index}\mathbf{A}_{\mathbf{R}} \\
&= \operatorname{index}\mathcal{P}(A)\,g\,\mathcal{P}(A) = \operatorname{index}\mathcal{P}(A)G^{-1}G\,g\,\mathcal{P}(A) \\
&= \operatorname{index}(G\mathcal{P}(A)G^{-1})G\,g\,\mathcal{P}(A) = \operatorname{index}\mathcal{P}(A_-)G\,g\,\mathcal{P}(A_+)\,,
\end{aligned}$$

with $A_+ = A$ and $A_- = A^*$. This ends the proof of Theorem 24.13.

□

At the end of this chapter we present a calculation which shows the difference between the indices of two problems of this type.

Theorem 24.14. *Under the assumptions of the preceding theorem, let $\mathbf{A}_i$ denote the problem $\mathbf{A}$ with boundary condition $s|_Y = G(g_i(s_+|_Y))$, where we assume that $E = S^+ \otimes V$ and g_i, $i = 1, 2$ are automorphisms of the auxiliary bundle V. Then we have*

$$\operatorname{index}\mathbf{A}_1 - \operatorname{index}\mathbf{A}_2 = \operatorname{index}\{(\mathrm{Id} - \mathcal{P}(\mathbf{A})) + g_2 g_1^{-1}\mathcal{P}(\mathbf{A})\}\,.$$

Proof. From Theorem 21.2 it follows that this difference of indices is equal to the index of the *boundary integral* $\mathbf{R}_2\mathcal{P}(\mathbf{A})\mathbf{R}_1^*$, where once again:

$$\mathbf{A} = \begin{pmatrix} A & 0 \\ 0 & A^* \end{pmatrix}, \quad \text{hence} \quad \mathcal{P}(\mathbf{A}) = \begin{pmatrix} \mathcal{P}(A) & 0 \\ 0 & \mathcal{P}(A^*) \end{pmatrix},$$

and

$$\mathbf{R}_1 \begin{pmatrix} s_+ \\ s_- \end{pmatrix} := s_- - G(g_1 s_+) \quad \text{and} \quad \mathbf{R}_2 \begin{pmatrix} s_+ \\ s_- \end{pmatrix} = s_- - G(g_2 s_+)\,.$$

Since G is unitary

$$\mathbf{R}_1^* = \begin{pmatrix} -g_1^{-1}G^{-1} \\ 1 \end{pmatrix}$$

and hence

$$\begin{aligned}\mathbf{R}_2\mathcal{P}(\mathbf{A})\mathbf{R}_1^* &= (\,-Gg_2 \quad 1\,)\begin{pmatrix}\mathcal{P}(A) & 0\\ 0 & \mathcal{P}(A^*)\end{pmatrix}\begin{pmatrix}-g_1^{-1}G^{-1}\\ 1\end{pmatrix}\\ &= Gg_2\mathcal{P}(A)g_1^{-1}G^{-1} + \mathcal{P}(A^*)\\ &= G\{g_2\mathcal{P}(A)g_1^{-1} + (\mathrm{Id} - \mathcal{P}(A))\}G^{-1}.\end{aligned}$$

This shows that

$$\begin{aligned}\operatorname{index}\mathbf{A}_1 - \operatorname{index}\mathbf{A}_2 &= \operatorname{index}\{(\mathrm{Id} - \mathcal{P}(A)) + g_2\mathcal{P}(A)g_1^{-1}\}\\ &= \operatorname{index}\{(\mathrm{Id} - \mathcal{P}(A)) + g_2 g_1^{-1}\mathcal{P}(A)\}\end{aligned}$$

since g_1 and therefore also g_1^{-1} commute with $\mathcal{P}(A)$ modulo compact operators. □

25. Cutting and Pasting of Elliptic Operators

We discuss the process of the cutting and pasting of elliptic operators and show that some operators, like the signature operator, are stable in a suitable sense under this operation.

Essentially this chapter shows how to obtain new operators from operator pieces under the process of cutting and pasting of elliptic operators. We distinguish two extreme situations:

(1) On the one hand, we have the *flexible* case: We may obtain complicated operators from simple pieces solely by cutting and pasting. We shall show later in Chapter 26 that the process is highly non-trivial even in the simplest possible case of Dirac operators on S^2.

(2) On the other hand, we have the *rigid* case: classes of operators which are invariant under cutting and pasting. We shall show in this chapter that the signature operator is in fact stable in a suitable sense under this operation.

Assume that we are in the situation of Chapter 24. We have an operator A on a closed partitioned manifold M which is essentially the *sum* of two operators $A_\pm$ on suitable pieces $X_\pm$. We use the symbolic notation (see Chapter 9 for a similar situation or Douglas & Wojciechowski [1989] for exactly this situation):

$$A = A_+ \cup A_- . \tag{25.1}$$

Assume once again that we have a unitary automorphism $\Phi : E|_Y \to E|_Y$ of the bundle $E|_Y$ (possibly with shift in the basis Y). Let a denote the principal symbol of the operator A. In a bicollar neighbourhood of Y, the symbol a has the form:

$$a(u, y; \nu, \zeta) = G(y)(i\nu + b(y; \zeta)) . \tag{25.2}$$

We assume a consistency of Φ and G, namely that for all $y \in Y$ and $\zeta \in T^*Y_y$:

$$\begin{aligned} \Phi(y)G(y) &= G(f(y))\Phi(y) \quad \text{and} \\ \Phi(y)b(y;\zeta) &= b\left(f(y); (f^{-1})^*(\zeta)\right)\Phi(y) . \end{aligned} \tag{25.3}$$

Here f is a diffeomorphism of Y covered by Φ. We introduce the notation $\Phi_E := \Phi$ and $\Phi_F := G\Phi_E G^{-1}$, and then we can reformulate (25.3) as the condition:

$$\Phi_F(y)a(0,y;\nu,\zeta) = a\left(0,f(y);(f^{-1})^*(\nu),(f^{-1})^*(\zeta)\right)\Phi_E(y)\,. \tag{25.4}$$

Now we can introduce a new manifold and new bundles:

$$\begin{aligned} M^f &:= X_+ \cup_f X_-\,, \\ E^\Phi &= (E|_{X_+}) \cup_{\Phi_E} (E|_{X_-})\,, \qquad \text{and} \\ F^\Phi &= (F|_{X_+}) \cup_{\Phi_F} (F|_{X_-})\,, \end{aligned} \tag{25.5}$$

with the obvious identifications. For instance, to get M^f, identify the point $(0,y)$ with the point $(0,f(y))$. Actually, to have everything smooth, we have to be much more careful and must work as in Chapter 9. We extend the manifold X_+ to $X_+ \cup ((-1,0] \times Y)$ and the manifold X_- to $X_- \cup ([0,1) \times Y)$, and then we identify (u,y) with $(u,f(y))$ on the collar N, where the normal variable $u \in (-1,1)$. Now we can define a new elliptic symbol in this situation:

$$a^\Phi : \pi^*(E^\Phi) \to \pi^*(F^\Phi)\,, \tag{25.6}$$

by the formula

$$a^\Phi(x;\xi) := a(x;\xi)\,.$$

Although the algebraic formula for the symbol is unchanged, the symbol generally behaves in a different way, since it acts between the sections of the new bundles over the new manifold M^f.

Choose an arbitrary elliptic operator A^Φ with principal symbol a^Φ. The problem is to find a formula for the difference:

$$\operatorname{index} A^\Phi - \operatorname{index} A\,. \tag{25.7}$$

Our solution of the problem is based on an explicit realization of the operator A^Φ. An alternative approach, which deals with a more complicated analytical situation and therefore uses the intricate machinery of Boutet de Monvel, was offered in Booß & Rempel [1981], [1982]. There, however, the explicit topological formula of Theorem 25.2 was not obtained for the difference (25.7).

The main point in our explicit realization of the operator A^Φ is the use of (25.3). We cannot directly paste $A_+ = A|_{X_+}$ to the operator $A_- = A|_{X_-}$, but must modify the tangential operator B. If we substitute B with the operator $B^{\Phi} = \Phi^{-1}B\Phi$, then everything works fine. The sections of the bundle E^{Φ} are couples (s_1, s_2) such that on some collar neighbourhood of Y we have

$$s_2(u, f(y)) = \Phi_E(y)s_1(u, y) \,. \tag{25.8}$$

We show that any operator, which is equal to A_- on X_- and which, restricted to the same collar on X_+, takes the form

$$G\left(\partial_u + \Phi_E^{-1}B\Phi_E\right) \,, \tag{25.9}$$

maps sections of E^{Φ} in sections of F^{Φ} which are just couples (s_3, s_4) satisfying the condition

$$s_4(u, f(y)) = G(f(y))\, \Phi_E(y)\, G^{-1}(y)\, s_3(u, y) \,. \tag{25.10}$$

Essentially we used this argument before in Chapter 9, where we presented our construction of the operator $A \cup A^*$. Let (s_1, s_2) denote a section of E^{Φ}. We have to show that (A_+s_1, A_-s_2) is a section of F^{Φ}. This is reduced to the equality:

$$\Phi_E\{G\left(\partial_u + \Phi_E^{-1}B\Phi_E\right) s_1\} = G(\partial_u + B)s_2 \,. \tag{25.11}$$

The left side of (25.11) is equal to

$$G\circ\partial_u(\Phi_E(s_1)) + (GB)(\Phi_E(s_1)) = G\,\partial_u\, s_2 + GBs_2 = A_-(s_2) \,. \tag{25.12}$$

Now the crucial point is the assumption (25.3) which tells us that the principal symbol of $\Phi_E^{-1}B\Phi_E$ is equal to the principal symbol of B. The second equality in (25.3) can be written as:

$$b(y;\zeta) = \Phi_E^{-1}\, b\left(f(y); (f^{-1})^*\zeta\right)\, \Phi_E \,. \tag{25.13}$$

Therefore there exists a family $\{B_t\}_{t\in[0,1]}$ of Dirac operators on Y, such that $B_0 = B$ and $B_1 = \Phi_E^{-1}\, B\, \Phi_E$. Take for instance:

$$B_t := \psi(t)\left(\Phi_E^{-1}B\Phi_E\right) + (1 - \psi(t))B \,, \tag{25.14}$$

where ψ is a smooth real valued function on the interval $[0, 1]$ equal to 0 in some neighbourhood of 0, and equal to 1 in a certain neighbourhood of 1. We introduce an operator A_+^{Φ} on X_+ in the following way:

$$A_+^{\Phi} := \begin{cases} A_+ & \text{on } X_+ \setminus N \\ G(y)(\partial_u + B_u) & \text{on } N. \end{cases} \tag{25.15}$$

Now, we can explicitly define an operator $A^\Phi : C^\infty(M^f; E^\Phi) \to C^\infty(M^f; F^\Phi)$ with the required principal symbol $\sigma_{A^\Phi} = a^\Phi$ by putting:

$$A^\Phi := \begin{cases} A_+^\Phi & \text{on } X_+ \\ A_- & \text{on } X_- . \end{cases} \tag{25.16}$$

To compute the index difference (25.7) observe that we can *localize* the problem as follows. The index of the Dirac operator is given by a local formula. This means that

$$\operatorname{index} A = \int_M \alpha(x) , \tag{25.17}$$

where $\alpha(x)$ is a density given at the point x by a complicated algebraic formula in terms of coefficients of the operator A in x (accounted for e.g. in Gilkey [1984], Getzler [1986], or Berline, Getzler & Vergne [1992]). It is a consequence of the locality of the index expressed in (25.17) that we have the following formula:

$$\operatorname{index} A^\Phi = \operatorname{index} A + \operatorname{index} T^\Phi . \tag{25.18}$$

In (25.18) the operator $T^\Phi : C^\infty(Y^f; W_1) \to C^\infty(Y^f; W_2)$ is given by the formula $G(\partial_u + B_u)$ and acts between sections of the bundles $W_1 := E|_N/\Phi$ and $W_2 := F|_N/G\Phi G^{-1}$ over the mapping torus $Y^f := (I \times Y)/f$. The index of the operator is given by the Atiyah-Singer formula. We are looking for the *desuspension* of this formula to the manifold Y.

With this purpose in mind, we must exploit the *spectral flow* which was introduced earlier in Chapter 17. Recall the following: We have a family of self-adjoint operators $\{B_u\}$ parametrized over S^1. It is explained that such families have only one topological invariant, $\operatorname{sf}\{B_u\}$. It was noticed in Atiyah, Patodi & Singer [1976] that the spectral flow is in fact the analytical index of the family $\{B_u\}$ with values in $K^1(S^1) = \mathbf{Z}$. The following result is a consequence of the index theorem for families:

Proposition 25.1. *Under the preceding assumptions we have*

$$\operatorname{sf}\{B_u\} = \operatorname{index} T^\Phi .$$

An elementary argument was offered in Vafa & Witten [1984]. We have included a proof of Proposition 25.1 in Chapter 17, see Theorem 17.13. To obtain the following result, we just use a formula presented in Theorem 17.17.

Theorem 25.2. *Assume that Φ is a unitary automorphism of $E|_Y$ which satisfies the consistency condition (25.3). Let $\mathcal{P}(A_+)$ denote the Calderón projector and $H(A_+)$ the Cauchy data space. Then the operator $\mathcal{P}^\Phi := \mathcal{P}(A_+)\,\Phi\,\mathcal{P}(A_+) : H(A_+) \to H(A_+)$ is a Fredholm operator and we have the following equalities:*

$$(25.19)\quad \begin{aligned} \operatorname{index} A^\Phi - \operatorname{index} A &= \operatorname{index} T^\Phi = \operatorname{sf}\{B_u\} \\ &= \operatorname{index}((\operatorname{Id} - \mathcal{P}(A_+)) - \Phi\mathcal{P}(A_+)) = \operatorname{index} \mathcal{P}^\Phi . \end{aligned}$$

Corollary 25.3. *The index of the general linear conjugation problem* $\mathrm{LCP}(A, \Phi)$ *(see Chapter 24) on the manifold M equals the index of the operator A^Φ.*

Let us give an important topological example of the situation in which cutting and pasting operations have no influence on the index of the corresponding operator:

Theorem 25.4. *Let* $\operatorname{sign} M$ *denote the signature, and $\chi(M)$ the Euler characteristic of a partitioned manifold $M = X_+ \cup X_-$ with an automorphism $f : Y \to Y$, where $Y = \partial X_+ = \partial X_-$. Then* $\operatorname{sign} M^f = \operatorname{sign} M$ *and $\chi(M^f) = \chi(M)$.*

Proof. As discussed in Chapter 7, the signature of a manifold is equal to the index of a special Dirac operator

$$D_M : C^\infty(M; \Omega^+) \to C^\infty(M; \Omega^-) .$$

Therefore we only have to show that

$$(25.20)\qquad \operatorname{index} D_M = \operatorname{index} D_{M^f} .$$

Fix a Riemannian structure g on M and let g_Y denote its restriction to the boundary (in fact any $Y_u := \{u\} \times Y$ carries the same metric g_Y). This defines the operator D_M. Under the cutting and pasting of the manifold M, we have to modify the Riemannian structure in order to get a metric on the manifold M^f. Replace the metric g_Y by a family of metrics:

$$(25.21)\qquad g_u := \psi(u) f^*(g) + (1 - \psi(u))g .$$

Now any metric g_u defines a boundary signature operator B_u and it follows from Theorem 25.2 that:

$$(25.22)\qquad \operatorname{sign} M^f - \operatorname{sign} M = \operatorname{sf}\{B_u\} .$$

As follows from Chapter 7, the kernel of any of the operators B_u consists of all harmonic forms on the manifold Y. Therefore the dimension of $\ker B_u$ is constant. This argument shows that there is no spectral flow, since there are no eigenvalues changing the sign. That concludes the proof for the signature operator.
We use the same argument for the operator which gives the Euler characteristic. □

The preceding example illustrates a situation in which the topological constraints prohibit any change of the index under cutting and pasting operations. On the other hand, in one of the most fundamental examples of the index theory, the index is completely determined by the cutting and pasting operations. This will be discussed in the next chapter.

26. Dirac Operators on the Two-Sphere

We show that the index of any elliptic operator on any closed Riemann surface or even-dimensional sphere is completely determined by the pasting of *trivial* pieces of the operator. Our example gives a full analysis of the index on S^2 and determines the index of the generalized Riemann-Hilbert problem.

At the end of Chapter 25 we discussed an important example of a situation in which the cutting and pasting of elliptic operators have no influence on the value of the index of the operator. Here we discuss another extreme case in which the index is completely determined by the pasting of the *trivial* pieces of the operator. The argument presented below gives the index of any elliptic operator on any closed Riemann surface and any even-dimensional sphere. Since the generalization is easy, we concentrate on the simplest and most important case of the two-dimensional sphere.

First we introduce the basic operator. It is not the Cauchy-Riemann operator $f \mapsto \partial f/\partial z\, dz$, as is usually assumed, but a total Dirac operator which acts on true spinors. Although we do not discuss K-homology groups here, we shall mention that the operator, which we construct, provides a generator of the group $K_0(D^2, S^1)$. As explained in Wojciechowski [1985] (see also Douglas & Wojciechowski [1989]), the cycles for the relative K_0-homology group of a manifold X with boundary Y are given by operators of the form $A \cup A^*$ on the double of X. We have already used the construction of the double of elliptic operators in Chapter 12.

In the simplest possible two-dimensional case, we consider the Cauchy-Riemann operator $\bar{\partial} : C^\infty(D^2) \to C^\infty(D^2)$, where $\bar{\partial} = \frac{1}{2}(\partial/\partial x + i\partial/\partial y)$. In polar coordinates out of the origin, this operator has the form $\frac{1}{2}e^{i\varphi}(\partial_r + (i/r)\partial_\varphi)$. Therefore, after some small smooth perturbations (and modulo the factor $\frac{1}{2}$), we assume that $\bar{\partial}$ has the following form in a certain collar neighbourhood of the boundary:

$$\bar{\partial} = e^{i\varphi}(\partial_r + i\partial_\varphi)\,. \tag{26.1}$$

Now we can apply the construction from Chapter 9 directly. Remember that in this case $G(\varphi) = e^{i\varphi}$. By H^k we denote the bundle, which is obtained from two copies of $D^2 \times \mathbf{C}$ by the identification $(z, w) = (z, z^k w)$ near the equator, $k \in \mathbf{Z}$. We apply the construction from Chapter 9 and obtain the operator

$$A_0 = \bar{\partial} \cup (\bar{\partial})^* : C^\infty(S^2; H^1) \to C^\infty(S^2; H^{-1}) .$$

Let us analyse the situation more carefully. We fix $N := (-\varepsilon, +\varepsilon) \times S^1$, a bicollar neighbourhood of the equator. The formally adjoint operator to $\bar{\partial}$ has the following form:

$$(\bar{\partial})^* = e^{-i\varphi}(-\partial_u + i\partial_\varphi + 1) \tag{26.2}$$

$(u = r - 1)$ in this cylinder. A section of H^1 is a couple (s_1, s_2) such that in N:

$$s_2(u, \varphi) = e^{i\varphi} s_1(u, \varphi) . \tag{26.3}$$

The couple $(\bar{\partial} s_1, (\bar{\partial})^* s_2)$ is a smooth section of H^{-1}. To show that, we check the equality $(\bar{\partial})^* s_2 = e^{-i\varphi} \bar{\partial} s_1$. We have in the neighbourhood N:

$$\begin{aligned}(\bar{\partial})^* s_2 &= (\bar{\partial})(e^{i\varphi} s_1) = e^{-i\varphi}(\partial_u + i\partial_\varphi + 1)(e^{i\varphi} s_1) \\ &= \partial_u s_1 + ie^{-i\varphi}\partial_\varphi(e^{i\varphi} s_1) + s_1 = (\partial_u + i\partial_\varphi) s_1 \\ &= e^{-i\varphi} e^{i\varphi}(\partial_u + i\partial_\varphi) s_1 = e^{-i\varphi}(\bar{\partial} s_1) .\end{aligned} \tag{26.4}$$

Here index $\bar{\partial} \cup \bar{\partial}^* = 0$. We produce an operator with a nontrivial index. Modify the operator A_0 in such a way that it will act from sections of the bundle H^2 to sections of the bundle $H^0 = S^2 \times \mathbf{C}$. The resulting operator A_1 has index equal to 1. The index can be computed by using Theorem 25.2. In this particular case, we can actually make a direct calculation. First we give the construction of the operator A_1:

Any section of H^2 is a couple (s_1, s_2) such that we have in N:

$$s_2(u, \varphi) = e^{2i\varphi} s_1(u, \varphi) .$$

Following the procedure from Chapter 25 we do not change the operator $(\bar{\partial})^*$, but modify the operator $\bar{\partial}$. We have to substitute the operator $\bar{\partial}$ with $\bar{\partial}_1$. In order to obtain a section of the trivial bundle, we must have the following equality:

$$\bar{\partial}_1(s_1) = (\bar{\partial})^* s_2 . \tag{26.5}$$

Calculations similar to those in (26.4) allow us to conclude that the operator $\bar\partial_1$ is equal to $\bar\partial$ outside of N. On N we define:

$$\bar\partial_1 := e^{i\varphi}(\partial_u + i\partial_\varphi - \psi(u)), \tag{26.6}$$

where $\psi(u) \geq 0$ is a smooth function equal 0 near $u = \varepsilon$ and 1 in a small neighbourhood of $u = 0$. Hence, on a smaller cylinder (on which $\psi(u) = 1$), we have the following equalities, for any couple (s_1, s_2), such that $s_2 = e^{2i\varphi}s_1$ on this cylinder:

$$\begin{aligned}(\bar\partial)^* s_2 &= (\bar\partial)^*(e^{2i\varphi}s_1) = e^{-i\varphi}(\partial_u + i\partial_\varphi + 1)(e^{2i\varphi}s_1) \\ &= e^{i\varphi}\partial_u s_1 + e^{-i\varphi}i\partial_\varphi(e^{2i\varphi}s_1) + e^{i\varphi}s_1 = e^{i\varphi}(\partial_u + i\partial_\varphi - 1)s_1 = \bar\partial_1(s_1).\end{aligned} \tag{26.7}$$

These elementary calculations show that the operator $A_1 := \bar\partial_1 \cup (\bar\partial)^*$ is well-defined. The only difference from the operator $\bar\partial \cup (\bar\partial)^*$ is the modification on the collar N. It follows from the discussion in Chapter 25 that

$$\operatorname{index} A_1 = \operatorname{index} A_1 - \operatorname{index} A_0 = \operatorname{index}[e^{i\varphi}(\partial_u + i\partial_\varphi - \psi(u))], \tag{26.8}$$

where the operator on the right side is a well-defined elliptic operator on $S^1 \times S^1$. We know that the index of an elliptic operator is unchanged under small continuous deformations, and we can calculate the index of the operator on the right side directly. After a small deformation we see that we must calculate the index of the elliptic operator $T := \partial_u + i\partial_\varphi - u$ acting on functions $f(u, \varphi)$ from $C^\infty(\mathbf{R} \times S^1)$ which satisfy the periodicity condition

$$f(u+1, \varphi) = e^{-i\varphi} f(u, \varphi). \tag{26.9}$$

Any such f can be represented by a series

$$f(u, \varphi) = \sum_{k \in \mathbf{Z}} f_k(u) e^{ik\varphi} \tag{26.10}$$

with $f_{k+1}(u) = f_k(u+1)$ because of

$$\begin{aligned}\sum_k f_k(u+1)e^{ik\varphi} = f(u+1,\varphi) \overset{!}{=} e^{-i\varphi} f(u,\varphi) &= e^{-i\varphi}\sum_k f_k(u)e^{ik\varphi} \\ &= \sum_k f_k(u)e^{i(k-1)\varphi} = \sum_k f_{k+1}(u)e^{ik\varphi}.\end{aligned}$$

Since

$$T(f_k(u)e^{ik\varphi}) = f_k'(u)e^{ik\varphi} - (k+u)f_k(u)e^{ik\varphi}$$

the equation $Tf = 0$ gives the following series of equations for f_k:

$$f_k'(u) = (k+u)f_k(u) \qquad \text{for all } k \in \mathbf{Z}.$$

As a result we get

$$f_k(u) = C_k e^{\frac{1}{2}(k+u)^2}.$$

Moreover, C_k does not depend on k because of

$$C_{k+1}e^{\frac{1}{2}(k+1+u)^2} = f_{k+1}(u) \overset{!}{=} f_k(u+1) = C_k e^{\frac{1}{2}(k+1+u)^2}.$$

This means that the coefficients $f_k(u)$ of the series (26.10) explode for $k \to \infty$, if $C = C_k \neq 0$. So there is no non-trivial solution of $Tf = 0$ with the required periodicity.

The same calculations for the operator $T^* = -\partial_u + i\partial_\varphi - u$ yield

$$\begin{aligned} T^*(f_k(u)e^{ik\varphi}) &= -e^{ik\varphi}\{f_k'(u) + (k+u)f_k(u)\} = 0 \\ &\iff f_k(u) = C_k e^{-\frac{1}{2}(k+u)^2} \end{aligned}$$

with C_k again independent of k because of $f_k(u+1) = f_{k+1}(u)$. Thus the solution of $T^*f = 0$ has the form

$$C\sum_k e^{ik\varphi}e^{-\frac{1}{2}(k+u)^2}$$

which converges nicely; hence the kernel of T^* with the required periodicity is one-dimensional. This gives the following formula:

$$\text{(26.11)} \qquad \operatorname{index} A_1 = -1.$$

More generally, this argument proves the following result:

Theorem 26.1. *Let A_k denote an operator over S^2 which is equal to $\bar\partial$ on $D^2 \setminus N$, $(\bar\partial)^*$ on the second copy of D^2, and $e^{i\varphi}(\partial_u + i\partial_\varphi - k\psi(u))$ on N. The operator A_k is in fact the operator $A_0 \otimes \mathrm{Id}_{H^k}$: $C^\infty(S^2; H^{k+1}) \to C^\infty(S^2; H^{k-1})$ and we have the following index theorem:*

$$\text{(26.12)} \qquad \operatorname{index} A_k = -k.$$

Remarks 26.2. (a) Alternatively, we can prove (26.11) by Proposition 25.1 which gives index $A_1 = \mathrm{sf}\{B_u = i\partial_\varphi - u\}$. We can also calculate the spectral flow of the family $\{B_u\}$ directly. We consider $\{B_u\}$ as a family of ordinary differential operators over the circle S^1, parametrized by $u \in S^1 = I/\{0,1\}$. We have a spectral decomposition of $L^2(S^1)$ by the system $\{e^{ik\varphi}\}_{k\in\mathbf{Z}}$ of eigenfunctions of

$$(i\frac{d}{d\varphi} - u)h = \lambda h, \qquad h \in C^\infty(S^1)$$

with corresponding eigenvalues $\{\lambda = -(k+u)\}_{k\in\mathbf{Z}}$, i.e. the spectrum of $\{B_u\}$ is given by the graph in Figure 26.1 and hence $\mathrm{sf}\{B_u\} = -1$. A further analysis of this example (with opposite signs) has been given in Example 17.8b.

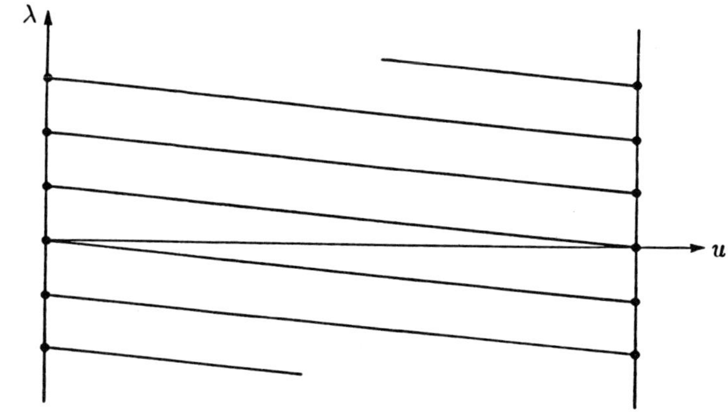

Fig. 26.1 The spectrum of $\{i\frac{d}{d\varphi} - u\}_{u\in I}$

(b) Direct generalization yields the corresponding index theorem for Dirac operators on even-dimensional spheres which gives the Atiyah-Singer index theorem on these spheres by stable homotopy. That result, together with the computation of the signature operator on complex projective spaces, was the basis of the first proof of the index theorem (see Palais [1965a]).
(c) Notice that the cutting and pasting procedure explained above both in general terms and specifically for the 2-sphere also provides a direct proof of the Atiyah-Singer index theorem for elliptic pseudo-differential operators on arbitrary closed Riemann surfaces.

Example 26.3. The Cauchy-Riemann operator $\bar{\partial}' : C^\infty(S^2) \to \Omega^{0,1}(S^2)$ is given by the formal formula

$$f \mapsto (\bar{\partial} f)\, d\bar{z}\,. \tag{26.13}$$

Now the bundle of $(0,1)$-forms is in fact isomorphic to the bundle H^2. The transition function is $z \mapsto 1/z$, which gives for the tangent bundle $z \mapsto -1/z^2$ as the glueing function on the equator, hence $TS^2 = H^{-2}$ and $T^*S^2 = H^2$, or directly $d\bar{z} \mapsto d(1/\bar{z}) = -(1/(\bar{z})^2)\, d\bar{z} = -z^2\, d\bar{z}$. So it becomes obvious that $\bar{\partial}' = (A_1)^*$. That gives the following well-known equation:

$$\operatorname{index} \bar{\partial}' = -\operatorname{index} A_1 = 1\,. \tag{26.14}$$

Now we explain how the index of the classical *Riemann-Hilbert problem* (24.25) fits into our construction. Recall that we have a fixed continuous (we assume g to be a smooth function) invertible function g on the circle, and we are looking for couples (f_+, f_-) which satisfy:

$$\begin{array}{c} f_+ \text{ is holomorphic in } D^2, \\ f_- \text{ is holomorphic in } \mathbf{C} \setminus D^2, \\ f_-(z) = g(z) f_+(z) \text{ on } S^1, \qquad \text{and} \\ f_-(\infty) = 0\,. \end{array} \tag{26.15}$$

It is obvious that f_+ has the form $\sum_{k \geq 0} a_k z^k$, and f_-, thanks to the normalization condition at ∞, has the form $\sum_{k<0} b_k z^k$. We want to compute the index of the Fredholm pair $(H_+\ ,\ gH_-)$ of subspaces in $L^2(S^1)$, where

$$H_+ := \mathcal{L}_{\mathbf{C}}\{z^k\}_{k \geq 0} \quad \text{and} \quad H_- := \mathcal{L}_{\mathbf{C}}\{z^k\}_{k<0}\,.$$

Actually, $H_- = (H_+)^\perp = z^{-1} H((\bar{\partial})^*)$, and we can transfer the Riemann-Hilbert problem into a linear conjugation problem for the operator $A_0 = \bar{\partial} \cup (\bar{\partial})^*$. In this corresponding linear conjugation problem we are looking for couples (s_+, s_-) such that

$$\bar{\partial} s_+ = 0\,, \qquad (\bar{\partial})^* s_- = 0 \qquad \text{and} \qquad s_-(z) = z g(z) s_+(z)\,. \tag{26.16}$$

The factor z reflects the fact that the couple (s_+, s_-) is a section of the bundle H^1. Therefore we have here the linear conjugation problem $\text{LCP}(A_0, g)$ of Definition 24.7. Then Theorem 24.9 and Proposition 24.12 give a formula for the index of the problem

$$\operatorname{index}_{\text{LCP}}(\bar{\partial}; g) = -\deg(g)\,. \tag{26.17}$$

Remark 26.4. Of course, in the linear conjugation problem we consider in $\bar{\partial}_1$ and $(\bar{\partial}_1)^*$ smooth perturbations of the true operators $\bar{\partial}$ and $(\bar{\partial})^*$. It follows from the formula for the Calderón projection given in Chapter 12 that the Calderón projections of the new operators are continuous perturbations of the original Calderón projections. Therefore they do not change their connected components in the Grassmannian, and the index of the corresponding Riemann-Hilbert problem (for the operator $\bar{\partial}_1$) is equal to the index of the original problem.

Formula (26.17) trivially extends to the situation of vector valued functions. The $f_\pm$ in this situation have values in $\mathbf{C}^N$ and satisfy (26.15) with $g : S^1 \to GL(N, \mathbf{C})$ and we can formulate the solution of the index problem for the vector-valued Riemann-Hilbert problem. The following theorem holds:

Theorem 26.5. *Under the assumptions made above, the vector-valued Riemann-Hilbert problem has an index equal to*

$$\begin{aligned}(26.18)\quad \operatorname{index}_{\mathrm{LCP}}(\bar{\partial}; g) &= \operatorname{index}_{\mathrm{LCP}}(A_0 \otimes \mathrm{Id}_{\mathbf{C}^N}; g)\\ &= (\mathrm{ch}[g])[S^1] = -(2\pi i)^{-1} \int_{S^1} \mathrm{tr}(g^{-1} dg)\,.\end{aligned}$$

Remark 26.6. See Mushelišvili [1962] for a general discussion of the Riemann-Hilbert problem from the classical point of view.

Bibliography

Agmon, S., Douglis, A. and Nirenberg, L.

(1) Estimates near the boundary for solutions of elliptic partial differential equations satisfying general boundary conditions I, *Comm. Pure Appl. Math. 12* (1959), 623–727.

Agranovič, M.S.

(1) Elliptic singular integro-differential operators, *Uspehi Mat. Nauk 20/5* (1965), 3–120 (Russian). English translation *Russian Math. Surveys 20*, no. 5/6, (1965), 1–121.

Agranovič, M.S. and Dynin, A.S.

(1) General boundary value problems for elliptic systems in an n-dimensional domain, *Dokl. Akad. Nauk. SSSR 146* (1962), 511–514 (Russian). English translation *Soviet Math. Dokl. 3* (1962/63), 1323–1327.

Alfsen, E.M.

(1) *Compact Convex Sets and Boundary Integrals*, Springer-Verlag, Berlin, 1971.

Alinhac, S.

(1) Non-unicité du problème de Cauchy, *Ann. Math. 117* (1983), 77–108.

(2) Unicité du problème de Cauchy pour des opérateurs du second ordre à symboles réels, *Ann. Inst. Fourier 34* (1984), 89–109.

Aronszajn, N.

(1) A unique continuation theorem for solutions of elliptic partial differential equations or inequalities of second order, *J. Math. Pures Appl. 36* (1957), 235–249.

Atiyah, M.F.

(1) Bott periodicity and the index of elliptic operators, *Quart. J. Math. Oxford (2) 19* (1968), 113–140.

(2) Topology of elliptic operators, *Amer. Math. Soc. Symposia in Pure Math. 16* (1970), 101–119.

(3) Eigenvalues of the Dirac operator, in: Hirzebruch, F. et al. (eds.), *Arbeitstagung Bonn 1984*, Lecture Notes in Math. 1111, Springer-Verlag, Berlin, 1985, pp. 251–260.

Atiyah, M.F. and Bott, R.

(1) The index problem for manifolds with boundary, in: *Coll. Diff. Analysis, Tata Institute, Bombay*, Oxford University Press, Oxford, 1964, pp. 175–186.

Atiyah, M.F., Bott, R. and Shapiro, A.

(1) Clifford modules, *Topology 3, Suppl. 1* (1964), 3–38.

Atiyah, M.F., Patodi, V.K. and Singer, I.M.

(1) Spectral asymmetry and Riemannian geometry, *Bull. London Math. Soc. 5* (1973), 229–234.
(2) Spectral asymmetry and Riemannian geometry. I, *Math. Proc. Cambridge Phil. Soc. 77* (1975), 43–69.
(3) Spectral asymmetry and Riemannian geometry. III, *Math. Proc. Cambridge Phil. Soc. 79* (1976), 71–99.

Atiyah, M.F. and Singer, I.M.

(1) The index of elliptic operators on compact manifolds, *Bull. Amer. Math. Soc. 69* (1963), 422–433.
(2) Index theory for skew-adjoint Fredholm operators, *Inst. Hautes Études Sci. Publ. Math. 37* (1969), 5–26.
(3) Dirac operators coupled to vector potentials, *Proc. Nat. Acad. Sci. U.S.A. 81* (1984), 2597–2600.

Baum, P. and Douglas, R.G.

(1) Toeplitz operators and Poincaré duality, in: Gohberg, I. C. (ed.), *Proceedings of the Toeplitz Memorial Conference, Tel Aviv 1981*, Birkhäuser, Boston, 1981, pp. 137–166.

Benn, I.M. and Tucker, R.W.

(1) *An Introduction to Spinors and Geometry with Applications in Physics*, Hilger, Bristol, 1987.

Berline, N., Getzler, E. and Vergne, M.

(1) *Heat Kernels and Dirac Operators*, Springer, Heidelberg, Berlin, 1992.

Berthier, A.-M. and Georgescu, V.

(1) Sur la propriété de prolongement unique pour l'opérateur de Dirac, *C. R. Acad. Sci. Paris Sér. I Math. 291* (1980), 603–606.

Bessaga, Cz. and Pelczynski, A.

(1) *Selected Topics in Infinite-Dimensional Topology*, Polish Scientific Publishers, Warsaw, 1975.

Birman, M. and Solomyak, A.

(1) On subspaces which admit pseudodifferential projections, *Vestnik Leningrad Univ. Nat. Mekh. Astronom. 82, no. 1* (1982), 18–25 (Russian).

Bismut, J.-M. and Cheeger, J.

(1) Invariants êta et indice des familles pour des variétés à bord, *C. R. Acad. Sci. Paris Sér. I Math. 305* (1987), 127–130.

(2) Families index for manifolds with boundary, superconnections and cones. I, *J. Funct. Anal. 89* (1990), 313–363.
(3) Families index for manifolds with boundary, superconnections and cones. II, *J. Funct. Anal. 90* (1990), 306–354.

Bismut, J.-M. and Freed, D.S.
(1) The analysis of elliptic families. II. Dirac operators, eta invariants, and the holonomy theorem, *Comm. Math. Phys. 107* (1986), 103–163.

Bitsadze, A.V.
(1) *Boundary Value Problems for Second Order Elliptic Equations*, Nauka, Moscow, 1966; (Russian). English translation North-Holland, Amsterdam, 1968.

Bojarski, B.
(1) The abstract linear conjugation problem and Fredholm pairs of subspaces, in: *In Memoriam I.N. Vekua* Tbilisi Univ., Tbilisi, 1979, pp. 45–60 (Russian).

Booß, B.
(1) *Elliptische Topologie von Transmissionsproblemen*, Bonner Math. Schriften 58, Bonn, 1972.
(2) *Topologie und Analysis - Eine Einführung in die Atiyah-Singer-Indexformel*, Springer, Berlin, 1977.

Booß, B. and Bleecker, D.D.
(1) *Topology and Analysis. The Atiyah-Singer Index Formula and Gauge-Theoretic Physics*, Springer, New York, 1985. (English translation of Booß [1977], revised and enlarged).

Booß, B. and Rempel, S.
(1) Découpage et recollage des opérateurs elliptiques, *C.R. Acad. Sci. Paris Sér. I Math. 292* (1981), 711–714.
(2) Cutting and pasting of elliptic operators, *Math. Nachr. 109* (1982), 157–194.

Booß, B. and Schulze, B.-W.
(1) Index theory and elliptic boundary value problems - remarks and open questions, in: Bojarski, B. (ed.), *Proc. Banach Semester "Partial Diff. Equ." (Warsaw 1978)*, Banach Center Publications Vol. 10, Warsaw, 1983, pp. 33–57.

Booß, B. and Wojciechowski, K.P.
(1) *Desuspension of splitting elliptic symbols, Tekster fra IMFUFA* No. 52, Roskilde, 1982.
(2) The index of elliptic operators on a mapping torus, *C.R. Math. Rep. Acad. Sci. Canada 7* (1985), 97–102.

(3) Desuspension of splitting elliptic symbols I, *Ann. Global Anal. Geom. 3* (1985), 337–383.
(4) Desuspension of splitting elliptic symbols II, *Ann. Global Anal. Geom. 4* (1986), 349–400.
(5) Pseudo-differential projections and the topology of certain spaces of elliptic boundary value problems, *Comm. Math. Phys. 121* (1989), 1–9.
(6) Dirac operators and manifolds with boundary, in: Delanghe, R. et al. (eds.), *Clifford Algebras and their Applications in Mathematical Physics*, Proc. 3rd Int. Conf. (Deinze, Belgium, May 1993), Kluwer Academic Pub., Dordrecht (to appear).

Borisov, N.V., Müller, W. and Schrader, R.

(1) Relative index theorems and supersymmetric scattering theory, *Comm. Math. Phys. 114* (1988), 475–513.

Boutet de Monvel, L.

(1) Boundary problems for pseudo-differential operators, *Acta Math. 126* (1971), 11–51.

Branson, T.P. and Gilkey, P.B.

(1) Residues of the eta function for an operator of Dirac type, *J. Funct. Anal. 108* (1992a), 47–87.
(2) Residues of the eta function for an operator of Dirac type with local boundary conditions, *Differ. Geom. Appl. 2* (1992b), 249–267.

Brown, L.G., Douglas, R.G. and Fillmore, P.A.

(1) Unitary equivalence modulo the compact operators and extensions of C^*-algebras, in: P.A. Fillmore (ed.), *Proceedings of a Conference on Operator Theory*, Lecture Notes in Math. 345, Springer-Verlag, Berlin, 1973, pp. 58–128.

Brüning, J. and Seeley, R.T.

(1) An index theorem for first order regular singular operators, *Amer. J. Math. 110* (1988), 659–714.

Bunke, U.

(1) Relative index theory, *J. Funct. Anal. 105* (1992a), 63–76.
(2) *On Constructions Making Dirac Operators Invertible at Infinity*, Humboldt Universität, Preprint, Berlin, 1992b.
(3) *A Glueing Formula for the eta-Invariant*, Humboldt Universität, Preprint, Berlin, 1993.

Calderón, A.P.

(1) Boundary value problems for elliptic equations, in: *Outlines of the Joint Soviet-American Symposium on Partial Differential Equations*, Novosibirsk, 1963, pp. 303–304.

(2) *Lecture Notes on Pseudo-Differential Operators and Elliptic Boundary Value Problems. I.* Consejo Nacional de Investigaciones Cientificas y Technicas, Inst. Argentino de Matematica, Buenos Aires, 1976.

Callan, C.G., Dashen, R. and Gross, D.J.

(1) Toward a theory of the strong interactions, *Phys. Rev. D (3) 17* (1978), 2717–2763.

Carleman, T.

(1) Sur un problème d'unicité pour les systèmes d'équations aux dérivées partielles à deux variables indépendentes, *Ark. Mat. Astr. Fys. 26 B, No 17* (1939), 1–9.

Carslaw, H.S. and Jaeger, J.C.

(1) *Conduction of Heat in Solids*, Oxford University Press, Oxford, 1959^{II}, reprinted Clarendon Press, Oxford, 1988.

Chazarain, J. and Piriou, A.

(1) *Introduction à la théorie des équations aux dérivées partielles linéaires*, Gauthier-Villars, Paris, 1981. English, North Holland, Elsevier, 1982.

Colton, D. and Kress, R.

(1) *Integral Equation Methods in Scattering Theory*, Wiley, New York, 1983.

Cordes, H.O.

(1) Über die eindeutige Bestimmtheit der Lösungen elliptischer Differentialgleichungen durch Anfangsvorgaben, *Nachr. Akad. Wiss. Göttingen Math.-Phys. Kl. IIa, Nr. 11* (1956), 239–258.

Costabel, M. and Stephan, E.

(1) Strongly elliptic boundary integral equations for electromagnetic transmission problems, *Proc. R. Soc. Edinb., A 109, no. 3/4* (1988), 271–298.

Costabel, M. and Wendland, W.L.

(1) Strong ellipticity of boundary integral operators, *J. reine angew. Math. 372* (1986), 34–63.

Delanghe, R., Sommen, F. and Souček, V.

(1) *Clifford Algebra and Spinor-Valued Functions: A Function Theory for the Dirac Operator*, Kluwer, Dordrecht, 1992.

Doetsch, G.

(1) *Einführung in Theorie und Anwendung der Laplace-Transformation*, Birkhäuser, Basel, 1970. English, Springer, Berlin, 1974.

Donaldson, S.K. and Kronheimer, P.B.

(1) *The Geometry of Four-Manifolds*, Clarendon Press - Oxford University Press, Oxford, 1990.

Douglas, R.G. and Wojciechowski, K.P.

(1) Analytic realization of relative K-homology on manifolds with boundary and pairing with K-cohomology, *Z. Anal. Anwendungen 8* (1989), 485–494.
(2) Adiabatic limits of the η-invariants. The odd-dimensional Atiyah-Patodi-Singer problem, *Comm. Math. Phys. 142* (1991), 139–168.

Dubrovin, B.A. Fomenko, A.T. and Novikov, S.P.

(1) *Modern Geometry - Methods and Applications I, II, III*, Nauka, Moscow, 1979 (Russian). English translation Springer, Berlin, 1984, 1985, 1990.

Duhamel, J.M.C.

(1) Mémoire sur la méthode générale relative au mouvement de la chaleur dans les corps solides plongés dans les milieux dont la température varie avec le temps, *J. Éc. polyt. Paris 14* (1833), Cah. 22, 20.

Eskin, G.I.

(1) *Boundary Value Problems for Elliptic Pseudodifferential Equations*, Nauka, Moscow, 1973 (Russian). English translation Transl. Math. Monogr., vol. 52, Providence, 1981.

Freed, D.S. and Uhlenbeck, K.K.

(1) *Instantons and 4-Manifolds*, Math. Sci. Res. Inst. Publ., Springer, New York, 1984.

Fuglede, B.

(1) *Om Hilbert-Schmidt-homotope projektioner*, Copenhagen University, Notes, 1976.

Fujii, K.

(1) A representation of complex K-groups by means of a Banach algebra, *Mem. Fac. Sci. Kyushu Univ. Ser. A 32* (1978), no. 2, 255–265.

Furutani, K. and Otsuki, N.

(1) Spectral flow and intersection number, *J. Math. Kyoto Univ. 33* (1993), 261–283.

Getzler, E.

(1) A short proof of the local Atiyah-Singer index theorem, *Topology 25* (1986), 111–117.

Gilkey, P.B.

(1) The boundary integrand in the formula for the signature and Euler characteristic of a Riemannian manifold with boundary, *Adv. Math.* *15* (1975), 334–360.

(2) *Invariance Theory, the Heat Equation, and the Atiyah-Singer Index Theorem*, Publish or Perish, Wilmington, 1984. (New edition in preparation).

(3) The geometrical index theorem for Clifford modules, in: Rassias, T.M. (ed.), *Topics in Mathematical Analysis*, World Scientific Press, Singapore, 1989, pp. 315–327.

(4) *On the Index of Geometrical Operators for Riemannian Manifolds with Boundary*, Preprint, University of Oregon, Eugene, 1991.

Gilkey, P.B. and Smith, L.

(1) The eta invariant for a class of elliptic boundary value problems, *Comm. Pure Appl. Math.* *36* (1983), 85–131.

(2) The twisted index problem for manifolds with boundary, *J. Differential Geom.* *18* (1983), 393–444.

Goldstone, J. and Jaffe, R.L.

(1) Baryon number in chiral bag models, *Phys. Rev. Lett.* *51* (1983), 1518–1521.

Gromov, M. and Lawson, H.B.

(1) Positive scalar curvature and the Dirac operator on complete Riemannian manifolds, *Inst. Hautes Études Sci. Publ. Math.* *58* (1983), 295–408.

Grubb, G.

(1) Boundary problems for systems of partial differential operators of mixed order, *J. Funct. Anal.* *26* (1977), 131–165.

(2) *Functional Calculus of Pseudo-Differential Boundary Problems*, Birkhäuser, Boston, 1986.

(3) Pseudo-differential boundary problems in L_p spaces, *Comm. Partial Differential Equations* *15* (1990), 289–340.

(4) *Heat Operator Trace Expansion and Index for General Atiyah-Patodi-Singer Boundary Problems*, Preprint, Copenhagen, 1991.

Heinz, E.

(1) Über die Eindeutigkeit beim Cauchyschen Anfangswertproblem einer elliptischen Differentialgleichung zweiter Ordnung, *Nachr. Akad. Wiss. Göttingen Math.-Phys. Kl. IIa* 1955/1, 1–12.

Hirzebruch, F.

(1) *Neue topologische Methoden in der algebraischen Geometrie*, Springer-Verlag, Berlin, 1956. English 1966.

Hitchin, N.

(1) Harmonic spinors, *Adv. Math. 14* (1974), 1-55.

Hörmander, L.

(1) Pseudo-differential operators and non-elliptic boundary problems, *Ann. of Math. (2) 83* (1966), 129-209.
(2) *The Analysis of Linear Partial Differential Operators III*, Springer, Berlin, 1985.

Jänich, K.

(1) Charakterisierung der Signatur von Mannigfaltigkeiten durch eine Additivitätseigenschaft, *Invent. Math. 6* (1968), 35–40

Jaffe, A. and Taubes, C.H.

(1) *Vortices and Monopoles*, Birkhäuser, Boston, 1980.

Jerison, D.

(1) Carleman inequalities for the Dirac and Laplace operators and unique continuation, *Adv. Math. 62* (1986), 118–134.

John, F.

(1) *Plane Waves and Spherical Means Applied to Partial Differential Equations*, Interscience Publ., New York, 1955.

Julg, P.

(1) Indice relatif et K-théorie bivariante de Kasparov, *C.R. Acad. Sci. Paris Sér. I Math. 307* (1988), 243–248.

Kalf, H.

(1) Non-existence of eigenvalues of Dirac operators, *Proc. Roy. Soc. Edinburgh Sect. A 89* (1981), 309–317.

Kaminker, J. and Wojciechowski, K.P.

(1) Index theory on $\mathbf{Z}_k$ manifolds and pseudodifferential Grassmannian, in: Arveson, W.B. et al. (eds.), *Proc. OATE 2 Conf. (Craiova, Romenia 1988)*, Longman Publ. House, London, 1992, pp. 88–92.

Karoubi, M.

(1) *K-Theory*, Springer, Berlin, 1978.

Karras, U., et al.

(1) *Cutting and Pasting of Manifolds, SK-Groups*, Publish or Perish Inc., Boston, 1973.

Kato, T.

(1) *Perturbation Theory for Linear Operators*, Springer, Berlin, 1976.

Kiskis, J.

(1) Fermion zero modes and level crossing, *Phys. Rev. D (3) 18* (1978), 3690–3694.

Klimek, S. and Wojciechowski, K.P.

(1) *Adiabatic Cobordism Theorems for Analytic Torsion and* η*-Invariant*, Preprint, Purdue (IUPUI), Indianapolis, 1992.

(2) η-Ivariants on manifolds with cylindrical end, *Differ. Geom. Appl. 3* (1993), 191–201.

Klingenberg, W.

(1) *Riemannian Geometry*, DeGruyter, Berlin, 1982.

Kobayashi, S. and Nomizu, K.

(1) *Foundations of Differential Geometry I, II*, Interscience, John Wiley, New York, 1963, 1969.

Koranyi, A.

(1) Harmonic functions on symmetric spaces, in: Boothby, W.M., and Weiss, G.L. (eds), *Symmetric Spaces - Short Courses Presented at Washington University*, Marcel Dekker, New York, 1972, pp. 379–412.

Koranyi, A. and Vagi, S.

(1) Rational inner functions on bounded symmetric domains, *Trans. Am. Math. Soc. 254* (1979), 179–193.

Kori, T.

(1) *Boundary Value Problems for the Dirac Operator on* S^4, Preprint, Waseda University, Tokyo, 1993.

Lawson, H.B. and Michelsohn, M.-L.

(1) *Spin Geometry*, Princeton University Press, Princeton, 1989.

Lesch, M. and Wojciechowski, K. P.

(1) *On the* η*-Invariant of Generalized Atiyah-Patodi-Singer Boundary Problems*, Report No. 278, Augsburg University, Augsburg, 1993.

Levi, E.E.

(1) Caratterische multiple e problema di Cauchy, *Ann. Mat. Pura Appl. (3) 16* (1909), 161-201.

Lions, J.L. and Magenes, E.

(1) *Problèmes aux limites non homogènes et applications I*, Editions Dunod, Paris, 1968. English, Springer, Berlin, 1972.

Lopatinskiĭ, Ya.B.

(1) On a method of reducing boundary problems for a system of differential equations of elliptic type to regular integral equations,

Ukrain Mat. Zurnal 5 (1953), 123–151 (Russian). English translation *Amer. Math. Soc. Transl. (2) 89* (1970), 149–183.

Lück, W.

(1) Analytic and topological torsion for manifolds with boundary and symmetry, *J. Differential Geom. 37* (1993), 263–322.

Mathai, V.

(1) Heat kernels and Thom forms, *J. Funct. Anal. 104* (1992), 34–46.

Mazzeo, R.R. and Melrose, R.B.

(1) *Analytic Surgery and the Eta Invariant*, Preprint, M.I.T., 1992.

McKean, H. and Singer, I.M.

(1) Curvature and the eigenvalues of the Laplacian, *J. Differential Geom. 1* (1967), 43–69.

Meister, E.

(1) *Randwertaufgaben der Funktionentheorie. Mit Anwendungen auf singuläre Integralgleichungen und Schwingungsprobleme der mathematischen Physik*, Teubner, Stuttgart, 1983.

Melrose, R.B.

(1) Transformation of boundary problems, *Acta Math. 147* (1981), 149–236.

(2) *The Atiyah-Patodi-Singer Index Theorem*, Preprint, M.I.T, 1992.

Melrose, R.B. and Mendoza, G.A.

(1) *Elliptic Boundary Problems on Spaces with Conic Points*, Journées équations aux dérivées partielles, Saint-Jean-de-Mont 1981, Exposé No. 4, 1981.

Mickelsson, J.

(1) *Current Algebras and Groups*, Plenum Press, New York, 1989.

Milnor, J.

(1) On spaces having the homotopy type of a CW-complex, *Trans. Amer. Math. Soc. 90* (1957), 272–280.

Minakshisundaram, S. and Pleijel, A.

(1) Some properties of the eigenfunctions of the Laplace operator on Riemannian manifolds, *Canad. J. Math. 1* (1949), 242–256.

Müller, W.

(1) L_2-index and resonances, in: *Proc. 21st Int. Taniguchi Symp. Kyoto 1987*, Lecture Notes in Math. 1339, Springer-Verlag, Berlin, 1988, pp. 203–211.

(2) *Eta Invariants and Manifolds with Boundary*, Preprint, Max-Planck-Inst. f. Math., Bonn, 1993.

Mushelišvili, N.I.

(1) *Singular Integral Equations*, Fizmatgiz, Moscow, 1962 (Russian). German translation Akademie-Verlag, Berlin, 1965; English translation of 1946 edition: Noordhoff, Groningen, 1953.

Nash, C.

(1) *Differential Topology and Quantum Field Theory*, Academic Press, London, 1991.

Nicolaescu, L.

(1) *The Maslov Index, the Spectral Flow, and Splittings of Manifolds*, Preprint, Michigan State Univ., East Lansing, 1993.

Otsuki, N. and Furutani, K.

(1) Spectral flow and Maslow index arising from Lagrangian intersections, *Tokyo J. Math. 14* (1991), 135–150.

Palais, R.S.

(1) *Seminar on the Atiyah-Singer Index Theorem*, Ann. of Math. Studies 57, Princeton University Press, Princeton, 1965a.
(2) On the homotopy type of certain groups of operators, *Topology 3* (1965b), 271–279.

Plamenevski, B. and Rozenblum, G.

(1) Pseudodifferential operators with isolated singularities in symbols: K-theory and index formulas, *Funkcional. Anal. i Priložen. 26/4* (1992), 45–56 (Russian). English translation *Functional Anal. Appl. 26/4*.

Pliś, A.

(1) A smooth linear elliptic differential equation without any solution in a sphere. *Comm. Pure Appl. Math. 14* (1961), 599–617.

Pressley, A. and Segal, G.B.

(1) *Loop Groups*, Clarendon Press - Oxford University Press, Oxford, 1986.

Protter, M.H.

(1) Unique continuation for elliptic equations, *Trans. Amer. Math. Soc. 95* (1960), 81–91.

Reed, M. and Simon, B.

(1) *Methods of Modern Mathematical Physics. I*, Academic Press, New York, 1972.
(2) *Methods of Modern Mathematical Physics. IV*, Academic Press, New York, 1978.

Rees, H.D.

(1) The η-invariant and Wall non-additivity, *Math. Ann. 267* (1984), 449–452.

Rempel, S. and Schulze, B.-W.

(1) *Index Theory of Elliptic Boundary Problems*, Akademie-Verlag, Berlin, 1982.

Roe, J.

(1) *Elliptic Operators, Topology and Asymptotic Methods*, Pitman, London, 1988a.

(2) An index theorem on open manifolds. I. II. *J. Differential Geom. 27* (1988b), 87–113, 115–136.

Roze, S.N.

(1) On the spectrum of the Dirac operator, *Theoret. Math Fiz. 2* (1970), 275-279 (Russian). English translation *Theoret. and Math. Phys.*

Ryaben'kii, V.S.

(1) Boundary equations with projections, *Uspehi Mat. Nauk 40:2* (1985), 121-149 (Russian). English translation *Russian Math. Surveys 40:2.*

Šapiro, Z.Ya.

(1) The first boundary problem for an elliptic system of differential equations, *Mat. Sbornik N.S. 28 (70)* (1951), 55–78 (Russian).

(2) On general boundary problems for equations of elliptic type, *Izvestiya Akad. Nauk SSSR, Ser. Mat. 17* (1953), 539–562 (Russian).

Sato, M.

(1) Soliton equations as dynamical systems on infinite dimensional Grassmann manifolds, *RIMS Kokyuroku 439* (1981), 30–40.

Sato, M., Miwa, J. and Jimbo, M.

(1) Holonomic quantum fields II: The Riemann-Hilbert problem, *Publ. RIMS, Kyoto Univ. 15* (1979), 201–278.

Schmidt, J.R.

(1) Chiral anomaly for the Dirac operator in an instanton background via the Atiyah-Patodi-Singer theorem, *Phys. Rev. D (3) 36* (1987), 2539–2544.

Schmidt, J.R. and Bincer, A.M.

(1) Chiral asymmetry and the Atiyah-Patodi-Singer index for the Dirac operator on a four-dimensional ball, *Phys. Rev. D (3) 35* (1987), 3995–4000.

Schulze, B.-W.

(1) Über eine Verheftung elliptischer Randwert-Probleme, *Math. Nachr. 95* (1979), 205–222.

(2) *Pseudo-Differential Operators on Manifolds with Singularities*, Teubner, Leipzig, and North Holland, Amsterdam, 1990.

Seeley, R.T.

(1) Integro-differential operators on vector bundles, *Trans. Amer. Math. Soc. 117* (1965), 167–204.

(2) Singular integrals and boundary value problems, *Amer. J. Math. 88* (1966), 781-809.

(3) Topics in pseudo-differential operators, in: *CIME Conference on Pseudo-Differential Operators (Stresa 1968)*. Ed. Cremonese, Rome, 1969, pp. 167–305.

Segal, G.B. and Wilson, G.

(1) Loop groups and equations of KdV type., *Inst. Hautes Études Sci. Publ. Math. 61* (1985), 5–65.

Shubin, M.A.

(1) *Pseudodifferential Operators and Spectral Theory*, Nauka, Moscow, 1978 (Russian). English translation Springer, Berlin, 1986.

Simon, B.

(1) Schrödinger semigroups, *Bull. Amer. Math. Soc. 7* (1982), 447–526.

Singer, I.M.

(1) Operator theory and periodicity, *Indiana Univ. Math. J. 20* (1971), 949–951.

(2) The η-invariant and the index, in: Yau, S.-T. (ed.), *Mathematical Aspects of String Theory*, World Scientific Press, Singapore, 1988, pp. 239-258.

Steenrod, N.

(1) *The Topology of Fibre Bundles*, Princeton University Press, Princeton, 1965^V.

Stein, E.M. and Weiss, G.

(1) *Introduction to Fourier Analysis on Euclidean Spaces*, Princeton University Press, Princeton, 1971.

Taubes, C.H.

(1) Casson's invariant and gauge theory, *J. Differential Geom. 31* (1990), 547–599.

Taylor, M.E.

(1) *Pseudodifferential Operators*, Princeton University Press, Princeton, 1981.

Treves, F.

(1) *Pseudodifferential and Fourier Integral Operators I,* Plenum Press, New York, 1980.

Vafa, C. and Witten, E.

(1) Eigenvalue inequalities for fermions in gauge theories, *Comm. Math. Phys. 95, 3* (1984), 257–274.

Vogelsang, V.

(1) Absence of embedded eigenvalues of the Dirac equation for long range potentials, *Analysis 7* (1987), 259–274.

Wall, C.T.C.

(1) Non-additivity of the signature, *Invent. Math. 7* (1969), 269–274.

Witten, E.

(1) Supersymmetry and Morse theory, *J. Differential Geom. 17* (1982), 661–692.

(2) Quantum field theory, Grassmannians, and algebraic curves, *Comm. Math. Phys. 113* (1988), 529–600.

Wojciechowski, K.P.

(1) *Spectral Flow and Some Applications to the Index Theory,* Doctoral Thesis, Warsaw, 1981 (Polish).

(2) Spectral flow and the general linear conjugation problem, *Simon Stevin 59* (1985a), 59–91.

(3) Elliptic operators and relative K-homology groups on manifolds with boundary, *C.R. Math. Rep. Acad. Sci. Canada 7* (1985b), 149–154.

(4) A note on the space of pseudodifferential projections with the same principal symbol, *J. Operator Theory 15* (1986), 207–216.

(5) On the Calderón projections and spectral projections of elliptic operators, *J. Operator Theory 20* (1988), 107–115.

(6) *Elliptic Boundary Problems and the Holonomy Theorem of Witten,* Preprint, IUPUI, Indianapolis, 1992.

(7) *On the Adiabatic Additivity of the η-Invariant.,* Preprint, IUPUI, Indianapolis, 1993.

Yoshida, T.

(1) Floer homology and splittings of manifolds, *Ann. of Math. 134* (1991), 277–323.

Index